SCIENCE AND SOUND IN NINETEENTH-CENTURY BRITAIN

Nineteenth-Century Science, Technology and Medicine: Sources and Documents

Forthcoming:

Scientific Methodology in Nineteenth Century Britain
4 volume set
Edited by Charles H. Pence

Women in Medicine in the Long Nineteenth Century
4 volume set
Edited by Claire Brock

Insects in Victorian Culture
4 volume set
Edited by John Clark

Electric Power in Victorian Britain
4 volume set
Edited by Nathan Kapoor

Women, Gender and Science in Victorian Culture
4 volume set
Leila McNeill & Anna Reser

For more information about this series, please visit www.routledge.com/Nineteenth-Century-Science-Technology-and-Medicine-Sources-and-Documents/book-series/NCSTM

SCIENCE AND SOUND IN NINETEENTH-CENTURY BRITAIN

Edited by
Edward J. Gillin

Volume I:
Sounds Experimental and Entertaining

Routledge
Taylor & Francis Group
LONDON AND NEW YORK

First published 2024
by Routledge
4 Park Square, Milton Park, Abingdon, Oxon OX14 4RN

and by Routledge
605 Third Avenue, New York, NY 10158

Routledge is an imprint of the Taylor & Francis Group, an informa business

British Library Cataloguing-in-Publication Data
A catalogue record for this book is available from the British Library

ISBN: 978-1-032-50077-5 set
ISBN: 978-1-032-50078-2 (volume I) hbk
ISBN: 978-1-032-61840-1 (volume I) ebk

DOI: 10.4324/b23406

Typeset in Times New Roman
by Apex CoVantage, LLC

Publisher's Note
References within each chapter are as they appear in the original complete work.

For the "Sound and Materialism in the Nineteenth Century" research group, based in the Faculty of Music, Cambridge (2016–2019) – David Trippett, Melissa Van Drie, Melle Kromhout, and Stephanie Probst.

CONTENTS

ACKNOWLEDGEMENTS

The completion of these volumes owes a great deal to David Lewis, Graeme Gooday, Simon Schaffer, Alexandra Hui, William Whyte, Fanny Gribenski, D'Maris Coffman, David Pantalony, Viktoria Tkaczyk, John Kelsey, Priti Parikh, Judy Stephenson, and the staff of London's Royal Society Library, Cambridge University Library, the National Archives, the Bodleian Library in Oxford, Oxford University Museum Library, Trinity College Cambridge Library, the British Library, and King's College London Library, as well as all of my colleagues at the Bartlett School of Sustainable Construction at University College London. This work is part of the Leverhulme Early Career Research Project "The State of Science: governing knowledge of nature in Victorian Britain", hosted at the University of Leeds and University College London. Thanks go to my family, Steve Spencer, Alexander Teague, Alfred Spencer, Louise Spencer, and Victoria, Warren, Bassi, and Reuben Putt-Gillin, and to my friends Tommy Grimshaw, Crosbie Smith, Harry Mace, Peter Roberts, Ben Allen, Charlie Brown, Christopher Yabsley, and Sid Sharma. These volumes are dedicated to my colleagues on the "Sound and Materialism in the Nineteenth Century" European Research Council research project, namely, David Trippett, Melissa Van Drie, Melle Kromhout, and Stephanie Probst.

CHRONOLOGY

1787 – Ernst Chladni publishes his *Entdeckungen über die Theorie des Klanges*
1789 – French Revolution
1793 – Outbreak of war between France and Britain
1815 – Battle of Waterloo
1818 – John Ross's Arctic Expedition
1820 – Hans Christian Ørsted unites magnetism and electricity
1821 – Michael Faraday observes electromagnetic rotation
1828 – Michael Faraday and Charles Wheatstone first collaborate to deliver lectures on sound at the Royal Institution
1830 – Publication of John Herschel's "Sound" in the *Encyclopaedia Metropolitana*
1831 – Michael Faraday experiments on "crispations"
1831 – James Clark Ross reaches the north magnetic pole
1831 – Michael Faraday observes electromagnetic induction
1831 – First meeting of the British Association for the Advancement of Science held in York
1831 – Two failed reform bills for extension of Britain's franchise
1830 – John Herschel publishes his *A Preliminary Discourse on the Study of Natural Philosophy*
1831 – Mary Somerville publishes her *Mechanism of the Heavens*
1831–36 – Charles Darwin circumnavigates the world on board HMS *Beagle*
1832 – Great Reform Act
1833 – William Whewell first coins the word *scientist*
1834 – Mary Somerville publishes her *On the Connexion of the Physical Sciences*
1834 – Palace of Westminster burns down
1834 – Stuttgart Conference on musical pitch appoints A440 as a standard
1837 – Ascent of Queen Victoria
1837 – First patent of the electric telegraph
1839 – Launch of the British magnetic scheme
1845 – Michael Faraday records the effect of a magnetic force on polarized light
1848 – Revolutions across Europe, including in France, the Netherlands, Denmark, the Austrian Empire, and Italy
1851 – Great Exhibition held in London

1851 – First submarine telegraphy cable between England and France
1853–56 – Crimean War
1857 – Herbert Spencer publishes his article "Progress: its Law and Cause"
1858 – Emperor Napoleon III launches an inquiry into French musical pitch standards
1859 – French government appoints the diapason normal at A435.
1859–60 – Society of Arts committee on musical pitch
1859 – Publication of Charles Darwin's *On the Origin of Species*
1861 – James Clerk Maxwell publishes his *On Physical Lines of Force*
1861–65 – American Civil War
1863 – Publication of Hermann Helmholtz's *Die Lehre von den Tonempfindungen als physiologische Grundlage für die Theorie der Musik*
1865 – James Clerk Maxwell publishes his article "A Dynamical Theory of the Electromagnetic Field"
1867 – Second Reform Act
1871 – Publication of Charles Darwin's *The Descent of Man, and Selection in Relation to Sex*
1875 – Alexander Ellis publishes his translation of Hermann Helmholtz's *On the sensations of tone as a physiological basis for the theory of music*
1876 – Alexander Graham Bell takes out a patent in the United States for his telephone
1877 – Thomas Edison invents his phonograph
1877 – Publication of John William Strutt's *The Theory of Sound*
1877 – Thomas Edison develops his telephone transmitter
1878 – First commercial telephone exchange
1884 – Third Reform Act
1894 – First gramophone record
1901 – Death of Queen Victoria
1914–18 – First World War

AN INTRODUCTION TO SCIENCE AND SOUND IN NINETEENTH-CENTURY BRITAIN

Throughout the nineteenth century, the soundscapes of industrial Britain underwent a radical transformation. The industrial harnessing of coal for steam-powered transportation and manufacturing, combined with rapid urbanization and the development of novel inventions and electroacoustic apparatus, presented British society with a range of unfamiliar sounds and fundamentally changed how it listened and what it thought about what it heard. Amid all this economic, social, cultural, and technological upheaval, a diverse network of natural philosophers, mathematicians, experimentalists, theologians, scientists, and physicists provided radical new understandings of what sound was, how it behaved, and its broader place in nature. Equally, acoustics was an object of innovation, with instrument makers and natural philosophers alike mobilizing sonorous knowledge to develop an array of marvels, from talking automata to microphones and recording devices, that promised a communications revolution that would completely reorganize the human experience of sound. For the built environment, managing echo and resonance proved a challenge for architects as they worked to deliver brilliant acoustics for churches, debating chambers, and theatres. Yet sound was more than a subject to be studied: it was also a medium through which knowledge of nature could be obtained. As much as the eye, the ear had an epistemological role, offering valuable insights into a diverse range of natural phenomena that could not be obtained through sight alone. *Science and Sound in Nineteenth-Century Britain* explores these complex and changing connections between science and sound during the nineteenth century. Through four volumes of primary sources, including works both celebrated and obscure, it places this relationship within its varying historical contexts to offer direction over how sound can, and should, feature in the broader history and philosophy of science.

SCIENCE AND SOUND IN HISTORY

The science of acoustics was not new in the nineteenth century, nor was it limited to the British Isles or, for that matter, Europe and North America. Sound, especially musical sound, and its relationship with mathematical principles has been a subject of study since at least the sixth century BC. The ancient Greek philosopher

Pythagoras (ca. 570–ca. 495 BCE) argued that the musical relations between varying pitches were reducible to ratios of vibrations that could be illustrated through the production of notes on oscillating strings: a string sounding one given note could, on being halved in length, produce the same note an octave higher.[1] In his *Harmonics*, Roman mathematician Claudius Ptolemy (ca. 100–ca. 170 CE) subsequently extended Pythagoras's mathematical treatment of sound and music and, during the medieval and early modern period, this science of harmony remained a subject of investigation and speculation. In particular, Europe's great churches and cathedrals provided natural philosophers with valuable sites of acoustic inquiry, provided unusual sonorous phenomena and insights into how sound behaved in a contained space. Gloucester Cathedral, for instance, included a fourteenth-century passageway that formed what was known as a "whispering gallery", in which faint sounds could be heard over an unusually extensive distance. Giovanni Pisano's (ca. 1250–ca. 1315) thirteenth-century cupola for Pisa's baptistery of St John produced a similar effect. These sonorous experiences took on philosophical value in the seventeenth century when new inductivist frameworks transformed such churches into places of acoustic examination. For Francis Bacon (1561–1626), working within an Aristotelian tradition in which a philosopher's first task was to question the role of the senses, understanding sound was vital to knowing how the external world was perceived.[2] In his *Sylva Sylvarum* (1626), Bacon made observations on echo, including those in Gloucester Cathedral, which was of particular interest to early modern natural philosophers, being one of the most elusive natural phenomena to observe.[3] On hearing of a brilliant echo that had been heard in the chapel of Charenton-le-Pont, near Paris, Bacon went to experience this phenomenon for himself. A depot for watermen conveying timbre to the capital, the chapel's isles of pillars were stacked high with wood, which caused spoken words to be repeated thirteen times.[4] Similarly, in Marin Mersenne's (1588–1648) *Harmonie universelle* (1636–1637), ecclesiastical buildings features as locations in which sound's propagation through echo and resonance could be examined.[5] Bacon's and Mersenne's inductivist approach to the investigation of sound fit within a broader cultivation of seventeenth-century natural philosophy. Following a Baconian experimental framework, the first members of the Royal Society of London, established in 1660, devoted considerable time to acoustic study. This included a range of experiments, from Robert Boyle's (1627–1691) attempts to measure the speed of sound to observations on echo and vibrating strings.[6] Most famously in his undergraduate manuscript, "Of Musick", written between 1664 and 1666, Isaac Newton (1642–1726/27) compared the seven notes of the diatonic scale to the seven perceived colours of the spectrum, effectively uniting the phenomena of light and sound. These connections between the optical and the aural would become increasingly central to eighteenth-century sonorous discourse, notably in the works of Swiss mathematicians Leonhard Euler (1707–1783) and Daniel Bernoulli (1700–1782).[7]

Particularly significant for the development of nineteenth-century acoustic science were the early experiments of German natural philosopher Ernst Chladni

(1756–1827). Representing one of the earliest frameworks for visualizing the relationship between frequency and musical tones, Chladni made invisible sound waves observable to the eye by distributing sand over a glass plate and then putting this into vibration with a violin bow. This oscillatory motion caused the sand to form patterns, "Chaldni figures", which varied in relation to the frequency of the pitch sounded: sand collected at the points of least vibration, revealing the movement of sound waves through solid bodies. Published in his *Entdeckungen über die Theorie des Klanges* (1787) and *Die Akustik* (1802), Chladni's experiments secured attention with scientifically and socially prominent audiences across Europe, most notably, French emperor Napoléon Bonaparte.[8] In Britain, however, Chladni's works would not become widely known until the late 1810s and 1820s. With Britain and France having been in almost constant conflict since 1793, the British Isles were largely isolated from the latest philosophical theories and scientific claims of Continental Europe until the final defeat of Napoléon in 1815. It is this moment of cultural exchange, in the peace that followed the decisive Battle of Waterloo, that marks the point of departure for *Science and Sound in Nineteenth-Century Britain.*

The cultivation of acoustic knowledge and the place of sound in scientific activity was, evidently, far from new in the nineteenth century. Yet, between the 1810s and 1900s, this historic relationship took on a very distinctive character in the British Isles. As ever, when analysing past scientific activity, it is to the contexts of knowledge production that historians should look, and this is no different to the history of acoustic science. As becomes clear throughout the following four volumes, a broad array of cultural contexts shaped the relationship between sound and science in nineteenth-century Britain. The rise of industrial mechanized manufacturing, railways, and steamships, combined with rapid urbanization and population growth, sustained unfamiliar contexts for sonorous examination, as did the escalation of European colonialism.[9] Along with encountering diverse societies and their musical cultures, the spread of empire provided exciting opportunities for sonorous experiments in new environments, notably in the rarefied atmospheres of mountainous and polar regions. Likewise, the post-Napoleonic peace of 1815 inaugurated an influx of philosophical discourse from the continent, including the very latest investigations into sound, and encouraged a proliferation of sonorous inquiry across the British Isles, especially in London. The vast church-building programmes of the mid-nineteenth century stimulated intense concern with architectural acoustics.[10] Politically too, the gradual expansion of the British electorate, with the franchise extended to include middle- and working-class men for the first time through the reform acts of 1832, 1867, and 1884 shaped scientific activity relating to sound.[11] More than ever before, it mattered who could hear political discourse in Parliament, with the House of Commons' debating chamber a leading venue of acoustic investigation, and the communication of MPs' speech through new media technologies a powerful inducer of sonorous innovation. Natural philosophers and theologians regularly conscripted sound into their accounts of nature and contributions to religious and philosophical controversies, including those surrounding evolution and natural selection that came to dominate the natural sciences in the second half of the century. Entertainers, similarly, took full advantage of sonorous phenomena and scientific understandings of sound to deliver exhibitions and spectacles to diverse social audiences, especially in the fashionable showrooms and lecture theatres of London's West End.[12]

Science and Sound in Nineteenth-Century Britain deals not only with sound as a body of knowledge but also with the role of sound as a sense in past scientific activity, although these were, of course, connected. The experience of sound was inseparable from the development of new acoustic knowledge. Frequently, sonorous encounters transcended the science of acoustics, with natural philosophers and experimentalists mobilizing the ear as a valuable organ with which to scrutinize a range of natural phenomena, from electromagnetism to heat and light. As Jonathan Sterne observed in his influential 2003 *The Audible Past: cultural origins of sound reproduction*, there has been a bias towards vision in traditional histories of knowledge production, as well as in the formation of perceptions over what it has historically meant to be "modern".[13] After all, if we think about distinctions between the "Dark Ages" and the "Enlightenment", these immediately invoke notions of light and vision. While successive histories since Steven Shapin's *The Scientific Revolution* (1996), have undermined the validity of such terms, they nevertheless continue to carry significant cultural capital with twenty-first-century audiences: to be reasoned, learned, and empirical is to be enlightened.[14] Following Stern's challenge to this ocular hegemony, historians of science have stressed the multi-sensory character of past scientific activity, so much so that critiques of a visual dominance in contemporary scholarship seems dated and unnecessary.

So what then does the examination of sound and, specifically, listening as an epistemological practice do for our understanding of nineteenth-century science? Or, rather, why does the ear matter as an organ of knowledge production? Well, as these four volumes demonstrate, in nineteenth-century Britain, the experience of sound shaped new ways of thinking about nature. Michael Faraday's (1791–1867) investigations into electromagnetism throughout the 1820s and 1830s accompanied his experimental demonstrations to make vibrating bodies visible, as did Charles Wheatstone's (1802–1875) development of Chladni's acoustic figures, which rendered the relationship between sonorous pitch and the oscillations of glass plates observable to the eye. It is telling that in the months running up to his discovery of electrometric induction in August 1831, effectively inducing electricity through the combination of movement and magnetism, Faraday was, in fact, engaged in an industrious experimental programme of rendering sonorous vibrations visible. As Faraday remarked in March 1832, when depositing his earliest thoughts on electromagnetism to the Royal Society's strong box, he was "inclined to compare the diffusion of magnetic forces from a magnetic pole, to the vibrations upon the surface of disturbed water, or those of air in the phenomena of sound" and was sure that future investigation would show that "the vibratory theory will apply to these phenomena, as it does to sound, and most probably to light".[15] Such analogous thinking extended beyond electromagnetic phenomena, with natural philosophers frequently employing sound and sonorous experiences in their examination of how invisible forces like heat and light might operate. Listening was also a productive means of communicating knowledge to diverse audiences. When explaining a phenomenon like resonance or the relationship between vibratory frequency and pitch, experimentalists and natural philosophers found dramatic or entertaining sonorous performances, especially those musical, of considerable value.

In turn, these experiments and demonstrations influenced contemporary accounts of broad natural phenomena, from the movement of heat and light to that of electricity and magnetism, in which sound provided a useful tool for conceptualizing the wave-like motions of invisible forces. In the works of natural philosophers such as John Herschel (1792–1871), Mary Somerville (1780–1872), and William Whewell (1794–1866), sound was crucial to their depictions of a universe in which all natural phenomena were united, acting to a few governing laws, set in motion by a divine Creator.[16] In this way, the study of sound was central to contemporary religious interpretations of nature and Creation. Religious and scientific authorities frequently drew on sonorous phenomena and the human power of hearing as evidence of benevolent design while the existence of music appeared as something that transcended the material world. Through its ethereal, aesthetic beauty, romanticized understandings of the musical attributed to it a metaphysical value, being above scientific explanation. It was, after all, difficult for scientific practitioners to account for why some sounds were pleasing to the ear and others were not. While not new to the nineteenth century, during the 1800s, German Romantic philosophy grew more entrenched in its rejection of reducing all visions of nature to physical and chemical principles: in such interpretations, music took on a spiritual character.[17] While Georg Wilhelm Friedrich Hegel (1770–1831) claimed music to be beyond mathematical understanding, music critic Ernst Theodor Amadeus Hoffman (1776–1822) lauded music's metaphysical power to deliver transcendental experiences.[18] This Romantic account of music, with its stronghold in the German lands, endured throughout the century thanks to the writings of music theorists and composers like Eduard Hanslick (1825–1904) and Richard Wagner (1813–1883).[19] Britain might not have boasted such prominent musical authorities, but it was home to a network of musicians, philosophers, and theological guides who shared in this belief in music's transcendental qualities.

Yet such romantic, often religious, interpretations became increasingly strained throughout the century. During the 1820s and 1840s, new geological claims over the great age of the Earth and nebular theory surrounding the formation of stars challenged traditional orthodoxies over Creation as told in Genesis. In a similar fashion, chemistry threatened to reduce all human thought and emotions to chemical reactions. And with Robert Chambers's *Vestiges of the Natural History of Creation* (1844), Herbert Spencer's "Progress: its Law and Cause" (1857), and Charles Darwin's *On the Origin of Species* (1859) encouraging increasing support for evolutionary accounts of nature, fears abounded that, before long, everything in the universe would be reduced to material, godless processes. Sound, and especially music, was not immune to such controversy. Authors, including Darwin and Spencer, were keen to reduce nature's sounds, as well as music, to evolutionary processes.[20] Be it the development of birdsong through natural selection or the advancement of musical culture through the progress of civilization, the notion that aesthetically pleasing sonorous phenomena delivered evidence of a Divine creator came under growing strain. Likewise, new experimental and mathematical accounts of sound, such as those of German physicist Hermann Helmholtz (1821–1894), Lord Rayleigh (1842–1919), and John Tyndall (1820–1893), threatened to completely redefine what society thought about what it heard.[21] Psychologists and physiologists also contributed to this fashioning of new understandings of the sensory perception of sound. Alexandra Hui has brilliantly shown that such thinking, uniting acoustic experiments with physiological and psychological accounts of sound was particularly prominent in late nineteenth-century Germany and Austria. In the Germanic lands, celebrated musical authorities, including Helmholtz, Gustav Fechner (1801–1887), Ernst Mach (1838–1916), Carl Stumpf (1848–1936), and Wilhelm Wundt (1832–1920) all dealt with the problem of how humans experienced music.[22] These authors undoubtedly shaped sonorous thought in Britain, particularly those of English psychologist Edmund Gurney (1847–1888). Along with these often-troubling theories over sound and music, the proliferation of new technologies also shaped a reorganization of society's relationship with the sonorous. When, in August 1878, *The Times* reported that within the past twelve months, Alexander Graham Bell (1847–1922) had designed an articulating electric telegraph, Thomas Edison (1847–1931) had realized his phonograph, and David Edward Hughes (1830–1900) had announced the invention of his microphone, the newspaper foresaw a sonorous revolution underway.[23] These devices shaped new practices of listening, initiated in what historian Roland Wittje described as the "age of electroacoustics".[24] Sound in 1900 was, clearly, very different to what it had been in 1800. As Emily Thompson observed in her now classic *The Soundscape of Modernity* (2002), these technological developments fundamentally redefined how audiences encountered sound. Before the rise of electroacoustics, the experience of sound was determined by the physical space in which it was heard and at the moment it was sounded. New communications and recording technologies, Thompson argued, divorced this relationship between space and sound, not only ushering in "modern" soundscapes but also unfamiliar ways of listening.[25]

It is clear then, that in industrial Britain, sound had changed a great deal between the start and end of the nineteenth century. This transformation spanned the world well beyond the British Isles, and yet a focus on this region is valuable to our understanding of the nineteenth-century relationship between science and sound. In the first instance, this is because of Britain's comparatively early industrialization, with it being the first country to mobilize fossil fuels for the mechanization of its economy. Britain was also home to a bustling scientific culture, with the likes of Faraday, Somerville, Rayleigh, Tyndall, Gurney, Alexander Ellis, John

Herschel, and Wheatstone all leading figures in the cultivation of sonorous knowledge, to say nothing of Darwin's and Spencer's claims over sound and evolution. In particular, it was London that was a centre of acoustic discussion, as well as the experience of sonorous phenomena as popular entertainment. And finally, Britain's growing network of overseas colonies and naval expeditions of discovery constituted the most ambitious imperial project of the age, stimulating multicultural exchanges of musical culture. Collectively, all of this put the British Isles at the forefront of the development of science's relationship with sound.

SOUND AND THE HISTORY OF SCIENCE

Recent histories of science have emphasized the multi-sensory nature of scientific activity. In her 2006 *Nervous Conditions*, Elizabeth Musselman stressed the epistemological value of combining visual ways of knowing, with those of temperature and feeling, such as in chemical research and the historic development of thermometers.[26] Making non-visual sensory experiences of nature visible to the eye has been a productive way of making and sharing knowledge of a wide range of natural phenomena. The historically significant role of sight in popularizing scientific work has also been well demonstrated, notably in Bernard Lightman's study of Victorian scientific visual culture, *Victorian Popularizers of Science* (2007), and Iwan Morus's extensive writing on mid-nineteenth-century London's scientific culture.[27] This historiographical examination of the role of senses in past scientific activity has, rightly, extended to questions of sound and hearing.[28] Myles Jackson's *Harmonious Triads* (2006) explored the intersections between nineteenth-century German musicians, instrument makers, and physicists; collectively, these helped fashion the very latest scientific understandings of sound and music. Peter Pesic's *Music and the Making of Modern Science* (2014) extended this analysis beyond modern Germany, offering an overview of the relationship between sound, specifically that musical, and scientific change since antiquity.[29] The 2013 *Osiris* special edition "Music, sound, and laboratory from 1750–1980" unpacked the close relationship between sound and science in the formation of new knowledge of nature since the mid-eighteenth century. Alexandra Hui, Julia Kursell, and Myles Jackson argued that sound took on a distinctive value in the development of European and American laboratory culture.[30] Instruments too, played an important part in the history of scientific culture, both through the sounds they produced and through their means of production.[31] Sound was of particular value for sharing truth claims over nature, as demonstrated in James Davies and Ellen Lockhart's *Sound Knowledge: music and science in London, 1789–1851* (2016). As they show, London's scientific and musical cultures united in the public display of new knowledge of the natural world: the city's audiences were encouraged to listen and acquire knowledge by their ears, as well as with their eyes.[32] Together, historians and philosophers of science have done much to illustrate the connected relationship between science and sound, as well as between past scientific and musical culture.

Science and Sound not only draws attention to the key texts that have shaped this existing scholarship but also looks to provide future direction to studies in the history of science and technology, as well as sound studies and broader cultural histories. It includes the most influential works of acoustic science, including John Herschel's *Encyclopaedia Metropolitana* entry on "Sound" (1830), Alexander Ellis's (1814–1890) English translation of Helmholtz's *On the Sensations of Tone* (1875), and John William Strutt's (1842–1919) *The Theory of Sound* (1877). However, *Science and Sound* goes beyond these elite, heavily mathematical treatises to deliver a wider understanding of the relationship between science and sound in nineteenth-century Britain. The scientific activity and discourse surrounding sound, as well as the use of sound in experimental investigation and demonstration, took place in diverse, often-unfamiliar contexts with broad audiences. Authors such as Mary Somerville and William Whewell were no less scientific in their treatment of sound than the likes of Herschel but directed their writing to readers who might not possess the mathematical skill of a university wrangler. Throughout the 1820s and 1830s, Charles Wheatstone and Faraday treated the pursuit of new sonorous knowledge to develop public demonstrations that were both informative and entertaining: instruments and apparatus that illustrated diverse sonorous phenomena taught philosophical lessons about nature to non-specialist publics. Sound also featured prominently in theological treatises and sermons, as well as controversial studies of nature that challenged religious orthodoxies through radical ideas of evolution, progress, and psychological analysis. Acoustics was, of course, an area of concern for architects and those invested in the building sciences, such as David Boswell Reid, John Scott Russell, and Thomas Roger Smith. And outside of the built environment too, sound was a subject of experimentation, notably for discovering officers of British naval expeditions who took advantage of the atmospheric conditions encountered in tropical and polar regions to acquire new sonorous measurements. Understanding nineteenth-century science and sound involves the study of a wide range of actors, audiences, and locations: to achieve this, historians must evaluate historical documents beyond the celebrated works of elite natural philosophers, scientists, and mathematicians, and consider sermons, philosophical treatises, architectural texts, expeditionary narratives, popular accounts of experiments, and non-specialist journals and newspapers.

Although the contents of each individual volume appear chronologically, *Science and Sound* is organized thematically. Volume 1, "Sounds Experimental and Entertaining", explores the experimental culture surrounding sound between the 1810s and 1840s, with a strong emphasis on London. Following Wheatstone's arrival in the West End at his uncle's Pall Mall workshop in 1816 and his promotion of Chladni's acoustic figures, the capital became a centre of sonorous inquiry. Wheatstone extended Chladni's experimental programme and introduced his techniques for visualizing sound to British audiences. From 1819, he applied his acoustic knowledge to the cultivation of a series of devices that mobilized sonorous phenomena for public entertainment. Wheatstone's "Enchanted Lyre" appeared to play music of its own accord while suspended from a ceiling by wire.[33]

This sensational apparatus left audiences mystified and intrigued at Wheatstone's acoustic ingenuity. More than this, the transmission of sound through linear conductors seemed to promise that a utopian age of rapid, long-distance communication was at hand: if sounds could be conveyed between far-removed rooms and buildings then, Wheatstone reasoned, why not between towns and cities, and across empires? With Cambridge mathematician Robert Willis (1800–1875) developing increasingly sophisticated speaking machines, building on the human-like automaton of the Hungarian inventor Wolfgang von Kempelen (1734–1804), this communications revolution coincided with a more general blurring of the boundaries between human and mechanical speech. Willis's mechanical works were of great interest to Wheatstone who was ever in search of new acoustic principles. Wheatstone combined Willis's work on the production of vowel sounds through reed-organ pipes with his own examination of instruments that sounded through the reciprocal vibration of air columns. In this, he took full advantage of the return of musical instruments from far beyond Europe, notably the Javanese Gender, which used resonating columns of air to augment its vibrating metallic plates.

In 1820s' London, Wheatstone brought together musical, colonial, mechanical, and philosophical resources in the fashioning of the very latest knowledge of sound. What becomes apparent through the publication of Wheatstone's acoustic work, as well as reviews of his public exhibitions, is the lack of boundary between science as knowledge production and science as popular culture. Take, for instance, Wheatstone's "Kaleidophone". As with his "Enchanted Lyre" and work on resonance, accounts of this invention appeared in both the technical and non-specialist press. This question of place of publication is significant to our broader understanding of what science actually was in Regency Britain: Wheatstone's acoustic experiments and apparatus featured as easily in elite scientific journals, such as in the *Philosophical Transactions of the Royal Society*, as they did in more general periodicals like *The Literary Gazette* and *Dublin Literary Gazette*. Nowhere was science's, and specifically acoustic science's, flexible character more obvious than at the Royal Institution. Sometime in the mid-1820s, Wheatstone became acquainted with Faraday and would draw the celebrated experimentalist's interests towards questions of sound. Between 1828 and 1832, Faraday and Wheatstone orchestrated a series of Royal Institution lectures illustrating sonorous phenomena, including the exhibition of Wheatstone's most famous instruments. In this way, marvels like the Enchanted Lyre and Kaleidophone transitioned from the public showrooms of London's West End to one of the city's most eminent bastions of metropolitan science. Sound's ability to permeate popular and elite scientific demonstrations was evident from Faraday and Wheatstone's use of musicians in their Royal Institution lectures. During the 1820s and 1830s, German musician Karl Eulenstein (1802–1890) won fame for his performances on the Jew's Harp, producing a wide range of tones through the manipulations of his oral cavity and the use of wax on this basic instrument. In 1828, Wheatstone included Eulenstein as a guest at the Royal Institution to illustrate the phenomena of resonance before drawing out implications concerning vibrating columns of air from this performance.[34] The mobilization of musical resources was not trivial but demonstrative of the value of sound in London's scientific culture: audiences without mathematical or scientific training could, by listening, experience and come to know natural phenomena otherwise unobtainable. The principles of resonance and of mathematical ratios between musical notes were rendered accessible through sonorous performances.

As Volume 1 shows, Faraday and Wheatstone's acoustic work was part of a wider effort to use sound to examine nature beyond purely sonorous questions. While French natural philosopher Félix Savart (1791–1841) applied similar techniques for measuring vibrations through plates to the examination of crystalline bodies and the history of their formation, experimentalists like Arthur Trevelyan (1802–1878) and J. P. Marrion endeavoured to measure the movements of heat and electromagnetism with mixed results. Little wonder, then, that when Scottish natural philosopher David Brewster (1781–1868) reviewed the state of scientific knowledge in the British Isles in 1832, he drew specific attention to the very latest advancements in acoustic knowledge. It was, specifically, the cultivation of new sonorous wonders and effects that had, Brewster claimed, the power to arouse superstition and wonder in the human mind. As he explained, among "the discoveries of modern science there are few more remarkable than those which relate to the production of harmonic sounds. We are all familiar with the effects of musical instruments. . . . We sit entranced under their magical influence".[35]

What becomes apparent throughout Volume 1 is the extent to which London, during the 1820s and 1830s, was home to an unusually intense and sustained culture of sonorous inquiry. This was, in fact, a brief moment but, as becomes evident in Volumes 2, 3, and 4, one that was significant. If we take the Royal Institution's fashionable Friday Night Lecture series as an index, we can see that London's acoustic moment peaked between 1828 and 1832, with the delivery of some thirteen papers addressing sound, followed by Wheatstone's extensive eight-lecture series at King's College London in 1835.[36] After this, it would not be until John Tyndall drew attention back to sound in the 1850s and 1860s that the subject returned to prominence with London audiences. Part of this eclipse was undoubtedly due to Wheatstone's growing interest in electric telegraphy, as an alternative to linear sound conductors as the most promising means of realizing rapid communication, as well as Faraday's prioritizing of electrical research following his observation of electromagnetic induction in 1831. Nevertheless, as Volume 2, "Philosophies and Epistemologies of Sound", shows, actors beyond London took the study of sound in new directions, although they often conscripted and extended Wheatstone and Chladni's claims in their accounts of sonorous phenomena. The most influential of these was undoubtedly John Herschel's article "Sound", published in 1830 in the *Encyclopaedia Metropolitana.* He followed this thorough mathematical treatise with an examination of the epistemological value of sound in his *A Preliminary Discourse on the Study of Natural Philosophy* which sought to establish an empirical framework for the production of true knowledge of nature. Yet while Herschel directed his writing towards mathematical and specialist audiences, Mary Somerville's *Mechanism of the Heavens* (1831) and *On the Connexion of the Physical Sciences* (1834) brought acoustic science to a far more general readership. Following David Brewster's review of the latter of these works, in which he encouraged Somerville to include Wheatstone's latest acoustic figure experiments, her second edition of *On the Connexion of the Physical Sciences* (1835) included a greatly expanded account of sound, complete with illustrations taken from Wheatstone's *Philosophical Transactions of the Royal Society* paper of 1833.

Both Herschel and Somerville were using sound in quite a specific way: they were not just providing accounts of how sound worked but using its behaviour to draw parallels with other invisible natural phenomena, notably those of heat and light. Sound, in this manner, provided a means of examining nature in a much broader fashion, and this had theological implications. To be able to use the laws of sound to draw out similar governing principles for other phenomena suggested evidence of an ordered, contrived universe which, in turn, suggested the existence of a divine architect, much in the style of William Paley's (1743–1805) classic design argument. Polymath and, from 1841, Trinity College Cambridge's domineering Master, William Whewell (1794–1866) shared these sentiments, employing sound, specifically the gifts of speech and hearing, as evidence of God within a distinctly Anglican framework. His *Astronomy and General Physics*, one of the eight Bridgewater Treatises commissioned between 1833 and 1836 to deliver evidence of God's power and wisdom through studies of nature, united terrestrial phenomena, such as those sonorous, with celestial motions and the astronomical laws of gravity. In his influential *History of the Inductive Sciences* (1837) and subsequent *Philosophy of the Inductive Sciences* (1840), Whewell would unite sound's moral value with what he perceived to be its epistemological worth. For Whewell, hearing was an invaluable way of knowing nature, and the human ability to detect mathematical principles by the ear, such as variations in tone and fractions of notes in terms of vibrations, further enhanced sound's theological, as well as scientific, worth.

Through Whewell's, Somerville's, and Herschel's works, we get a real sense of the broad consensus that existed in mid-nineteenth-century Britain that maintained acoustic science to be of both epistemological and theological significance. A look at some of the sermons and musical tracts of the period adds weight to this assertion: to popular, non-specialist audiences, religious and musical guides alike applied scientific reasoning to their understanding of sound and, specifically, music while promoting its divinity and metaphysical quality. However, as Volume 2 demonstrates, by the 1850s, such visions of nature were becoming increasingly strained. Although Robert Chambers's (1802–1871) anonymously published *Vestiges of the Natural History of Creation* had raised troubling questions over the universe's evolution and the Earth's deep history as early as 1840, it was Charles Darwin's (1809–1882) *On the Origin of Species* in 1859 and subsequent *The Descent of Man, and Selection in Relation to Sex*, published in 1871, that would provoke immense alarm with Britain's religious audiences, raising profound anxieties over Man's place in nature, as well as God's role in Creation. In the latter of these works, Darwin would pay specific attention to the subject of sound, examining music as an evolutionary function of sexual selection across a range of different species. It was the English philosopher Herbert Spencer (1820–1903), however, who would do the most to unite matters sonorous and musical within an account of nature stressing evolutionary progress. For Spencer, the development of music fit within his vision of nature and society as governed by the laws of progression from the simple to the complex. Spencer taught that language and music were both consistent with this idea of progress. First in his influential essay "Progress: its law and cause", published in 1857, and then his more focused "The origin and function of music", Spencer explained how music had evolved over time, from speech to the simple sounds of ancient societies and tribes, to what he contended to be the more sophisticated works and practices of modern European culture. As late as 1891, Spencer still engaged in discourse surrounding the development of music, in terms of melody and harmony, and its place in nature.

So what we see throughout the nineteenth century, is an unravelling of sound and music as sources of evidence for divine creation and theological assurance. Instead, through the works of Darwin and Spencer, acoustic science became a deeply problematic source of controversy in a way that was, to a considerable extent, consistent with the growth of religious uncertainty throughout the natural sciences. Volume 3, "Sound in Context", shifts away from the examination of sound's philosophical and epistemological value and focuses on the shifting contexts of acoustic science. Throughout the nineteenth century, a diverse range of technical challenges, cultural encounters, and musical controversies shaped the cultivation of new knowledge and understanding of sound. Given that this was a moment of rapid colonial expansion, it is hardly surprising that the expanding geographical remit of Europe's imperial powers provided opportunities for sonorous examinations in unfamiliar environments. With sound's velocity contingent on its atmosphere, Britain's overseas territories offered a wide range of climatic conditions in which to experiment, from the heat of India to the extreme cold of the polar regions. The communication of sound through differing media was an important phenomenon to measure because of the role of sonorous time signals in oceanic navigation. To calculate a ship's longitude at sea required an accurate chronometer giving the time of a known location, from which a relative position could be calculated using the sun. To obtain this time before setting sail relied on visual communication via time balls and clocks, as well as the sound of time guns, but the latter of these travelled slowly and was relative to atmospheric conditions. From Madras to Barbados, Edinburgh to Cape Town, Britain's maritime empire relied on networks of timekeeping that initiated a series of sonorous measurements around the globe. While Britain possessed no colonial entrepôts in the polar regions, the series of post-1818 expeditions in search of a North-West Passage through the Arctic, uniting the Atlantic and Pacific Oceans, proved a rich source of scientific activity. Naval officers like William Edward Parry (1790–1855), Henry

Foster (1797–1831), and James Clark Ross (1800–1862) frequently performed experiments, comparing the communication of sound through a polar atmosphere with that of light. With time, such techniques became a valuable means of measuring sudden drops in temperature. These understandings had philosophical value but, as with timekeeping practices, also held practical worth for navigation, with sound signalling a subject of interest for both Faraday and Tyndall, as scientific advisers to Trinity House, responsible for the nation's lighthouses. At moments of poor visibility, reliable sound signalling was essential for safe navigation, but this depended on understanding how sonorous phenomena acted in varying atmospheric conditions.

Imperial expansion, polar exploration, and global oceanic navigation provided rich contexts for sonorous knowledge formation but, arguably, understanding sound was of even greater urgency for those in the business of architecture. Architecture has had a long connection with the development of acoustic knowledge, with Europe's great medieval cathedrals and churches representing some of the largest enclosed spaces in which sound could be experienced. Likewise, attention to sound's behaviour has been of concern to theatre builders since at least the time of Roman engineer Virtuvius's (ca. 80–70 BCE– ca. 15 BCE) *De Architectura*, constituting the earliest known treatise to address architectural acoustics. In nineteenth-century Britain, however, this challenge of managing sonorous communication in a built environment took on unprecedented urgency. During the late-eighteenth century and 1800s, British architects had been concerned primarily with securing good hearing in theatres, with Benjamin Wyatt's (1775–1852) new Drury Lane Theatre, opened in 1812, being an especially celebrated example. As historian Viktoria Tkaczyk has shown, this was very much part of a wider European expansion in theatre culture, in which theatre-building was prominent in acoustic knowledge formation.[37] Yet, by the 1820s and 1830s, British architectural acoustics had become quite distinctive, and this was largely due to two construction projects of national extent. The first of these was the ambitious church-building programme that Parliament initiated with the 1818 Church Building Act, providing funding for new places of Anglican worship in Britain's rapidly expanding industrial towns and cities. With neoclassical stone porticoes being expensive, the favoured style for these constructions quickly became Gothic, with a large amount of more economical brickwork. Yet this presented an unusual acoustic challenge: Gothic was highly appropriate for Roman Catholic worship, largely consisting of Latin chant and mass, with the style's division of space through vaulting, decoration, transepts, and columns often proving of great value in terms of resonance. However, Gothic was less well suited to worship in a Protestant tradition, such as in the Church of England, in which the clear spoken word of the sermon, in a local vernacular, was of crucial importance for spiritual and moral guidance. Since the Reformation, Protestant church builders had favoured neoclassical constructions, not just stylistically but also due to the perception that smooth ceilings and walls communicated spoken sounds efficiently.[38] Certainly this was the opinion of Christopher Wren (1632–1723), England's great seventeenth-century church builder and architect of St Paul's Cathedral.[39] For nineteenth-century church builders, however, uniting elaborate Gothic interiors with effective spoken communication was exceedingly difficult, stimulating considerable acoustic experiment, such as Rev John Blackburn's development of sounding boards for church pulpits, including his own in Attercliffe parish church: experiments that even the Royal Society's *Philosophical Transactions* thought of suitable significance for publication in 1828. In his *The Acoustics of Public Buildings*, which in 1861 represented the most extensive acoustic treatise to date, architect Thomas Roger Smith (1830–1903) drew heaving on some forty years of industrious church construction to provide advice for future architectural endeavours.

Along with British church building, the second architectural project of significance was the reconstruction of the Palace of Westminster, home to the nation's Houses of Parliament.[40] Since Wren's reworking of the House of Commons debating chamber during the 1690s, this room had been subject to continual acoustic scrutiny; it was, after all, essential for MPs to be able to hear and be heard in the physical centre of the nation's political discussion. Historically, few places have proven so difficult in which to secure satisfactory acoustics and it is no exaggeration to say that, throughout the nineteenth century, no other building was subject to such intensive sonorous investigation as the Commons' chamber. With the medieval building destroyed in a devastating fire in 1834, architect Robert Smirke (1780–1867) erected a temporary chamber, ready by 1835 and, from 1836, Charles Barry (1795–1860) oversaw the design and construction of a permanent new House of Commons building, which came into use between 1850 and 1851. In both of these structures, acoustics took on immense importance, with scientific inquiry accompanying almost constant controversy. Politicians and civil servants alike saw the reconstruction project as a great opportunity to ensure a new debating chamber, as well as a replacement House of Lords building, with satisfactory acoustics. To this end, a select committee of 1835, charged with investigating parliamentary accommodation, appointed Edinburgh chemist David Boswell Reid (1805–1863), to enhance the atmosphere and acoustics of the Commons' temporary chamber. Reid's conviction was that improved ventilation and air purity would advance the communication of speech in any room, and to this end, he undertook some twenty years of chemical and acoustic experiments at Westminster, first in the temporary accommodation before then endeavouring to apply this work to Parliament's permanent chambers.[41] In 1844, Reid published an account of these inquires, complete with a substantial chapter on architectural acoustics. When Thomas Roger Smith reviewed the state of acoustic knowledge in 1861, Reid's work was one of the few sources he cited as worthy of attention, along with Wyatt's account of his Drury Lane Theatre, and naval architect John Scott Russell's (1808–1882) writings on theatre and lecture-room seating.

Scientific practitioners specializing in sound were increasingly prominent in architecture and construction, but they also endeavoured to engage with musical audiences. Yet such encounters often proved revealing of the limits of empirical science as an authoritative framework for knowledge production. This became especially apparent throughout the debates surrounding the regulation of musical pitch during the 1850s and 1860s and which would persist well into the twentieth century.[42] The standardization and the formation of agreed units of measurement have often been associated with the development of modern scientific culture, especially with the rise of research laboratories as accepted venues of scientific activity that accompanied the expansion of Europe's electrical industries and telegraphic communications networks.[43] However, musical pitch proved a difficult measure to unify: the frequency at which a musical A or C should be set was an aesthetic problem as much as it was a mathematical challenge. Across Europe, there was a consensus that the pitch at which orchestras tuned their instruments was escalating in the search for "brilliancy", threatening the character of historical compositions and the longevity of vocalists' voices. Concurrently, different musical institutions employed different frequencies, ensuring a lack of uniformity throughout the world of musical practice. Following French emperor Napoléon III's appointment of a *diapason normal* in 1859, setting a standard A above middle C to 435 full vibrations, London's Society of Arts organized its own committee to establish a British standard. However, within this committee, there was a huge discrepancy between those favouring a mathematical standard of C above middle C set to 512 vibrations per second, based on the square root of a string measured to vibrate once a second, and those preferring a higher frequency based on the perceived aesthetic superiority of elevated pitches: a higher-pitched C was generally regarded as giving a more pleasing sound to duller, lower frequencies. In the end, in 1860, the Society of Arts recommended a C at 528, much to the consternation of John Herschel, who had campaigned extensively for a C512 on mathematical grounds.[44] Nevertheless, even the Society of Arts' proposed standard failed to gain weight with Britain's musical communities, with the lack of pitch uniformity a source of concern throughout the 1860s and 1870s. Following his 1880 *The History of Musical Pitch*, the English mathematician and phonetician Alexander Ellis campaigned for pitch reform but met a similar fate to the Society of Arts' earlier scheme of 1859–60, with musical standardization proving divisive until the mid-twentieth century. Scientific authorities were often opposed to musical practitioners in matters beyond pitch. Herschel's efforts to standardize C accompanied a wider ambition to reform harmonic scales by dividing an octave into 1,000 equal intervals, while musician and engineer William Pole (1814–1900) proposed to reorganize the practices of organ tuners, applying musical rules in place of the traditional method by ear alone. As with pitch, such utopian visions failed to impress musicians.

Volume 3's focus on the varying contexts of science and sound, be they colonial, architectural, or musical, demonstrates how, in nineteenth-century Britain, this relationship was both immensely productive, while simultaneously illustrative of the limits of scientific knowledge and authority in practical matters. For all these shortcomings, Volume 4, "Sound Transformed", details the ascent of new scientific ways of thinking about sound and music that emerged in the second half of the nineteenth century. This was a moment when mathematicians and physicists subjected sound to increasingly mathematical, psychological, physiological, and technological examination. German physicist Hermann Helmholtz's publication of *Die Lehre von den Tonempfindungen als physiologische Grundlage für die Theorie der Musik* in 1863 established that while dissonance was the result of a rapid succession of beats on the ear, a vibration's pleasing sonorous qualities were due to overtones: a pure sound would be dull and flat so to be euphonically satisfying required combinations of overtones and fundamental tones. This was a physical explanation for distinguishing between pleasant and unpleasant sounds. Helmholtz's claims mobilized the latest technological devices, notably his "Helmholtz resonator", for producing sonorous knowledge. Alexander Ellis's 1875 translation of this influential work, published as *On the Sensations of Tone as a Physiological Basis for the Theory of Music*, brought Helmholtz's theories to English-speaking audiences, but the impact of this work, in fact, preceded this. John Tyndall's *Sound: a course of eight lectures delivered at the Royal Institution of Great Britain*, published in 1867, had very much been inspired by Helmholtz's treatise, while works like George Biddell Airy's (1801–1892) *On Sound and Atmospheric Vibrations, with the mathematical elements of music* (1868) represented a broad renewed interest in mathematical treatments of sonorous phenomena, not been witnessed since the 1820s and 1830s. The Cambridge mathematician John Strutt, Baron Rayleigh from 1873, published his *The Theory of Sound* in 1877, representing the most authoritative, thorough mathematical account of sound to be published in English during the nineteenth century. Rayleigh's treatise would remain central to university mathematical training well into the twentieth century. And with English psychologist Edmund Gurney's (1847–1888) *The Power of Sound*, first published in 1880, moving beyond material examinations of musical phenomena, to subject music itself to scientific scrutiny in terms of its agency over human emotions and sentiment, Helmholtz's physical accounts of sound and music were now firmly accompanied by a range of physiological, mathematical, and psychological understandings. Though often subtle, this marked a huge shift from the mid-century consensus, implicit and explicit in works such as Herschel's, Somerville's, and Whewell's, that sound was evidence of design and music of a transcendent, metaphysical character. For religious audiences, this was a moment of concern, exasperated by the growing cultural capital of evolutionary interpretations of sound and music. All of this was accompanied by the proliferation of a host of radical new electroacoustic technologies, including microphones, telephones, radios, and recording devices, that seemed to materialize, demystify, and dehumanize the production and communication of sound.

With works such as those of Helmholtz, Rayleigh, and Gurney, as well as the evolutionary texts of Spencer and Darwin, it becomes clear that sound was, in 1900, a very different scientific entity to what it had been in 1800. Throughout the four volumes of *Science and Sound*, we are dealing with a transformation of the relationship between science and sound. Most obviously, we see the growing conception that sound, as an object of empirical, scientific examination, as well as music and the existence of euphonically pleasing sounds, fit within the domain of scientific explanation. Yet we also see the reorganization of sound's place in nature, from a phenomenon within a divinely ordered vision of the universe to a product of evolutionary biology: from something sacred to a matter of mathematical, physiological, and psychological reduction. Sound was a subject of scientific study as well as being of epistemological worth, but it was also of shifting philosophical and metaphysical value, which transcended the boundaries between the material and the spiritual.

NOTES

1 On ancient harmony, see Andrew Barker, *The Science of Harmonics in Classical Greece*, (Cambridge: Cambridge University Press, 2007); David Creese, *The Monochord in Ancient Greece Harmonic Science* (Cambridge: Cambridge University Press, 2010).

2 Penelope M. Gouk, "Acoustics in the early Royal Society, 1660–1689," *Notes and Records of the Royal Society* 36, no. 2 (1982), pp. 155–75, 156.

3 Leendert van der Miesen, "Studying the Echo in the Early Modern Period: between the academy and the natural world," in V. Tkaczyk and L. van der Miesen, ed., "Sonic things: knowledge formation in flux," *Sound Studies* 6, no 2 (2020), pp. 196–214.

4 Francis Bacon, *Sylva Sylvarum: or a naturall historie* (London: William Lee, 1635), p. 67.

5 Marin Mersenne, *Harmonie universelle contenant la théorie et la pratique de la musique* (1636), facsimile, ed. F. Lesure (Paris: Éditions du Centre National de la Recherche Scientifique, 1963), p. 25.

6 Gouk, "Acoustics in the early Royal Society", p. 160.

7 Peter Pesic, *Music and the Making of Modern Science* (Cambridge, MA: MIT Press, 2014), pp. 9–20, 121–31, and 151–60; on early modern acoustics see, Penelope Gouk, *Music, Science and Natural Magic in Seventeenth-Century England* (New Haven, CT: Yale University Press, 1999), pp. 224–57. For the preceding paragraph on early modern acoustics, I am indebted to Fanny Gribenski's insights.

8 On Chladni, see Myles W. Jackson, *Harmonious Triads: physicists, musicians, and instrument makers in nineteenth-century Germany* (Cambridge, MA: MIT Press, 2006), pp. 13–44; on Chladni and his audiences, see Viktoria Tkaczyk, "The Making of Acoustics around 1800, or How to Do Science with Words," in Mart Helen Dupree and Sean B. Franzen (eds.), *Performing Knowledge, 1750–1850* (Berlin and Boston: De Gruyter, 2015), pp. 27–56.

9 On nineteenth-century soundscapes, see John M. Picker, *Victorian Soundscapes* (Oxford: Oxford University Press, 2003), pp. 4–6; on the encounter of new musical instruments through colonial expansion, see James Q. Davies, "Instruments of Empire", in James Q. Davies and Ellen Lockhart (eds.), *Sound Knowledge: music and science in London, 1789–1851* (Chicago: University of Chicago Press, 2016), pp. 145–74.

10 On architectural acoustics, see Graeme Gooday, "Architectural Acoustics: Thomas Roger Smith and the science of hearing buildings in nineteenth-century Britain", in Edward Gillin and H. Horatio Joyce (eds.), *Experiencing Architecture in the Nineteenth Century: buildings and society in the modern age* (London: Bloomsbury, 2019), pp. 101–14; Viktoria Tkaczyk, "Listening in circles. Spoken drama and the architects of sound, 1750–1830", *Annals of Science*, 71, no. 3 (2014), pp. 299–334.

11 On public speech, including political, in Victorian culture, see Joseph S. Miesel, *Public Speech and the Culture of Public Life in the Age of Gladstone* (New York: Columbia University Press, 2001); on the importance of analysing past auditory landscapes for understanding the historical exertion of political power, see Sophia Rosenfeld, "On

being heard: a case for paying attention to the historical ear", *American Historical Review*, 116, no. 2 (April 2011), pp. 316–34, 318–19; on speech in parliamentary culture, see Ludovic Marionneau and Josephine Hoegaerts, "Throwing one's voice and speaking for others: performative vocality and transcription in the assemblées of the long nineteenth century", *Journal of Interdisciplinary Voice Studies* 6, no. 1 (2021), pp. 91–108.

12 James Q. Davies and Ellen Lockhart, "Fantasies of Total Description", in James Q. Davies and Ellen Lockhart (eds.), *Sound Knowledge: music and science in London, 1789–1851*, (Chicago: University of Chicago Press, 2016), pp. 1–26; also see Roger Parker and Susan Rutherford, "Introduction. London Voices, 1820–1840: A 'Luminous Guide'", in Roger Parker and Susan Rutherford (eds.), *London Voices, 1820–1840: vocal performers, practices, histories* (Chicago: University of Chicago Press, 2019), pp. 1–14.

13 Jonathan Sterne, *The Audible Past: cultural origins of sound reproduction* (Durham, NC: Duke University Press, 2003), p. 3.

14 As Shapin put it in his much-quoted opening, "There was no such thing as the Scientific Revolution, and this is a book about it." See Steven Shapin, *The Scientific Revolution*, (Chicago: University of Chicago Press, 1996), p. 1.

15 "Letter 557: Faraday to John George Children and sealed note, 12 Mar., 1832", in Frank A. J. L. James (ed.), *The Correspondence of Michael Faraday, Vol. 2: 1832–December 1840. Letters 525–1333* (London: Institution of Electrical Engineers, 1993), pp. 26–27.

16 See Edward J. Gillin, *Sound Authorities: scientific and musical knowledge in nineteenth-century Britain* (Chicago: University of Chicago Press, 2022), pp. 83–120.

17 Myles W. Jackson, *Harmonious Triads: physicists, musicians, and instrument makers in nineteenth-century Germany* (Cambridge, MA: MIT Press, 2006), pp. 75–78; also see Jamie James, *The Music of the Spheres: music, science, and the natural order of the universe* (London: Little, Brown and Company, 1993), pp. 3–4.

18 G. W. F. Hegel, *Hegel's Aesthetics: lectures on fine art* T. N. Knox (Trans.), 2 vols. (Oxford: Clarendon Press, 1988), p. 891; E. T. A. Hoffmann in, David Charlton (ed.), *E. T. A. Hoffmann's Musical Writings: Kreisleriana, the poet and the composer, music criticism* (Cambridge: Cambridge University Press, 1989), pp. 237 and 239. Thanks to David Trippett for discussions on German Romantic philosophy and materialism.

19 Eduard Hanslick, *On the Musically Beautiful: a contribution towards the revision of the aesthetics of music*, Geoffrey Payzant (Trans.), (Indianapolis, IN: Hackett Publishing Company, 1986), pp. 1–2; David Trippett, *Wagner's Melodies: aesthetics and materialism in German musical identity* (Cambridge: Cambridge University Press, 2013), p. 356.

20 On evolutionary thinking and music, see Bennett Zon, "The "Non-Darwinian" Revolution and the Great Chain of Musical Being", in Bernard Lightman and Bennett Zon (eds.), *Evolution and Victorian Culture* (Cambridge: Cambridge University Press, 2014), pp. 196–226; Bennett Zon, '"Spiritual" Selection: Joseph Goddard and the Music Theology of Evolution', in Martin V. Clarke (ed.), *Music and Theology in Nineteenth-Century Britain* (Farnham: Ashgate, 2012), pp. 215–35.

21 On Helmholtz and sound, see Michel Meulders, *Helmholtz: from enlightenment to neuroscience* (Cambridge, MA: MIT Press, 2010), pp. 153–78 and 179–99; Benjamin Steege, *Helmholtz and the modern listener* (Cambridge: Cambridge University Press, 2012); Timothy Lenoir, "Helmholtz and the materialities of communication", *Osiris* 9 (1994), pp. 184–207; Robert T. Beyer, *Sounds of Our Times: two hundred years of acoustics* (New York: Springer-Verlag, 1999), pp. 55–82.

22 Alexandra Hui, *The Psychophysical Ear: musical experiments, experimental sounds, 1840–1910* (Cambridge, MA: MIT Press, 2013), pp. xiii–xvii.

23 (Anon.), "Recent Developments in Acoustical Science", 19th Aug. 1878, *The Times*, (London, England), Issue 29337, p. 4.

24 Roland Wittje, *The Age of Electroacoustics: transforming science and sound* (Cambridge, MA: MIT Press, 2016), p. 4.
25 Emily Thompson, *The Soundscape of Modernity: architectural acoustics and the culture of listening in America, 1900–1933* (Cambridge, MA: MIT Press, 2002), pp. 18–19; for comparable studies of historical change and soundscapes, see James G. Mansell, *The Age of Noise in Britain: hearing modernity* (Urbana: University of Illinois Press, 2017); Alain Corbin, *Village Bells: sound and meaning in the 19th-century French countryside*, (Tran.) Martin Thom (Basingstoke: Macmillan Publishers Ltd, 1999).
26 Elizabeth Green Musselman, *Nervous Conditions: science and the body politic in early industrial Britain* (New York: State University of New York Press, 2006), pp. 17–19.
27 Bernard Lightman, *Victorian Popularizers of Science: designing nature for new audiences*, (Chicago: University of Chicago Press, 2007); Iwan Morus, "Seeing and believing science", *Isis* 97, no. 1, (Mar. 2006), pp. 101–10; Iwan Morus, "Worlds of wonder: sensation and the Victorian scientific performance", *Isis* 101, no. 4 (Dec. 2010), pp. 806–16; Iwan Morus, Simon Schaffer, and Jim Secord, "Scientific London", in Celina Fox (ed.), *London – World City, 1800–1840* (New Haven, CT: Yale University Press, 1992), pp. 129–42.
28 Mark M. Smith, *Sensing the Past: seeing, hearing, smelling, tasting, and touching in history*, (Berkeley: University of California Press, 2007), esp. pp. 41–58; for example, see Jutta Schickore, *The Microscope and the Eye: a history of reflections, 1740–1870* (Chicago: University of Chicago Press, 2007).
29 Myles W. Jackson, *Harmonious Triads: physicists, musicians, and instrument makers in nineteenth-century Germany* (Cambridge, MA: MIT Press, 2006); Peter Pesic, *Music and the Making of Modern Science* (Cambridge, MA: MIT Press, 2014).
30 Alexandra Hui, Julia Kursell, and Myles W. Jackson, "Music, sound, and the laboratory from 1750 to 1980", *Osiris*, 28, no. 1 (Jan. 2013), pp. 1–11, 5; on sound and laboratories, see Henning Schmidgen, "Silence in the Laboratory: the history of soundproof rooms", in Julia Kursell (ed.), *Sounds of Science – Schall im Labor (1800–1930)* (Berlin: Max Planck Institute for the History of Science, 2008), pp. 47–61; Julia Kursell, "'A Gray Box': the phonograph in laboratory experiments and fieldwork, 1900–1920", in Trevor Pinch and Karin Bijsterveld (eds.), *The Oxford Handbook of Sound Studies* (Oxford: Oxford University Press, 2012), pp. 176–97.
31 Thomas L. Hankins and Robert J. Silverman, *Instruments and the Imagination* (Princeton, NJ: Princeton University Press, 1995), pp. 86–108; John Tresch and Emily I. Dolan, "Towards a new organology: instruments of music and science", *Osiris*, 28, no. 1 (Jan. 2013), pp. 278–98; on musical and scientific instruments, see Emily I. Dolan, *The Orchestral Revolution: Haydn and the technologies of timbre* (Cambridge: Cambridge University Press, 2013), pp. 11–3; on instruments in nineteenth-century Paris, see David Pantalony, *Altered Sensations: Rudolph Koenig's acoustical workshop in nineteenth-century Paris* (New York: Springer, 2009).
32 James Q. Davies and Ellen Lockhart, "Fantasies of Total Description", in James Q. Davies and Ellen Lockhart (eds.), *Sound Knowledge: music and science in London, 1789–1851*, (Chicago: University of Chicago Press, 2016), pp. 1–26, 3.
33 Melissa Dickson, "Charles Wheatstone's Enchanted Lyre and the Spectacle of Sound", in James Q. Davies and Ellen Lockhart (eds.), *Sound Knowledge: music and science in London, 1789–1851* (Chicago: University of Chicago Press, 2016), pp. 125–44.
34 (Anon.), "Proceedings of the Royal Institution: February 15th", *The Quarterly Journal of Science, Literature and Art* 25 (London, 1828), pp. 173–74.
35 David Brewster, *Letters on Natural Magic addressed to Sir Walter Scott*, Bart, (New York: Harper & Brothers, 1842), p. 167.
36 King's College London Archives (KCL), Wheatstone 5, untitled note in "Various: 8 lectures on sound" (ca. 1835).

37 Viktoria Tkaczyk, "Listening in circles. Spoken drama and the architects of sound, 1750–1830", *Annals of Science* 71, no. 3 (2014), pp. 299–334.
38 Edward J. Gillin and Fanny Gribenski, "Science in the church: the sacred spaces of sonorous experiment and the formation of modern acoustics", *Isis*, 114, no. 3 (Sep., 2023), pp. 537–558; also see Anna Kvíčalová, *Listening and Knowledge in Reformation Europe: hearing, speaking and remembering in Calvin's Geneva* (Cham: Palgrave Macmillan, 2019).
39 Stephen Wren, *Parentalia: or, memoirs of the family of the Wrens; Viz. of Mathew Bishop of Ely, Christopher Dean of Windsor, &c. but chiefly of Sir Christopher Wren, late surveyor-general of the Royal buildings, president of the Royal Society, &c. &c.* (London: T. Osborn, 1750), p. 320.
40 Edward J. Gillin, *The Victorian Palace of Science: scientific knowledge and the building of the Houses of Parliament* (Cambridge: Cambridge University Press, 2017), pp. 30, 56–59, and 140–43.
41 On Reid at Parliament, see Edward J. Gillin, *The Victorian Palace of Science: scientific knowledge and the building of the Houses of Parliament* (Cambridge: Cambridge University Press, 2017), pp. 121–83.
42 For the history of pitch standardization, see Fanny Gribenski, *Tuning the World: the rise of 440 hertz in music, science, and politics, 1859–1955* (Chicago: University of Chicago Press, 2023).
43 Bruce J. Hunt (1997), "Doing Science in a Global Empire: cable telegraphy and electrical physics in Victorian Britain", in Bernard Lightman (ed.), *Victorian Science in Context* (Chicago: University of Chicago Press), pp. 312–33.
44 Edward J. Gillin and Fanny Gribenski, "The politics of musical standardization in nineteenth-century France and Britain", *Past & Present*, no. 251 (May 2021), pp. 153–87.

Editorial Headnote

1. Charles Wheatstone, "New Experiments on Sound", *Thomson's Annals of Philosophy* vi (London, England, 1823), pp. 81–90.

During the 1820s, instrument maker and natural philosopher Charles Wheatstone (1802–1875) transformed the place of sound in London's scientific culture. Originally from Gloucester, Wheatstone went to live in his father's premises, a musical instrument shop at 128 Pall Mall, right in the midst of London's fashionable West End. Living there from 1806 onward, Wheatstone became well acquainted with the principles of musical instrument design and construction. In 1816, his uncle employed Wheatstone as an apprentice at his nearby workshop at 436 The Strand, continuing the young philosopher's education in the business of musical instrument manufacturing.[1] In this employment, Wheatstone amazed polite London society with his mechanical marvels, aimed "to amuse and puzzle the gaping groups that gathered at the shop window".[2] During the late 1810s, Wheatstone purchased a copy of Italian chemist Alessandro Volta's (1745–1827) work on electrical experiments and subsequently began his own scientific investigations, replicating and extending German natural philosopher Ernst Chladni's (1756–1827) acoustic plate experiments. While Chladni had observed the movement of grains of sand on a vibrating plate, Wheatstone trialled this arrangement with fine powders and, later on, water, revealing the wave-like motions of sonorous impulses. These experiments would constitute the basis for his first scientific contribution, "New Experiments on Sound", which Wheatstone published in *Thomson's Annals of Philosophy* in 1823. This marked the earliest significant study of the phenomena that Chladni had first written about back in 1787, to be published in the British Isles, bringing the principles of acoustic figures to the attention of English-speaking readers for the first time. Wheatstone had first brought this experimental arrangement to London audiences in 1819, delivering acoustic figure demonstrations around the West End. His ensuing article divided sound into transversal and longitudinal vibrations, with those being, respectively, sonorous transmissions at right angles to the axis of a solid body and those moving in the direction of the body's axis. Wheatstone explained how the locations of rest in a vibrating solid body produced "nodal lines", where sand or water accumulated, this being the point of least vibration. Wheatstone favoured the use of water over Chladni's original use of sand as he claimed that this medium produced greater accuracy in terms of revealing the patterns resulting from the mathematical principles of various vibrating frequencies. Wheatstone termed these fluid motions "crispations", promoting his experimental techniques as a powerful means of visualising the transmission of sound. Wheatstone later found out that the Danish chemist Hans Christian Ørsted (1777–1851) had already adopted similar practices, employing fluids rather than sand, in 1813. On discovering that his work was not as original as he had first thought, Wheatstone contacted Ørsted, sharing his experimental results. The results of Ørsted's acoustic figures had been previously published in Lieber's *History of natural philosophy* (1813), with Ørsted observing secondary nodal lines which he believed represented more complex sub-vibrations.

NOTES

1 Don Metz, "The scientific instruments of Charles Wheatstone and the blending of science, art, and culture", *Interchange* 46, (2015), pp. 19–29; Brian Bowers, *Sir Charles Wheatstone FRS, 1802–1875* (London: Institution of Electrical Engineers, 2001), pp. 5–7 and 9–10.

2 Quoted in, Joseph Stratford, *Gloucestershire Biographical Notes* (The "Journal" Office: Gloucester, 1887), p. 190.

1

NEW EXPERIMENTS ON SOUND

Charles Wheatstone

Source: Charles Wheatstone, "New Experiments on Sound", *Thomson's Annals of Philosophy* vi (London, England, 1823), pp. 81–90.

ON THE PHONIC MOLECULAR VIBRATIONS

BEFORE I enter on the immediate subject of this article, it may be necessary to exhibit a general view of those bodies which, being properly excited, make those sensible oscillations which have been thought to be the proximate causes of all the phenomena of sound. These bodies, to avoid many circumlocutions otherwise inevitable, I have termed Phonics.

Linear Phonics.

Transversal,	***Longitudinal,***
Making their oscillations at right angles to their axis. 1. Capable of tension, or variable rigidity: chords or wires. 2. Permanently rigid: rods, forks, rings, &c.	Making their oscillations in the direction of their axis. 1. Columns of aeriform fluids or liquids; cylindric and prismatic rods.

Superficial Phonics

1. Capable of tension: extended membranes.
2. Permanently rigid: laminæ, bells, vases, &c.

Solid Phonics

1. Volumes of aeriform fluids.

The sensation of sound can be excited by any of these bodies when they oscillate sufficiently rapidly, either entire or divided into any number of parts in equilibrium with each other. The laws of these subdivisions differ in the various phonics according to their form and mode of connexion or insulation; and the velocities of the oscillations, or degrees of tune, depend on the form, dimensions, mode of

DOI: 10.4324/b23406-1

connexion, mode of division, and elasticity of the body employed. The points of division in linear phonics are called nodes; and the boundaries of the vibrating parts of elastic surfaces are termed nodal lines. The parts at which the oscillatory portions have their greatest excursions are named centres of vibration; these are always at the greatest mean distances from the nodal points or lines.

These mechanical oscillations are not, however, themselves the immediate causes of sound; they are but the agents in producing in the bodies themselves, and in other contiguous substances, isochronous vibrations of certain particles varying in magnitude according to the degree of tune. I convinced myself of this important fact by the following simple experiments: – I took a plate of glass capable of vibrating in several different modes, and covered it with a layer of water; on causing it to vibrate by the action of a bow, a beautiful reticulated surface of vibrating particles commenced at the centres of the vibrating parts, and increased in dimensions as the excursions were made larger. When a more acute sound was produced, the centres consequently became more numerous, and the number of coexisting vibrating particles likewise increased; but their magnitudes proportionably diminished. The sounds of elastic laminæ are generally supposed to be owing to the entire oscillations of the simple parts, as shown by Chladni when, by strewing sand over the sonorous plates, he observed the particles repulsed by the vibrating parts accumulate on the nodal lines and indicate the bounds of the sensible oscillations. Did no other motions exist in the plate but these entire oscillations, the water laid on its surface would, on account of its cohesion to the glass, show no peculiar phenomena; but the appearances above described clearly demonstrate that the oscillating parts consist of a number of vibrating particles of equal magnitudes, the excursions of which are greatest at the centres of vibration, and gradually become less as they recede further from it, until they become almost null at the nodal lines.

To multiply these surfaces, and to observe whether the magnitudes of these particles vary in different media, in a glass vessel of a cylindric form I superposed three immiscible fluids of different densities – namely mercury, water, and oil. On producing the sounds corresponding with each mode of division, I observed a number of vibrating parts, agreeing with the sound and showing similar appearances to the plate, formed on the surfaces of each of the fluids; not the least agitation appeared in the uniform parts. I afterwards inserted this glass in another vessel of water in order to observe the vibrations of the external surface, and found the same results as in the interior, though the levels of the surfaces were different.

The most accurate method to observe these phenomena is by employing a metallic plate of small dimensions, which must be fixed horizontally in a vice at one end, and covered on its upper side with a surface of water. On causing it to oscillate entirely by means of a bow, a regular succession of these vibrating corpuscles will appear arranged parallel to the two directions of the plate; and if the action of the bow be rendered continuous, their absolute number might be counted with the aid of a micrometer. Diminishing the oscillating part of the plate to one half of its length, the double octave to the preceding was heard, agreeably to the established rule that the velocities of the oscillations are inversely as the squares of the lengths: four vibrating corpuscles then occupied the space before occupied by one; and the absolute number was double to that in the former instance; but the absolute number of these corpuscles have no influence whatever on the degree of tune, which entirely depends on their relative magnitude in the same substance. Theory shows us that in plates of this description alteration of breadth does not affect the degree of tune: let us therefore reduce this half of the plate to half its breadth, and we shall find the note remain the same; but the absolute number of the corpuscles will in this case be equal to that in the entire plate. Let us now take two plates of equal lengths and breadths, but one double in thickness to the other; the rule is, that the velocities of the oscillations are as the thicknesses of the plates; we shall, therefore, in the thicker plate see a double number of particles to that of the other occupying the same extent of surface. The last circumstance in which two plates may differ is their specific rigidity; and in this respect it will be found that two plates of exactly equal dimensions, and covered with the same number of vibrating corpuscles of equal magnitudes, but of different substances, differ in sound; therefore the absolute magnitudes of the particles cannot be assumed as a standard of tune, unless regulated by the specific rigidity.

Unassisted by any means of actual admeasurement, the above are but the proximate results sensible to the eye; more extended and accurate experiments are necessary to confirm the results with mathematical certainty. As the absolute magnitudes of these particles will, I imagine, be hereafter a most useful element for calculation, I will here indicate the most effectual way I am acquainted with to arrive at this knowledge. A thick metallic slip of considerable length and breadth, bent similarly to a tuning-fork, and fixed at its curved part in a vice, is very easily excited by friction; and a more considerable surface of regularly arranged vibrating particles is seen than in most other superficies: any description of common exciter may be employed. When this bent plate is excited by percussion, the particles, before their disappearance, will assume an apparent rotatory motion, on account of the force exerted and its susceptibility of continuing the vibrations. Employing a parallelopipedal rod, the appearances of the higher modes of subdivisions are particularly neat; the entire vibrating parts between the nodes form ellipses, and the semipart at the free end a regular half of the same figure. It is important to remark that the crispations of the water only appear on the sides in the plane of oscillation; the other two sides, on one of which the exciter must be applied, do not show similar appearances.

I have also rendered the phonic molecular vibrations visible when produced by the longitudinal oscillations of a column of air. The following were the means employed: – I placed the open end of the head of a flute or flagiolet on the surface of a vessel of water; and on blowing to produce the sound, I observed similar crispations to those described above, forming a circle round the end of the tube, and afterwards appearing to radiate in right lines; on the harmonics of the tube being sounded, the crispations were correspondently diminished in magnitude. These phenomena will be more evident if the tube be raised a little from the surface of the liquid, and a thin connecting film be left surrounding it; the vibrating particles will then occupy a greater space, and be more sensible.

The existence of the molecular vibrations being now completely established, it becomes a critical question in what manner the sensible oscillations induce these vibrating particles. I do not know whether what I am now going to adduce will be admitted as the right explanation; but it is certainly analogous, so far as the superficial and transversal linear oscillations are concerned. A flexible surface, covered with a coat of resinous varnish, being made to assume any curve, the cohesion of the varnish will be destroyed in certain parts, and a number of cracks will be observed more regularly disposed as the force inducing the curve has been more regularly applied; when the original position of the surface is restored the cracks will be imperceptible, but will again appear at every subsequent motion. Be this as it may, these particles are invariable concomitants of the sensible oscillations; and there is no reason to suppose otherwise than that their vibrations are isochronous with them. To avoid confusion, I have restricted the word vibrations to the motions of the more minute parts, and the term oscillations to those of the sensible divisions. We may reasonably suppose that the molecular vibrations pervade the entire substance of a phonic; their excursions, however, are not the same in all parts; and they can only be rendered visible when these excursions are large. They may be so few in number as to be entirely inaudible, as in their transmission through linear conductors; but however few, when they are properly directed they induce the mechanical divisions of sonorous bodies, each of which will give birth to numerous vibrating corpuscles whose excursions are greater, and the sound will be rendered audible. Dr. Savart has well investigated the modes of division in surfaces put in motion by communicated vibrations. All those phonics whose limited superficies preclude them from exciting in themselves a sufficient number of vibrating corpuscles, when isolated, produce scarcely any perceptible sound – as extended chords, tuning-forks, &c.; but those whose superficies or solidities are more extended, as bells, elastic laminæ, columns of air, &c., produce sufficient volume of sound without accessory means.

Loudness of sound is dependent on the excursions of the vibrations; volume, or fulness of sound, on the number of coexisting particles put in motion. Thus the tones of the Æolian harp, on account of the number of subdivisions of the strings, are remarkably beautiful and rich without possessing much power; and the sounds of an Harmonica-glass, in which a greater number of particles are excited than by any other means, are extraordinarily so, united, according to the method of excitation, with considerable intensity: their pervading nature is one of the greatest peculiarities of these sounds.

The following is a recapitulation of the various properties of sound which are attributable to modifications of the vibrating corpuscles: –

The tune	Depends on the	velocities of the vibrations.
The time		continuance of the vibrations.
The intensity		excursions of the vibrations.
The richness, or volume		number of coexisting vibrations.
The quality (timbre)		magnitudes of the vibrating corpuscles.

It has often been thought necessary to admit the existence of more minute motions than the sensible oscillations, in order to account for many phenomena in the production of sound. Perault, in his 'Essai du Bruit,' insisted on their necessity more than any other author I have read: he imagined that the vibrations have a much greater velocity than the oscillations which cause them; but the experiment he adduced to prove this is far from conclusive; he mistook for these vibrations the oscillations of the subdivisions of the long string he employed. Other distinguished philosophers have had ideas of a similar nature; and Chladni thinks their existence necessary to account for the varieties of quality. I, however, conceived I was the first who had indicated these phenomena by experiment, until a few days ago, repeating them, together with the others which form the subject of this paper, in the presence of Prof. Oersted, of Copenhagen, he acquainted me with some similar experiments of his own. Substituting a very fine powder, Lycopodion, instead of the sand used by Chladni for showing the oscillations of elastic plates, this eminent philosopher found the particles not only repulsed to the nodal lines, but at the same time accumulated in small parcels on and near the centres of vibration; these appearances he presumed to indicate more minute vibrations, which were the causes of the quality of the sound. Subsequently he confirmed his opinion by observing the crispations of water or alcohol on similar plates, and showed that the same minute vibrations must take place in the transmitting medium, as they were equally produced in a surface of water when the sounding plate was dipped into a mass of this fluid. These experiments were inserted in Lieber's 'History of Natural Philosophy,' 1813.

RECTILINEAL TRANSMISSION OF SOUND

As the laws of the communication of the phonic vibrations are more evident in linear conductors, I shall confine the present article to a summary of their principal phenomena.

In my first experiments on this subject, I placed a tuning-fork, or a chord extended on a bow, on the extremity of a glass or metallic rod five feet in length, communicating with a sounding-board: the sound was heard as instantaneously as when the fork was in immediate contact; and it immediately ceased when the rod was removed from the sounding-board or the fork from the rod. From this it is evident that the vibrations, inaudible in their transmission, being multiplied by meeting with a sonorous body, become very sensibly heard. Pursuing my investigations on this subject, I have discovered means for transmitting, through rods of much greater lengths and of very inconsiderable thicknesses, the sounds of all musical instruments dependent on the vibrations of solid bodies, and of many descriptions of wind instruments. It is astonishing how all the varieties of tune, quality, and audibility, and all the combinations of harmony, are thus transmitted unimpaired, and again rendered audible by communication with an appropriate receiver. One of the practical applications of this discovery has been exhibited in London for about two years, under the appellation of "The Enchanted Lyre". So perfect was the illusion in this instance from the intense vibratory state of the reciprocating instrument, and from the interception of the sounds of the distant exciting one, that it was universally imagined to be one of the highest efforts of ingenuity in musical mechanism. The details of the extensive modifications of which this invention is susceptible, I shall reserve for a future communication; the external appearance and effects of the individual application above mentioned have been described in the principal periodical journals.

The transmission of the vibrations through any communicating medium as well as through linear conductors is attended by peculiar phenomena. Pulses are formed similar to those in longitudinal phonics; and consequently the centres of vibration and the nodes are reproduced periodically at equal distances: in this we observe an analogous disposition with regard to light. I had intended to include in this paper all the analogical facts I have observed illustratory of the identity of the causes of these two principal objects of sensation; but want of time and the danger of delay, now the subject is occupying so much the attention of the scientific world, have induced me hastily to collect the present experiments, and to defer the others for a future opportunity.

The thicknesses of conductors materially influence the power of transmission; and there is a limit of thickness, differing for the different degrees of tune, beyond which the vibrations will not be transmitted. The vibrations of acute sounds can be transmitted through smaller wires than those of grave sounds. A proof of this is easy. Attach a tuning-fork to one end of a very small wire, and apply the other end to the ear or a sounding-board. On striking the fork rather hard, two coexisting sounds will be produced: that which is more acute will be distinctly heard; but the other will not be transmitted. If the vibrations of a tuning-fork be conducted through a piece of brass wire of the size and thickness of a large needle, the sound, imperfectly transmitted, will become more audible by the pressure of the fingers on the conducting wire; but if a steel wire of the same length and thickness be employed, the sound will be unaltered by any pressure, because steel has a greater specific elasticity than brass.

POLARIZATION OF SOUND

Hitherto I have only considered the vibrations in their rectilineal transmission; I shall now demonstrate that they are peculiarly affected when they pass through conductors bent in different angles. I connected a tuning-fork with one extremity of a straight conducting-rod, the other end of which communicated with a sounding-board: on causing the tuning-fork to sound, the vibrations were powerfully transmitted, as might be expected from what has already been explained; but on gradually bending the rod, the sound progressively decreased, and was scarcely perceptible when the angle became a right one; as the angle was made more acute the phenomena were produced in an inverted order: the intensity gradually increased as it had before diminished; and when the two parts were nearly parallel, it became as powerful as in the rectilineal transmission. By multiplying the right angles in a rod, the transmission of the vibrations may be completely stopped.

To produce these phenomena, however, it is necessary that the axis of the oscillations of the tuning-fork should be perpendicular to the plane of the movable angle; for if they be parallel with it they will be still considerably transmitted. The following experiment will prove this. I placed a tuning-fork perpendicularly on the side of a rectilinear rod; the vibrations were therefore communicated at right angles. When the axis of the oscillations of the fork coincided with the rod, the intensity of the transmitted vibrations was at its maximum; in proportion as the axis deviated from parallelism the intensity of the transmitted vibrations diminished; and, lastly, when it became perpendicular, the intensity was at its minimum. In the second quadrant the order of the phenomena was inverted, as in the former experiment, and a second maximum of intensity took place when the axis of the oscillations had described a semicircumference and had again become parallel, but in an opposite direction. When the revolution was continued, the intensity of the transmitted vibrations was varied in a similar manner; it progressively diminished as the axis of the oscillations deviated from being parallel with the rod, became the least possible when it arrived at the perpendicular, and again augmented until it remained at its first maximum, which completed its entire revolution.

The phenomena of polarization may be observed in many corded instruments: the cords of the harp are attached at one extremity to a conductor which has the same direction as the sounding-board; if any cord be altered from its quiescent position, so that its axis of oscillation shall be parallel with the bridge or conductor, its tone will be full; but if the oscillations be excited so that their axis shall be at right angles with the conductor, its tone will be feeble. By tuning two adjacent strings of the harp unisons with each other, the differences of force will be sensible to the eye in the oscillations of the reciprocating string according to the direction in which the other is excited.

It now remains to explain the nature of the vibrations which produce the phenomena the existence of which has been proved by the preceding experiments. The vibrations generally assume the same direction as the oscillations which induce them: in a longitudinal phonic the vibrations are parallel to its axis; in a transversal phonic they are perpendicular to this direction; a circular or an elliptic form can be also given to the vibrations by causing the oscillations to assume the same forms. Any vibrating corpuscle can induce isochronous vibrations of similar contiguous corpuscles *in the same plane*, either parallel with, or perpendicular to, the direction of the original vibrations; and the polarization of the vibrations consists in the similarity of their directions, by which they propagate themselves equally in the same plane: therefore the vibrations being transmitted through linear conductors, it is the plane in which the vibrations are made that determines their transmission or nontransmission when the direction is altered. A longitudinal or a transversal vibration may be transmitted two ways to a conductor bent at right angles: their axis may be in that direction, as to be in the same plane with the right angle, in which case the former will be transversally, or the latter longitudinally transmitted in the new direction; or their axis may be perpendicular to the plane of this new direction, under which circumstances neither can be communicated.[1] In explaining the polarization of light, there is no necessity to suppose that the reflecting surfaces act on the luminous vibrations by any actual attracting or repulsing force, causing them to change their axes of vibrations; the directions of the vibrations in different planes, as I have proved exist in the communication of sound, are sufficient to explain every phenomenon relative to the polarization of light.

Let us suppose a number of tuning-forks oscillating in different planes and communicating with one conducting-rod: if the rod be rectilinear, all the vibrations will be transmitted; but if it be bent at right angles, they will undergo only a partial transmission; those vibrations whose planes are perpendicular, or nearly so, to the plane of the new direction will be destroyed. The vibrations are thus completely polarized in one direction while passing through the new path, and on meeting with a new right angle they will be transmitted or not, accordingly as the plane of the angle is parallel with, or perpendicular to, the axes of the vibrations. In this point of view, the circumstances attending the phenomena are precisely the same as in the elementary experiment of Malus on the polarization of light.

Double refraction is a consequence of the laws of polarization, by which a combination of vibrations having their axes in different planes, after travelling in the same direction, are separated into two other directions, each polarized in one plane only. That this well known property of light has a correspondent in the communication of phonic vibrations I shall now demonstrate. When two tuning-forks, sounding different notes by a constant exciter, and making their oscillations perpendicularly to each other, have their vibrations transmitted at the same time through one rod, at the opposite extremity of which two other conductors are attached at right angles, and when each of these conductors is parallel with one of the axes of the oscillations of the forks, on connecting a sounding-board with either conductor, those vibrations only will be transmitted through it which are polarized in the same plane with the angle made by the two rods through which the vibrations pass; either sound may be thus separately heard, or they may both be heard in combination by connecting both the conductors with sounding-boards.

The phenomena of diffraction regarding only the form of the surfaces, or the superficies over which the vibrations extend, are, by the conformation of the organs of hearing, not of any consequence to the perception of sound, though the same phenomena, when the chromatic vibrations are concerned, are very evident to the eye. They, however, undoubtedly take place equally in both instances, and may be well explained by the theory already laid down. Each separate vibration propagating itself in the plane of its vibrating axis, a number of vibrations in different planes, after passing through an aperture, naturally expand themselves transversely as well as rectilineally, and thereby occupy a greater space than they would were they only longitudinally transmitted.

I have still to indicate a new property of the phonic vibrations; but whether it is analogous to any of the observed phenomena of light I am yet ignorant. When the source of the vibrations is in progressive motion, the vibrations emanating from it are transmitted when the conductor is rectilineal and parallel with the original direction; and they are destroyed when the conductor is perpendicular to the direction, though the axis of vibration and the conductor being in both instances *in the same plane*, would transmit the vibrations were the phonic stationary. These circumstances are proved by the following experiments: – When a tuning-fork, placed perpendicularly to a rod, communicating at one or both extremities with sounding-boards, and caused to oscillate with its vibrating axis parallel with the rod, moves along the rod, preserving at the same time its perpendicularity and parallelism, the vibrations will not be transmitted while the movement continues; but the transmission will take place immediately after it has remained motionless. When the tuning-fork moves on the upper edge of a plane perpendicular to a sounding-board, the vibrations rectilineally transmitted will not be influenced by the progressive motion.

NOTE

1 I have just seen a paper by M. Fresnel, entitled "Considérations Mécaniques sur la Polarisation de la Lumière," in which this eminent philosopher had previously arrived at the same conclusions with respect to light as I have proved in this communication respecting sound. The important discoveries of Dr. Thomas Young, followed by those of M. Fresnel, have recently reestablished the vibratory theory of light; and new facts are every day augmenting its probability. The new views in acoustical science which I have opened in this paper will, I presume, give additional confirmation to the opinions of these eminent philosophers; and I hope, when I resume the subject, to be enabled to account for the principal phenomena of coloration, with regard to their acoustic analogies, in a way calculated to establish the permanent validity of the theory.

Editorial Headnote

2. (Anon.), "The Enchanted Lyre", *The Literary Gazette* 5 (London, 1821), p. 586.

In 1823, Charles Wheatstone inherited his uncle's shop, having already taken over his family's property at 128 Pall Mall. Although building something of a philosophical reputation for himself through his popularization and development of Chladni's acoustic figure experiments, Wheatstone's primary occupation remained the designing and building of mechanical and musical marvels, intended to entertain and bewilder London's fashionable clientele. Without a doubt, his greatest triumph was his "Enchanted Lyre" arrangement, first displayed in 1821. This spectacle consisted of an antique lyre, suspended from the ceiling of his shop, apparently playing music of its own volition. Wheatstone replicated this arrangement in a shop in the Royal Opera Arcade in 1822, charging a five-shilling entry fee for an hour's musical entertainment.[1] This was not just a musical spectacle but a philosophical wonder, with audiences and commentators at a loss as to how the floating instrument produced such beautiful music, seemingly free of human manipulation. In 1821, a correspondent from the *Literary Gazette* reviewed Wheatstone's marvel, reporting the music to be aesthetically pleasing, but confessed to complete confusion about how the instrument maker had achieved such a mysterious spectacle. As the correspondent put it, Wheatstone "disclaimed mechanism altogether", instead asserting that the Enchanted Lyre's performance was the "result of a new combination of powers".[2] In fact, Wheatstone's arrangement was remarkably simple. The lyre was connected either by a thin wire or deal wood to the ceiling, and from there it was in connection to the sounding board of a piano, some rooms away, linked by more wooden rods. As a pianist performed on the piano, sonorous vibrations travelled through the rods and wires, or what Wheatstone termed "linear conductors", all the way to the lyre, causing it to play in sympathy. At the centre of Wheatstone's apparatus were his own philosophical investigations into the communication of sounds through solid bodies, including his observations on Chladni's acoustic figures. Yet the value of the Enchanted Lyre was the way in which Wheatstone was able to transform his scientific studies of sonorous phenomena into a fashionable public exhibition. In 1820s' London, the boundaries between elite science and popular entertainment were extremely vague. Wheatstone, through his mechanical ingenuity, took full advantage of the opportunities this context provided.[3]

NOTES

1 Bowers, *Sir Charles Wheatstone*, pp. 8–9.
2 (Anon.), "The Enchanted Lyre", *The Literary Gazette*, 5 (1821, London), p. 586.

3 Melissa Dickson, "Charles Wheatstone's Enchanted Lyre and the spectacle of sound", in James Q. Davies and Ellen Lockhart (eds.), *Sound Knowledge: music and science in London, 1789–185* (Chicago: University of Chicago Press, 2016), pp. 125–44; also see, Bernard Lightman, "Refashioning the Spaces of London Science: elite epistemes in the nineteenth century", in David N. Livingstone and Charles W. J. Withers (eds.), *Geographies of Nineteenth-Century Science* (Chicago: Chicago University Press, 2011), pp. 25–50; Iwan Morus, Simon Schaffer, and Jim Secord, "Scientific London", in Celina Fox (ed.), *London – World City, 1800–1840* (New Haven, CT: Yale University Press, 1992), pp. 129–42.

2

THE ENCHANTED LYRE

(Anon.)

Source: (Anon.), "The Enchanted Lyre", *The Literary Gazette* 5 (London, 1821), p. 586.

We have been much gratified with an exhibition in Pall Mall of an instrument under the denomination of the Enchanted Lyre, the invention of a Mr. Wheatstone. The exhibition-room presents a work of handsome construction in the form of an ancient lyre, suspended from the ceiling. Its horns terminate in mouths resembling bugles. Its centre is covered on both sides with plates of a bright metallic lustre, and there is an ornamented key-hole, like that of a time-piece, which admits of its being wound up, which is evidently a mere *ruse*, as the instrument certainly does not utter melodious sounds in consequence of that operation. Round it there is a slight hoop-rail, perhaps five feet in diameter, which is supported by equally slight fixtures in the floor.

The inventor disclaims mechanism altogether, (though he winds up the machine,) and asserts that the performance of the Enchanted Lyre is entirely the result of a new combination of powers. Be that as it may, its execution is both brilliant and beautiful. The music *seems to proceed from it;* the tones are very sweet; the expression soft or powerful, and the whole really charming. We listened to Steibelt's battle-piece with unfeigned pleasure, and were equally delighted with several other compositions of simple melody and of more difficult harmony. Mr. Wheatstone professes to be able to give a concert, producing, by the same means, an imitation of various wind and stringed instruments: in the mean time the lovers of music will have a treat in hearing the Enchanted Lyre go through a half hour's entertainment.

DOI: 10.4324/b23406-2

Editorial Headnote

3. Thomas Busby, "The Acoucryptophone, or Enchanted Lyre", *Concert Room and Orchestra Anecdotes of Music and Musicians, ancient and modern*, Vol. I (London, 1825), pp. 9–10.

Thomas Busby (1754–1838) was most famous for his three-volume *Concert Room and Orchestra Anecdotes of Music and Musicians, Ancient and Modern*, published in 1825. This text was full of gossip and conjecture concerning contemporary music and musicians, often causing something of a sensation in polite British society. Busby had himself shown great musical promise from an early age, having been born in Westminster in 1754 and, beginning in 1769, serving a five-year apprenticeship with composer Jonathan Battishill (1738–1801). An organist and vocalist, Busby took his Doctorate in Music from the University of Cambridge in 1801 and proceeded to author a series of moderately successful compositions before launching several unsuccessful journals publishing the very latest musical news from Continental Europe.[1] Included among these enterprises was the *Monthly Musical Journal* which represented England's first musical periodical. Through these endeavours, Busby earned himself a reputation as a leader of musical taste in Georgian and Regency Britain. In all his work, Busby promoted music as a subject capable of scientific study, by which he meant that it could be defined by empirical laws, just as any other natural phenomena. This not only concerned the science of sound but also composition, the principles of harmony and melody, and musical performance. In his 1818 *A Grammar if Music*, he asserted that "[m]usic is a science having sounds for its element", and continued to explain how, "[b]y the theory of Music, we are enabled to measure the intervals between sonorous impressions; and so to understood the principles of *melody* and *harmony*".[2] In 1825, Busby reviewed Charles Wheatstone's Enchanted Lyre in his *Concert Room and Orchestra Anecdotes of Music and Musicians*, having been attracted by the spectacle's uniting of musical and scientific principles. For Busby, this was no mere commercial stunt but a carefully crafted scientific exhibition of considerable acoustic learning. Busby first witnessed Wheatstone's Enchanted Lyre in 1822, reporting how London's fashionable elites were excited in reaction to "one of the most beautiful and striking experiments that has ever been witnessed, in the philosophy of sound".[3] Busby thought that Wheatstone was entitled to immense credit for his manipulation of natural phenomena in this way and admired the natural philosopher's sense of drama in the manner in which he presented his apparatus to his audiences. Wheatstone appeared, for "the sake of illusion", to wind up his Enchanted Lyre with a key, adding much entertainment to an already-exhilarating acoustic experience.[4] In Busby's text, we get a sense of the diverse audiences that Wheatstone's spectacle attracted. This was a mechanism that transcended specialist scientific viewers and connected with musical, as well as popular and non-specialist, audiences.

NOTES

1 Philip Olleson, "Busby, Thomas (1754–1838)", *Oxford Dictionary of National Biography* (Oxford: Oxford University Press, 2004); online edition, May 2011, www.oxforddnb.com/view/article/4158, accessed 17 July 2022.
2 Thomas Busby, *A Grammar of Music: to which are prefixed observations explanatory of the properties and powers of music as a science. And of the general scope and object of the work* (London: John Walker, 1818), p. xi.
3 Thomas Busby, "The Acoucryptophone, or Enchanted Lyre", *Concert Room and Orchestra Anecdotes of Music and Musicians, Ancient and Modern*, Vol. I (London, 1825), pp. 9–10.
4 (Anon.), "The Enchanted Lyre and the Diaphonicon", *The New Monthly Magazine and Literary Journal* 6 (London, 1822), pp. 200–1; compare with a similar review in (Anon.), "Enchanted Lyre", *The Circulator of Useful Knowledge, Amusement, Literature, Science, and General Information* (London, 1825), p. 284.

3

THE ACOUCRYPTOPHONE, OR ENCHANTED LYRE

Thomas Busby

Source: Thomas Busby, "The Acoucryptophone, or Enchanted Lyre", *Concert Room and Orchestra Anecdotes of Music and Musicians, Ancient and Modern*, Vol. I, (London, 1825), pp. 9–10.

This perfectly novel musical invention, which, during its public exhibition in London, in 1822, excited so much surprise among the votaries of the harmonic art, forms one of the most beautiful and striking experiments that has ever been witnessed, in the philosophy of sound. It constitutes one of the practical applications of a series of original and interesting acoustical investigations, and reflects much credit on the science and talents of Mr. C. Wheatstone, its ingenious inventor. The description of the exhibition of this musical curiosity is briefly as follows: –

The *form* of an antique lyre of large dimensions was freely suspended from the ceiling by a silken cord. The instrument was not really furnished with any wires, but only a representation of them in steel rods. The company being seated, the inventor applied a key to a small aperture in the body of the instrument, giving it a few turns representative of the act of winding it up, and music was instantly heard, as if proceeding from the suspended lyre. As the sounds continued, they varied in their effect, and seemed to form a combination of the visible lyre with a conical piano-forte and a dulcimer. By these united tones, a variety of difficult compositions were executed; and the whole performance occupied about an hour.

This singular experiment was founded on the general principle by which sound is conducted, and the application of which is certainly capable of being carried to a much greater extent; and might be made to include even the mixed and multifarious tones of a complete orchestra.

DOI: 10.4324/b23406-3

4

ENCHANTED LYRE

(Anon.)

Source: (Anon.), "Enchanted Lyre", *The Circulator of Useful Knowledge, Amusement, Literature, Science, and General Information* (London, 1825), p. 284.

WE have lately witnessed a private exhibition of this singular instrument, in which its astonishing powers were called into action with peculiar effect. The enchanted lyre is the invention of Mr. Wheatstone, and consists of an ornamental instrument similar to that usually borne by the poet Orpheus. It is suspended to the ceiling of the apartment by a metal wire, and at a certain signal given by the exhibitor, the whole instrument is thrown into vibration – the most harmonious tones are produced – and an effect perfectly magical is the result. The most singular part of this apparatus remains, however, to be noticed; for it is confidently asserted by its ingenious inventor, that fifty such instruments might be constructed to produce the same effects by the execution of a single performer. This instrument is admirably adapted for those places of public amusement in which an extensive area is to be enlivened by the cheering strains of instrumental harmony. Such, indeed, is the capabilities of the enchanted lyre, that we expect, ere long, to have music laid on as the gas and water-pipes are now furnished through the streets, and at a probably still lower rate.

DOI: 10.4324/b23406-4

Editorial Headnote

5. Charles Wheatstone, "Description of the Kaleidophone, or Phonic Kaleidoscope", *Quarterly Journal of Science, Literature, and Art* I (London, England, 1827), pp. 344–51.

The Enchanted Lyre was Charles Wheatstone's most celebrated mechanical contrivance, but he also enjoyed notable fame for his Kaleidophone. This device consisted of a circular board on which Wheatstone placed three steel rods, perpendicularly, suspending various media. One rod had a spherical bead at its summit, while another had a moving plate, and the third was topped by a four-sided prism. There was a fourth rod, bent at a right angle with a bead at its end. When exhibiting his Kaleidophone, Wheatstone would put each rod, in turn, into vibration with a bow or hammer while reflecting a ray of light on each of the various media. As the rods vibrated, the reflecting light traced out the oscillations, rendering the sonorous visible by patterns that were symmetrical and resembled a mechanical engine in operation. In this way, Wheatstone provided a sonorous equivalent of Scottish natural philosopher David Brewster's (1781–1868) 1817 Kaleidoscope, which created aesthetically pleasing patterns by an arrangement of two mirrors, in a tube, positioned so as to create repeated reflections. The Kaleidophone was also, as Wheatstone was keen to emphasize, an effective device for uniting scientific knowledge of sound with easily accessible amusement. He claimed that the Kaleidophone rendered "obvious to the common observer what has hitherto been confined to the calculations of the mathematician; it presents another proof, that however remote from common observation the operations of nature may be, the most beautiful order and symmetry prevail through all".[1] As with his Enchanted Lyre, Wheatstone held public demonstrations of the Kaleidophone in his shop at 436 The Strand, informing audiences that it had been natural philosopher Thomas Young (1773–1829) who had first revealed the principle by reflecting light off vibrating piano strings. Young was more familiar to British audiences for his work surrounding Egyptology, doing much to decipher Egyptian hieroglyphs, including those on the Rosetta Stone. On the continent, however, Young had a considerable reputation as a specialist in the study of light. In particular, it was Young's claims over light's movement resembling a wave, rather than a projection of particles, or "corpuscles", that secured him attention from Europe's foremost mathematicians. In Britain, however, such a rejection of Isaac Newton's (1643–1727) corpuscular theory of light received a far more sceptical response, at first at least.[2] What made Wheatstone's Kaleidophone especially interesting to London audiences was its potential for illustrating the principles to which musical instruments, such as pianos, operated. The Kaleidophone was also comparable to Chladni's acoustic figures in its ability to make visible the otherwise invisible motions of a sound wave. Wheatstone's Kaleidophone might, at first, appear almost trivial, as he himself was well aware when he reminded his audiences that his device was no mere toy. Yet, just as with his Enchanted Lyre, Wheatstone's Kaleidophone provided a rich example of how seamlessly popular entertainment and philosophical investigations of sound merged. Sound, particularly when in combination with some sort of visualizing media, be it sand on an oscillating plate or light from a vibrating bead, was an exceptionally versatile phenomenon for popularizing the latest knowledge of nature.

NOTES

1 Charles Wheatstone, *The Scientific Papers of Sir Charles Wheatstone, D.C.L., F.R.S* (London: Taylor and Francis, 1879), p. 21.
2 See Geoffrey N. Cantor, *Optics After Newton: theories of light in Britain and Ireland, 1704–1840* (Manchester: Manchester University Press, 1983), pp. 159–72.

5

DESCRIPTION OF THE KALEIDOPHONE, OR PHONIC KALEIDOSCOPE

Charles Wheatstone

Source: Charles Wheatstone, "Description of the Kaleidophone, or Phonic Kaleidoscope", *Quarterly Journal of Science, Literature, and Art* I (London, England, 1827), pp. 344–51.

Description of the Kaleidophone, or Phonic Kaleidoscope; a new Philosophical Toy, for the Illustration of several Interesting and Amusing Acoustical and Optical Phenomena.

[Communicated by Mr. C. Wheatstone.]

The application of the principles of science to ornamental and amusing purposes contributes, in a great degree, to render them extensively popular; for the exhibition of striking experiments induces the observer to investigate their causes with additional interest, and enables him more permanently to remember their effects. I shall not, therefore, need an apology for presenting the tyro in science with another combination of philosophy with amusement, in addition to those already extant.

But this instrument possesses higher claims to attention; for it exemplifies an interesting series of natural phenomena, and renders obvious to the common observer what has hitherto been confined to the calculations of the mathematician; it presents another proof, that however remote from common observation the operations of nature may be, the most beautiful order and symmetry prevail through all.

In the property of "creating beautiful forms," the Kaleidophone resembles the celebrated invention of Dr. Brewster, from which its name is modified; but to the instrument itself, and its mode of action, it is almost superfluous to say there is no similarity. Previously to entering into an explanation of its construction and effects, the following brief summary may suffice to give a general idea of the nature of the experiments it is intended to illustrate.

These experiments principally consist in subjecting to ocular demonstration the orbits or paths described by the points of greatest excursion in vibrating rods, which in the most frequent cases, those of the combinations of different modes of vibration, assume the most diversified and elegant curvilinear forms.[1] The entire track of each orbit is rendered simultaneously visible by causing it to be delineated by a brilliantly luminous point, and the figure being completed in less time than the duration of the

DOI: 10.4324/b23406-5

visual impression, the whole orbit appears as a continuous line of light. As besides the changes which result from the combinations of the primitive with the higher modes of vibration, the figures of the orbits are affected by the form of the rod, by the extent of the excursions of the vibrations, by the mode of producing the motions, and by many other circumstances, a great variety of pleasing and regular forms is obtained. This variety is also enhanced by giving the same motions to a number of symmetrically disposed luminous points, the mutual intersections of the orbits of which produce innumerable elegant forms; and the appearances may be still more variegated by occasionally causing these points to reflect differently-coloured lights.

The apparatus for exhibiting these experiments consists of a circular board about nine inches in diameter, into which are

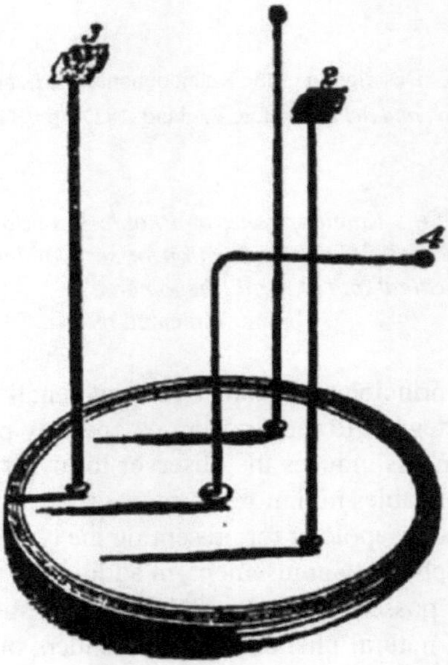

perpendicularly fixed, at equal distances from the circumference and from each other, three steel rods each about a foot in length. The first rod is cylindrical, about 1–10th of an inch in diameter, and is surmounted by a spherical bead[2] which concentrates and reflects the light which falls upon it. The second is a similar rod, upon the upper extremity of which is placed a plate moving on a joint, so that its plane may be rendered either horizontal, oblique, or perpendicular; this plate is adapted to the reception of the objects, which consist of beads differently coloured and arranged on pieces of black card in symmetrical forms. The third is a four-sided prismatic rod, and a similar plate is attached to its extremity for the reception of the same objects. Another rod is fixed at the centre of the board; this is bent to a right angle, and is furnished with a bead similarly to the first-mentioned rod. A small nut and screw is fixed to the board near the lower end of the first rod, in order by pressing upon it to render occasionally its rigidity unequal. A hammer, softened by a leather covering, is employed to strike the rods; and a violin-bow is necessary to produce some varieties of effect.

I shall now proceed to describe the different appearances which the rods present when in action, and to give directions for the production of the different effects, following the order in which the rods have been previously mentioned.

No. 1. On causing the entire rod to vibrate, so that its lowest sound be produced,[3] as it is seldom that the motions of a cylindrical rod can be confined to a plane, the vibrations will almost always be combined with a circular motion. When the pressure on the fixed end is exerted on two opposite points, and the rod put in motion in the direction of the pressure, the following progression in the changes of form will be distinctly observed: the track will commence as a line, and almost immediately open into an ellipse, the lesser axis of which will gradually extend as the larger axis diminishes, until it becomes a circle; what was before the lesser will then become the larger axis, and thus the motions will alternate until, from their decreasing magnitudes, they cease to be visible. In the case just described the ellipses make a right angle with each other; but by altering the direction of the motion, so as to render it oblique to the direction of the pressure, they may be made to intersect under any required angle, and when this angle = 0 the motion will be merely vibratory.

Every single sound formed by the subdivisions of the rod will present similar appearances, but the excursions will be smaller as the sound is higher, or, which is the same thing, as the number of the vibrations increases.

In the most simple case of the co-existence of two sounds, shown by putting the entire rod in motion, and producing also a higher sound by the friction of a bow; the original figure will appear waved or indented, and as unity is to the number of

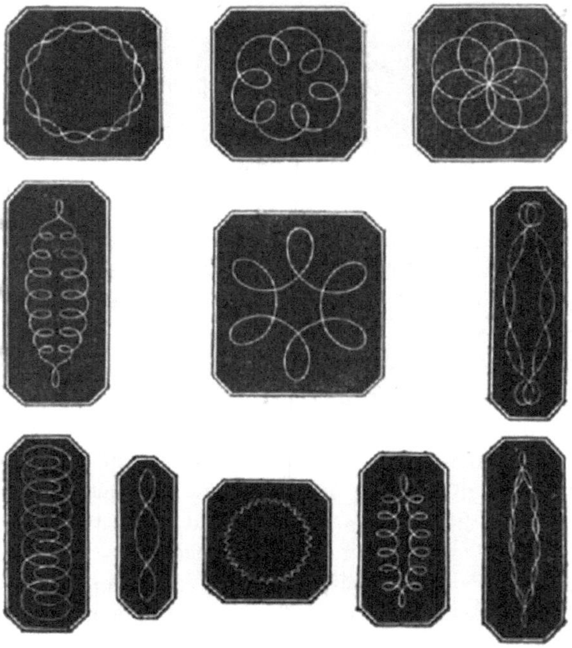

indentations, so will the number of vibrations in the lower sound be to the number in the higher sound. On varying the mode of excitation, by striking the rod in different parts and with different forces, very complicated and beautiful curvilinear forms may be obtained: some of these are represented by the opposite figures.

Placing the hand on the lower part of the rod, below the place at which it is excited, the excursions of the motions will rapidly decrease and exhibit spiral figures.

To obtain the figures with brilliancy and distinctness, a single light only should be employed, as that of the sun, a lamp, or a candle; rays of light proceeding from several points, as from a number of candles, or from the reflection of the clouds, occasion the track to be broad and indistinct; but double lights may be employed with effect, provided they be of equal intensity and symmetrically placed; each bead will then describe two similar figures. The appearances, in a bright sunshine, are remarkably vivid and brilliant.

No. 2. Although very beautiful and varied forms may be produced from the motion of a single point, yet the compound figures which are presented by objects, formed by a number of points, offer appearances still more pleasing to the eye.

An object being placed horizontally on the plate, and the rod being put in motion, the mutual intersections of the points, each describing a similar figure, present to the eye complicated, yet symmetrical figures, resembling elegant specimens of engine-turning.

When the plate is horizontal, the figures are all in one plane, but if it be inclined or perpendicular, the curves being then made in parallel planes, gives the idea of a solid figure, and in some cases the appearances are particularly striking.

Complementary colours alone should be employed in the objects; for these harmonising together, give greater pleasure to the eye than an injudicious combination of discordant tints: the intensities should be occasionally varied, and colourless light intermingled with the different shades.

No. 3. When this prismatic rod is put in motion, in the direction of either of its sides, the points move only rectilineally, but when the motion is applied in an oblique direction, a variety of compound curves is shown: this rod is principally employed to exhibit the optical phenomena, which will be afterwards mentioned.

No. 4. When a rod is straight, the curve produced by any point describing its motion, is always in the same plane; but in a rod bent to any angle, the two parts moving most frequently in different directions, curves are produced whose parts do not lie in the same plane. A few trials will soon indicate the best way of applying the motion, so as to cause the two parts to vibrate in different directions.

Before the conclusion of this subject, I will avail myself of the opportunity of observing, that the application of this mode of experimenting may be extended to the delineation of every description of curvilinear and angular motion, when the amplitudes of the tracks are not too great; and by this means, it is not improbable that the experimental knowledge of many interesting principles in science may be facilitated.

ON THE DUPLICATION AND MULTIPLICATION OF OBJECTS

When dark objects are substituted for luminous ones, their tracks become nearly invisible, and from the longer duration of the visual impression at the limits of vibration, the images are multiplied in proportion to the number of points at which they are retarded. Place horizontally on the rod No. 2, a word printed or written on a piece of card; in the lowest mode of vibration, at the opposite limits of the excursions, two legible images of the word will be distinctly seen, and but an indistinct shade, occasioned by the tracks of the letters, will appear in the intermediate space: the vibratory motion is imperceptible to the eye; the images will, therefore, appear stationary in this respect, but the diminution of the excursions will cause them to approximate very slowly and gradually towards the centre: this diminution operates so gradually, as to allow the images to superpose each other completely at each recurring vibration, without producing any intermingling or confusion.

On placing the object perpendicularly, the two images will appear in parallel planes, the furthest image appearing through the first apparent surface. Small pictures have a singular effect applied in this manner.

When other sounds co-exist with the fundamental, the images are multiplied, but they become fainter as their number increases: these multiplied images are equally visible, whether the vibrations be rectilineal, elliptical, or circular.

NOTES

1 We are indebted to Dr. T. Young for the first observation of these phenomena; the following account of his experiments is quoted from the *Philosophical Transactions* for 1800. "Take one of the lowest strings of a square piano-forte, round which a fine silvered wire is wound in a spiral form; contract the light of a window, so that when the eye is placed in a proper position, the image of the light may appear small, bright, and well defined, on each of the convolutions of the wire. Let the chord be now made to vibrate, and the luminous point will delineate its path like a burning coal whirled round, and will present to

the eye a line of light, which, by the assistance of a microscope, may be very accurately observed. According to the different ways by which the wire is put in motion, the form of this path is no less diversified and amusing than the multifarious forms of the quiescent lines of vibrating plates discovered by Professor Chladni; and it is, indeed, in one respect even more interesting, as it appears to be more within the reach of mathematical calculation to determine it."

The extremely limited extent of the excursions of a vibrating chord prevents its motions from being distinctly observed by the naked eye, but as the rods employed in the present experiments can extend their excursions to nearly two inches, and as the means employed greatly increase the intensity of the light, the phenomena are exhibited in a far more evident manner.

2 The only beads well adapted for this purpose are made of extremely thin glass silvered on the interior surface and about one-sixth of an inch in diameter; they are to be obtained at the shops under the name of steel beads. The protuberances at the apertures must be removed or blackened; otherwise, the reflections from them will render the images confused. To produce the coloured tracks, these beads must be coated with transparent colours, such as are ordinarily used for painting on glass; the light will then be reflected through the coloured surface, but in beads made of coloured glass, the reflection being made from the external surface shows only white light. The bead is cemented into a small brass cup screwed to the top of the wire.

3 The most simple mode of vibration of a rod vibrating transversely, when one of its extremities is fixed and the other is free, is that in which the entire rod makes its vibrations alternately on each side of the axis, which is nowhere intersected by the curve, but only touched at the fixed end. This gives the gravest sound which can be produced from the rod. In the other modes of vibration, the axis is intersected by the curve 1, 2, 3, or more times. The best means to command the production of these sounds is to touch a node of vibration lightly with the finger and put a vibrating part in motion by a violin bow. In the second sound, the number of vibrations is to that of the first, as $5^2:2^2$, or as 25:4; the difference of the sounds is, therefore, two octaves and an augmented fifth. Separating the first sound from the series, the number of the vibrations of all the others will be to one another as the squares of the numbers 3, 5, 7, 9, &c.; the third, in which there are three nodes, will therefore exceed the second by an octave and an augmented fourth; in the fourth, the acuteness will be augmented by nearly an octave; in the fifth, by nearly a major sixth, &c. To reduce to the same pitch all the proportions of the sounds which such a rod is capable of producing, I shall regard the sound corresponding with the most simple motion as the C one octave lower than the lowest of the piano-forte; the proportions of the sounds will then be –

Numbers of nodes	0	1	2	3	4	5
Sounds	$\overline{C}$	G♯^{2}	D^4	D^5–	B♭^{5}	F^3+
Numbers, the squares of which correspond with these sounds }	(2)	(5) 3	5	7	9	11

The possible series of sounds, regarding the fundamental as unity, will therefore be – 1, 6¼, , $17\frac{13}{36}$, 56¼, &c.; $34\frac{1}{36}$ or expressed in integral numbers – 36, 225, 625, 1225, 2025, &c. – Chladni, *Traité d'Acoustique,* p. 91.

Editorial Headnote

6. (Anon.), "The Kaleidophone, or phonic Kaleidoscope", *Mechanics' Magazine* 8 (London, England, 11 Aug. 1827), pp. 49–52.

On 11 August 1827, the *Mechanics' Magazine* published an account of Charles Wheatstone's Kaleidophone. Indeed, this popular instrument for displaying the oscillations of vibrating steel rods at different frequencies made the front cover, complete with a quote by Alexander Pope (1688–1744) on the divinity and beauty of nature. The *Mechanics' Magazine* included a picture of the Kaleidophone itself, as well as the patterns produced by the various media as traced out by the movement of rays of light. Wheatstone was able to produce an array of visually impressive patterns by altering the frequency at which he vibrated the Kaleidophone's rods, with higher speeds of oscillation producing smaller excursions of reflected light. The *Mechanics' Magazine* was confident that Wheatstone's latest invention had the potential to open up new lines of scientific investigation into the nature of sound. The journal's reporting of Wheatstone's Kaleidophone was significant in that it identified the instrument maker's work as worthy of specialist, learned attention, rather than simply as a source of popular entertainment. John Knight and Henry Lacey had launched the journal in 1823, shortly after British physician George Birkbeck's (1776–1841) establishment of the Mechanics' Institute in Liverpool. Running until 1893, the magazine's editors published on a weekly basis, with new issues released every Saturday morning. Birkbeck had already been influential in establishing a mechanics' institute in Edinburgh in 1821 and, in November 1823, would collaborate with Utilitarian philosopher Jeremy Bentham (1747/8–1832), Whig statesman Henry Brougham (1778–1868), and reform-minded politician John Hobhouse (1786–1869) to set up the London Mechanics' Institute. These institutions were intended to provide classes, libraries, and experimental apparatus for the technical education of working-class adults, otherwise excluded from formal university education. Birkbeck's role was crucial in these projects and would have his name given, posthumously, to the London Mechanics' Institute when it was renamed Birkbeck Literary and Scientific Institution in 1866.[1] Lacey and Knight would work closely with London's Mechanics' Institute in their editing and publishing of the *Mechanics' Magazine*. The journal made knowledge of the very latest mechanical, scientific, and engineering inventions, such as the Kaleidophone, accessible to diverse audiences, particularly labouring artisans who benefitted from the careful use of illustrations. Machines were of considerable interest, especially new designs of steam engines. Along with the rise of mechanics' institutions across the British Isles, the *Mechanics' Magazine* was part of a broad enterprise to educate and morally reform the nation's working classes, with individuals like Birkbeck, Brougham, and Bentham convinced that scientific learning was essential for social improvement. The featuring of Wheatstone's Kaleidophone in the *Mechanics' Magazine* provides evidence of the range of ways in which sonorous knowledge and acoustic marvels could be communicated to a wide selection of social audiences, from fashionable West End elites to artisanal readers across the British Isles.[2]

NOTES

1 See John George Godard, *George Birkbeck, the Pioneer of Popular Education: a memoir and a review* (London: Bemrose & Sons, 1884).

2 On audiences, see Iwan Rhys Morus, "Worlds of Wonder: sensation and the Victorian scientific performance", *Isis* 101, no. 4 (Dec. 2010), pp. 806–16; Bernard Lightman, *Victorian Popularizers of Science: designing nature for new audiences* (Chicago: University of Chicago Press, 2007); also see Bernard Lightman, "The visual theology of Victorian popularisers of science: from reverent eye to chemical retina", *Isis* 91, no. 4 (Dec. 2000), pp. 651–80.

6

THE KALEIDOPHONE, OR PHONIC KALEIDOSCOPE

(Anon.)

Source: (Anon.), "The Kaleidophone, or phonic Kaleidoscope", *Mechanics' Magazine* 8 (London, England, 11 Aug. 1827), pp. 49–52.

THE KALEIDOPHONE, OR PHONIC KALEIDOSCOPE

In the "Philosophical Transactions for 1800," Dr. T. Young first brought the following remarkable phenomena into notice: – "Take one of the lowest strings of a square pianoforte, round which a fine silvered wire is wound in a spiral form; contract the light of a window, so that when the eye is placed in a proper position, the image of the light may appear small, bright, and well defined, on each of the convelutions of the wire. Let the chord be now made to vibrate, and the luminous point will delineate its path like a burning coal whirled round, and will present to the eye a line of light, which, by the assistance of a microscope, may be very accurately observed. According to the different ways by which the wire is put in motion, the form of this path is no less diversified and amusing, than the multifarious forms of the quiescent lines of vibrating plates, discovered by Professor Chladni; and it is, indeed, in one respect even more interesting, as it appears to be more within the reach of mathematical calculation to determine it."

In consequence, however, of the difficulty of distinguishing with the naked eye the quickly changing movements of the vibrating chord, and of the necessarily limited extent of these movements, even when accurately observed with "the assistance of a microscope," the experiments of Dr. Young have been seldom repeated, either with a view to information or pleasure. It has at length occurred to Mr. C. Wheatstone, that, by employing rods instead of chords, and by illuminating intensely the delineating points of these rods, the vibrations might be not only greatly extended, but made, in all their symmetrical variety, clearly and distinctly visible. And both these ideas he has successfully realized in the construction of an instrument, to which he has given the appropriate name of "Kaleidophone" (from καλος, beautiful, and φωνή, sound or vibration).

DOI: 10.4324/b23406-6

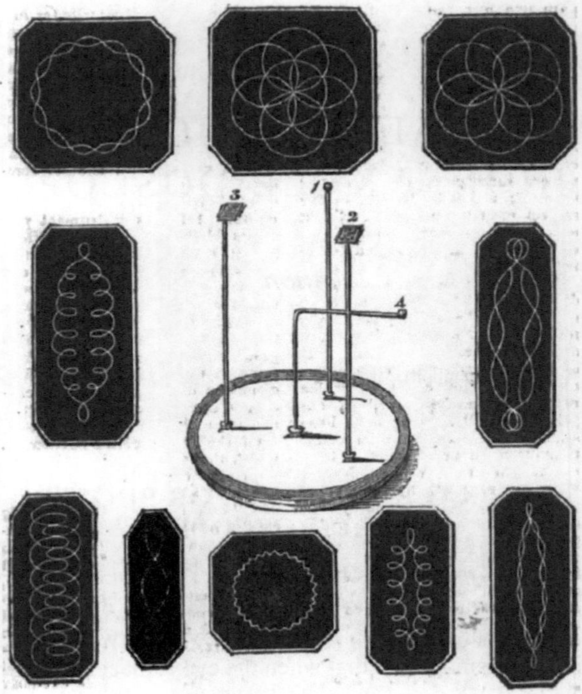

In the property of "creating beautiful forms," the Kaleidophone resembles the celebrated invention of Dr. Brewster, from which its name is modified; but between the two instruments and their modes of action there is no similarity whatever.

Mr. Wheatstone introduces his account of the instrument (Quarterly Journal of Science, New Series, No. 2) by the following just observations: –

> "The application of the principles of science to ornamental and amusing purposes, contributes in a great degree to render them extensively popular; for the exhibition of striking experiments induces the observer to investigate their causes with additional interest, and enables him more permanently to remember their effects. I shall not, therefore, need an apology for presenting the tyro in science with another combination of philosophy with amusement, in addition to those already extant. But this instrument possesses higher claims to attention; for it exemplifies an interesting series of natural phenomena, and renders obvious to the common observer what has hitherto been confined to the calculations of the mathematician; it presents another proof, that however remote from common observation the operations of nature may be, the most beautiful order and symmetry prevail through all."

Of the nature of the phenomena which the instrument is intended to illustrate, the following general idea is given: –

> "These experiments principally consist in subjecting to ocular demonstration the orbits of paths described by the points of greatest excursion in vibrating rods, which, in the most frequent cases (those of the combinations of different modes of vibration), assume the most diversified and elegant curvilinear forms. The entire track of each orbit is rendered simultaneously visible, by causing it to be delineated by a brilliantly luminous point; and the figure being completed in less time than the duration of the visual impression, the whole orbit appears as a continuous line of light: as, besides the changes which result from the combinations of the primitive with the higher modes of vibration, the figures of the orbits are affected by the form of the rod, by the extent of the excursions of the vibrations, by the mode of producing the motions, and by many other circumstances, a great variety of pleasing and regular forms is obtained. This variety is also enhanced by giving the same motions to a number of symmetrically disposed luminous points, the mutual intersections of the orbits of which produce innumerable elegant forms; and the appearances may be still more variegated by occasionally causing these points to reflect differently-coloured lights."

The apparatus for exhibiting these experiments (see the central figure in our front page) consists of a circular board about nine inches in diameter, into which are perpendicularly fixed, at equal distances from the circumference and from each other, three steel rods, each about a foot in length. The first rod is cylindrical, about 1–10th of an inch in diameter, and is surmounted by a spherical bead[1] which concentrates and reflects the light which falls upon it. The second is a similar rod, upon the upper extremity of which is placed a plate moving on a joint, so that its plane may be rendered either horizontal, oblique, or perpendicular; this plate is adapted to the reception of the objects, which consist of beads differently coloured, and arranged on pieces of black card in symmetrical forms. The third is a four-sided prismatic rod, and a similar plate is attached to its extremity for the reception of the same objects. Another rod is fixed at the centre of the board; this is bent to a right angle, and is furnished with a bead similarly to the first-mentioned rod. A small nut and screw is fixed to the board near the lower end of the first rod, in order, by pressing upon it, to render occasionally its rigidity unequal. A hammer, softened by a leather covering, is employed to strike the rods; and a violin-bow is necessary to produce some varieties of effect.

The different appearances which the rods present when in action, and the modes of producing the different effects, are thus described by Mr. Wheatstone: –

> "No. 1. On causing the entire rod to vibrate, so that its lowest sound be produced, as it is seldom that the motions of a cylindrical rod can be

confined to a plane, the vibrations will almost always be combined with a circular motion. When the pressure on the fixed end is exerted on two opposite points, and the rod put in motion in the direction of the pressure, the following progression in the changes of form will be distinctly observed: the track will commence as a line, and almost immediately open into an ellipse, the lesser axis of which will gradually extend as the larger axis diminishes, until it becomes a circle; what was before the lesser will then become the larger axis; and thus the motions will alternate, until, from their decreasing magnitudes, they cease to be visible. In the case just described, the ellipses make a right angle with each other; but by altering the direction of the motion, so as to render it oblique to the direction of the pressure, they may be made to intersect under any required angle, and when this angle = 0 the motion will be merely vibratory.

"Every single sound formed by the subdivisions of the rod will present similar appearances, but the excursions will be smaller as the sound is higher, or, which is the same thing, as the number of the vibrations increases.

"In the most simple case of the co-existence of two sounds, shown by putting the entire rod in motion, and producing also a higher sound by the friction of a bow, the original figure will appear waved or indented; and as unity is to the number of indentations, so will the number of vibrations in the lower sound be to the number in the higher sound. On varying the mode of excitation, by striking the rod in different parts and with different forces, very complicated and beautiful curvilinear forms may be obtained: (some of these are represented by the figures surrounding the instrument in the prefixed engraving).

"Placing the hand on the lower part of the rod, below the place at which it is excited, the excursions of the motions will rapidly decrease and exhibit spiral figures.

"To obtain the figures with brilliancy and distinctness, a single light only should be employed, as that of the sun, a lamp, or a candle; rays of light proceeding from several points, as from a number of candles, or from the reflection of the clouds, occasion the track to be broad and indistinct; but double lights may be employed with effect, provided they be of equal intensity and symmetrically placed; each bead will then describe two similar figures. The appearances, in a bright sunshine, are remarkably vivid and brilliant.

"No. 2. Although very beautiful and varied forms may be produced from the motion of a single point, yet the compound figures which are presented by objects formed by a number of points, offer appearances still more pleasing to the eye.

"An object being placed horizontally on the plate, and the rod being put in motion, the mutual intersections of the points, each describing a similar figure, present to the eye complicated yet symmetrical figures, resembling elegant specimens of engine-turning." (See the second of the figures in our front page).

"When the plate is horizontal, the figures are all in one plane; but if it be inclined or perpendicular, the curves being then made in parallel planes, give the idea of a solid figure, and in some cases the appearances are particularly striking.

"Complementary colours alone should be employed in the objects; for these harmonizing together, give greater pleasure to the eye than an injudicious combination of discordant tints: the intensities should be occasionally varied, and colourless light intermingled with the different shades.

"No. 3. When this prismatic rod is put in motion, in the direction of either of its sides, the points move only rectilineally, but when the motion is applied in an oblique direction, a variety of compound curves is shown.

"No. 4. When a rod is straight, the curve produced by any point describing its motion, is always in the same plane; but in a rod bent to any angle, the two parts moving most frequently in different directions, curves are produced whose parts do not lie in the same plane. A few trials will soon indicate the best way of applying the motion, so as to cause the two parts to vibrate in different directions."

Mr. W. remarks, in conclusion, "that the application of this mode of experimenting may be extended to the delineation of every description of curvilinear and angular motion, when the amplitudes of the tracks are not too great; and that, by this means, it is not improbable that the experimental knowledge of many interesting principles in science may be facilitated."

NOTE

1 "The only beads well adapted for this purpose are made of extremely thin glass silvered on the interior surface, and about one-sixth of an inch in diameter; they are to be obtained at the shops under the name of steel beads. The protuberances at the apertures must be removed or blackened, otherwise the reflections from them will render the images confused. To produce the coloured tracks, these beads must be coated with transparent colours, such as are ordinarily used for painting on glass; the light will then be reflected through the coloured surface; but in beads made of coloured glass, the reflection being made from the external surface, shows only white light. The bead is cemented into a small brass cup screwed to the top of the wire."

[illegible]

[illegible] [illegible] [illegible]

When [illegible] [illegible]

[illegible] particularly valid [illegible].

[illegible]

No. 1. When this [illegible]

No. 2. When [illegible]

[illegible] lower [illegible] in either [illegible] direction.

We [illegible] conclude [illegible] that [illegible] application [illegible]

[illegible]

[illegible]

7

EXPERIMENTS ON AUDITION

Charles Wheatstone

Source: Charles Wheatstone, "Experiments on Audition", in *The Scientific Papers of Sir Charles Wheatstone, D.C.L., F.R.S* (London: Taylor and Francis, 1879), pp. 30–35. Originally from the *Quarterly Journal of Science*, 1827.

[From the *Quarterly Journal of Science*, 1827, pt. ii.]

THE recent valuable experiments of Savart and of Dr. Wollaston have added to our stock of information several important and hitherto unnoticed phenomena relating to audition; but notwithstanding the investigations of these distinguished experimentalists, and though the physiology of the ear has been an object of unceasing attention for many centuries, yet we are far from possessing a perfect knowledge of the functions of the various parts of this organ. The description of new facts illustrative of this subject cannot, therefore, be devoid of interest; and though I do not anticipate that the observations contained in this communication will lead to any important results, their novelty may claim for them some attention from the readers of your Journal.

§ 1.

If the hand be placed so as to cover the ear, or if the entrance of the meatus auditorius be closed by the finger without pressure, the perception of external sounds will be considerably diminished, but the sounds of the voice produced internally will be greatly augmented: the pronunciation of those vowels in which the cavity of the mouth is the most closed, as *e, ou*, &c., produce the strongest effect; on articulating smartly the syllables *te* and *kew*, the sound will be painfully loud.

Placing the conducting stem of a sounding tuning-fork on any part of the head when the ears are closed as above described, a similar augmentation of sound will be observed. When one ear remains open, the sound will always be referred to the closed ear; but when both ears are closed, the sound will appear louder in that ear the nearer to which it is produced. If, therefore, the tuning-fork be applied above the temporal bone near either ear, it will be apparently heard by that ear to which it is adjacent; but on removing the hand from this ear (although the fork remains in the same situation), the sound will appear to be referred immediately to the opposite ear.

In the case of the vocal articulations, the augmentation is accompanied by a reedy sound, occasioned by the strong agitations of the tympanum. When the air in the meatus is compressed against this membrane by pressing the hand *close* to the ear, or when the eustachian tube is exhausted by the means indicated by Dr. Wollaston, the reedy sound is no longer heard, and the augmentation is considerably diminished. The ringing noise which simultaneously accompanies a very intense sound proceeds from the same cause, and may be prevented by the same means. This ringing may be produced by applying the stem of a sounding tuning-fork to the hand when covering the ear, or by whistling when a hearing-trumpet is placed to the ear. As a proof that the resulting augmentation, which, when great, excites the vibrations of the tympanum, is owing to the reciprocation of the vibrations by the air contained within the closed cavity, it may be mentioned, that when the entrance of the meatus is closed by a fibrous substance, as wool, &c., no increase is obtained.

If the meatus and the concha of the ear be filled with water, the sounds above mentioned will be referred to the cavity containing the water, in the same way as when it contained air and was closed by the hand; it will be indifferent whether any partition be interposed between the cavity and the external air, as the water is equally well insulated by a surface of air as by a solid body.

§ 2.

The preceding experiments have shown that sounds immediately communicated to the closed meatus externus are very greatly augmented; and it is an obvious inference, that if *external* sounds can be communicated so as to act on the cavity in a similar manner, they must receive a corresponding augmentation. The great intensity with which sound is transmitted by solid rods, at the same time that its diffusion is prevented, affords a ready means of effecting this purpose, and of constructing an instrument which, from its rendering audible the weakest sounds, may with propriety be named a Microphone.

Procure two flat pieces of plated metal, each sufficiently large to cover the external ear, to the form also of which they may be adapted; on the outside of each plate, directly opposite the meatus, rivet a rod of iron or brass wire about 16 inches in length and one eighth of an inch in diameter, and fasten the two rods together at their unfixed extremities, so as to meet in a single point. The rods must be so curved, that when the plates are applied to the ears, each rod may at one end be perpendicularly inserted into its corresponding plate and at the other end may meet before the head in the plane of the mesial line. The spring of the rods will be sufficient to fix the plates to the ears; but for greater security ribands may be attached to each rod near its insertion in the plate, and be tied behind the head.

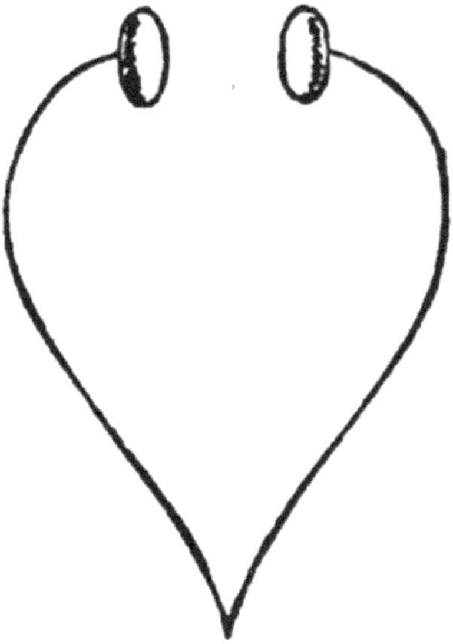

A more simple instrument may be constructed to be applied to one ear only, by inserting a straight rod perpendicularly into a similar plate to those described above.

The Microphone is calculated only for hearing sounds when it is in immediate contact with sonorous bodies: when they are diffused by their transmission through the air, this instrument will not afford the slightest assistance.

It is not my intention in this place to detail all the various experiments which may be made with this instrument; a few will suffice to enable the experimenter to vary them at his pleasure.

1. If a bell be rung in a vessel of water, and the point of the microphone be placed in the water at different distances from the bell, the differences of intensity will be very sensible. 2. If the point of the microphone be applied to the sides of a vessel containing a boiling liquid, or if it be placed in the liquid itself, the various sounds which are rendered may be heard very distinctly. 3. The instrument affords a means of ascertaining, with considerable accuracy, the points of a sonorous body at which the intensity of vibration is the greatest or least: thus, placing its point on different parts of the sounding-board of a violin or guitar, whilst one of its strings is in vibration, the points of greatest and least vibration are easily distinguished. 4. If the stem of a sounding tuning-fork be brought in contact with any part of the microphone, and at the same time a musical sound be produced by the voice, the most uninitiated ear will be able to perceive the consonance or dissonance of the two sounds; the roughness of discords and the beatings of imperfect consonances, are thereby rendered so extremely disagreeable, and form so evident a contrast to the agreeable harmony and smoothness of two perfectly consonant sounds, that it is impossible that they can be confounded.

§ 3.

Apply the broad sides of two sounding tuning-forks, both being unisons, to the same ear; on removing one fork to the opposite ear, allowing the other to remain, the sensation will be considerably augmented.

It is well known, that when two consonant sounds are heard together, a third sound results from the coincidences of their vibrations; and that this third sound, which is called the grave harmonic, is always equal to unity when the two primitive sounds are represented by the lowest integral numbers. This being premised, select two tuning-forks the sounds of which differ by any consonant interval excepting the octave: place the broad sides of their branches, while in vibration, close to one ear, in such a manner that they shall nearly touch at the acoustic axis; the resulting grave harmonic will then be strongly audible, combined with the two other sounds; place afterwards one fork to each ear, and the consonance will be heard much richer in volume, but no audible indications whatever of the third sound will be perceived.

§ 4.

Very acute sounds, such as the chirping of the *Gryllus campestris*, &c., are rendered inaudible by exhausting the air from the Eustachian tube, and thereby producing a tension of the membrane of the tympanum. The different thicknesses or tensions of this membrane may therefore occasion that diversity of the limits of audibility, with regard to the acute sounds, which Dr. Wollaston has pointed out as existing in different individuals; if so, it would be desirable to ascertain this limit in individuals in whom the tympanum is perforated, or destroyed.

§ 5.

When the auricula is brought forward, all *acute* sounds are rendered much more intense, but no sensible difference is perceived with regard to the grave sounds. The *higher* tones of glass staccados, or of an octave flute, the ticking of a watch, all kinds of sibilant sounds, &c. are thus greatly augmented: the experiment is easily tried by whistling very shrill notes. A still greater augmentation of the acute sounds is obtained, by placing the hands formed into a concave behind the ears, and by bending downwards the upper part of the auricula, so as to obtain a more complete cavity.

§ 6.

I will conclude with the following observation: I had, in consequence of a cold, a very slight pain in my left ear; on sounding the regular notes of the piano-forte, C^3 and C^4 were much louder than the others, and the loudness was much increased by placing the hand in the manner above described to the left ear. When it was pressed close, or when the Eustachian tube was closed, the intensities of all the notes were equalized. I attribute this affection to the diminished tension of the membrana tympani, which was again increased by the operation described.

Editorial Headnote

8. Charles Wheatstone, "On the Resonances, or Reciprocated Vibrations of Columns of Air", *Quarterly Journal of Science* III (London, England, 1828), pp. 175–83.

Between 1828 and 1832, Charles Wheatstone collaborated with celebrated experimentalist Michael Faraday (1791–1867) to deliver some fourteen lectures at the Royal Institution on subjects relating to sound. A matter of particular concern was the phenomenon of resonance, or reciprocating vibrating columns of air, which formed the basis of a paper Wheatstone published in the *Quarterly Journal of Science* in 1828. Although Faraday had delivered the contents of this paper through a series of Royal Institution lectures between February and April 1828, due to Wheatstone's anxieties surrounding public speaking, this was very much the product of the instrument maker's research. Rich in experiments conducted under Wheatstone's direction, Faraday demonstrated how the laws of resonance operated in a range of musical instruments, including flutes, of which he himself was an accomplished performer. Faraday and Wheatstone continued to demonstrate how it was the principles of resonance that had been utilized for musical effect in the Javanese "Gender", an instrument consisting of a series of plates suspended over tubes of bamboo. When the plates were vibrated, these pipes augmented the sound, with their lengths measured to reciprocate the lowest tone of each plate.[1] Sir Stamford Raffles (1781–1826) had brought the Gender back to London from Java on his final voyage, and it was, by 1828, in the possession of his widow, Lady Sophia Raffles (1786–1858). Faraday subsequently secured the loan of this instrument, which combined exotic excitement with philosophical understanding for the Royal Institution's fashionable clientele.[2] Of similar sensation were the performances of Karl Eulenstein (1802–1890), a German musician who had secured fame across Europe for his mastery of the Jew's Harp.[3] Wheatstone recruited Eulenstein for his Royal Institution lectures to help illustrate the relationship between musical sounds and the behaviour of vibrating columns of air in the cavity of a mouth. As Wheatstone explained, "when the number of vibrations of the contained volume of air is any multiple of the original vibrations of the tongue, a sound is produced, corresponding to the modification of the oral cavity".[4] Eulenstein's performances were both musical and theatrical, but Wheatstone transformed them into a subject of scientific discourse. Subsequent lectures would see Wheatstone extend this philosophical treatment of musical sounds, including with the assistance of a vocalist by the name of Mannin, who possessed the unusual "power of producing two simultaneous sounds from the mouth".[5] As with Eulenstein, Mannin's vocal performance was followed by Wheatstone's account of how the phenomenon was achieved. As much as these were, undoubtedly, Wheatstone's lectures, the role of Faraday in delivering them should not be underestimated. As correspondence from the *Dublin Literary Gazette* reflected in 1830, "What an extraordinary being is this Michael Faraday! . . . He is the improvisatore of the scientific world – the organ of new inventers – the mouthpiece of dumb mechanic geniuses. Do you wish to make known your discoveries in any art or science? Talk to Mr Faraday for ten minutes, and he will announce your labours to the public ear in the most captivating form imaginable".[6] In comparison, Wheatstone appeared to the correspondent as "a young man of anxious, spectacle-on-nose appearance", reluctant to speak but instead watching on from the side of the Royal Institution's lecture theatre.

NOTES

1 On resonance, see Myles W. Jackson, "Charles Wheatstone: musical instrument making, natural philosophy, and acoustics in early nineteenth-century London", in James Q. Davies and Ellen Lockhart (eds.), *Sound Knowledge: music and science in London, 1789–1851* (Chicago: University of Chicago Press, 2016), pp. 101–24, 113–15.

2 James Q. Davies, "Instruments of Empire", in James Q. Davies and Ellen Lockhart (eds.), *Sound Knowledge: music and science in London, 1789–1851* (Chicago: University of Chicago Press, 2016), pp. 145–74.

3 On Eulenstein, see (Anon.), "Mr Eulenstein's Jew's Harp", *Liverpool Mercury ect*, no. 1121 (26 Oct. 1832, Liverpool, England); (Anon.), "Eulenstein", *Bristol Mercury*, no. 2242 (6 Apr. 1833, Bristol, England); (Anon.), *A Sketch of the Life of C. Eulenstein, the celebrated performer on the Jews' Harps*, (London: James J. Welsh, 1833); Neuffer, "To Eulenstein, Musician on the Jew's Harp", in Leonard Fox (ed.), *The Jew's Harp: a comprehensive anthology*, (Lewisburg, PA: Bucknell University Press, 1988), pp. 162–13.

4 Wheatstone, *The Scientific Papers of Sir Charles Wheatstone*, pp. 43–44.

5 (Anon.), "Musical Sound", *Arcana of Science and Art* (1830, London, England), pp. 65–67, 66.

6 (Anon.), "Royal Institution", *The Dublin Literary Gazette, or Weekly Chronicle of Criticism, Belles Lettres, and Fine Arts* vol. 1 (Dublin, Ireland, 1830), pp. 395–96, 395.

8

ON THE RESONANCES, OR RECIPROCATED VIBRATIONS OF COLUMNS OF AIR

Charles Wheatstone

Source: Charles Wheatstone, "On the Resonances, or Reciprocated Vibrations of Columns of Air", *Quarterly Journal of Science* III (London, England, 1828), pp. 175–83.

An elastic body may be made to assume a vibratory state in two ways; either, immediately, by any momentary impulse, which, altering the natural positions of its particles, allows them afterwards to return by a succession of isochronous oscillations to their former state; or, secondarily, by means of an immediately sounding body, which causes it to reciprocate to the latter, when certain conditions, on which depends its susceptibility of vibrating in such a manner, are fulfilled. This reciprocation, to which, when the effect is referred to, the term *resonance* is applied, is effected by means of the undulations which are produced in the air, or in any fluid or solid medium, by the periodical pulses of the original vibrating body; these undulations being capable of putting in motion all bodies whose pulses are coincident with their own, and, consequently, with those of the primitive sounding body. – Galileo observed that a heavy pendulum might be put in motion by the least breath of the mouth, provided the blasts be often repeated, and keep time exactly with the vibrations of the pendulum; and this remark affords a correct explanation of the phenomenon.

Some of the most obvious cases of resonance are, – the vibrations of a string when another tuned in unison with it is made to vibrate; the resounding of a drinking-glass to the sound of the voice, or of a musical instrument; the reciprocated vibrations of a sounding-board, communicating immediately with a vibrating string or tuning-fork, &c. In the last mentioned instance, though the string and the fork are the original vibrating bodies, the audible sound is dependent on the resonance of the sounding-board.

As all these effects are well known, it is unnecessary to dilate upon them here, and I may uninterruptedly proceed to the immediate object of the present paper, viz., the investigation of the laws of the resonance or phonic reciprocation of columns of air.

DOI: 10.4324/b23406-8

§ 1. If one of the branches of a vibrating tuning-fork be brought near the embouchure of a flute, the lateral apertures of which are stopped so as to render it capable of producing the same sound as the fork, then the feeble and scarcely audible sound of the fork will be augmented by the rich resonance of the column of air within the flute.[1] The sound will be found greatly to decrease by closing or opening another aperture; for the alteration of the length of the column of air in such case renders it no longer proper to reciprocate perfectly the sound of the fork. This experiment may be easily tried on a concert flute, with a C tuning-fork. To ensure success, it is necessary to remark, that when a flute is blown into with the mouth, the under lip partly covering the embouchure, renders the sound about a semitone flatter than the sound when the embouchure is entirely uncovered; and as the latter must be unison to that of the tuning-fork, it is necessary, in most cases, to finger the flute for B when a C tuning-fork is employed.

A similar effect may be produced by substituting, for the column of air in the flute, the alterable volume of air contained within the cavity of the mouth. I have found the sounds of tuning-forks reciprocated most intensely by placing the tongue, &c. in the position for the nasal continuous sound of *ng* (in song), and then altering the aperture of the lips until the loudest sound is obtained.

§ 2. A column of air may also reciprocate a sound originally produced by a wind instrument, as the following experiment will show. Place two concert flutes on a table, parallel to, and at a short distance from each other; on the one which is nearer, sound C sharp (all the lateral apertures being open), and draw out the tube of the second flute, so that it shall be about a semitone flatter, to make it equivalent to the flattening of the first flute by the partial closing of the embouchure by the lip; a material difference will then be distinguished in the intensity of the tone by alternately closing and opening the first hole of the more distant instrument, thereby rendering it incapable or capable of reciprocating the original sound. That this effect is occasioned solely by the transmission of the sonorous undulations, and not by any wind actually blown into the second flute, is evident from the difference being in intensity and not in pitch.[2]

This experiment may be varied by placing the fipple of a flageolet at a short distance from the embouchure of a flute, provided, of course, that the columns of air, both in the flageolet and the flute, be capable of producing the same note.

§ 3. A cylindric or prismatic column of air, in a tube open at both ends, may vibrate not only in its entire length, but also in any number of aliquot parts, and in all cases the number of vibrations is inversely as the length of a single vibrating part. As a column of air is capable of reciprocating every sound which, according to its different modes of vibration, it is itself capable of producing; supposing $1=C^1$ to represent the lowest sound of the tube, it will without any change in its length, reciprocate sounds whose relations are $\begin{matrix} 1 & 2 & 3 & 4 & 5 & 6 & 7 & 8 \\ C^1 & C^2 & G^2 & C^3 & E^3 & G^3 & Bb^3 & C^4 \end{matrix}$ *&c.*

The harmonic subdivisions of a column of air in a tube closed at one end are different; a semi-vibrating part always exists near the closed end, but between two nodes, or a node and the open end, complete vibrating parts, as in an open tube, exist. The fundamental sound above mentioned, of an open tube, is given by a tube closed at one end, of one-half its length; the series, corresponding with the subdivisions, compared with the above, is $\begin{matrix} 1 & 3 & 5 & 7 & 9 \\ C^1 & G^2 & E^3 & Bb^3 & D^4 \end{matrix}$ &c. and these sounds it can consequently reciprocate.

§ 4. Any one among several simultaneous sounds may be rendered separately audible. Thus, if two vibrating tuning-forks, differing in pitch, be held over a closed tube, furnished with a moveable piston, either sound may be made to predominate by altering the piston, so as to enable the column of air to reciprocate the sound required. The same result may be obtained by selecting two bottles (which may be tuned with water), each corresponding to the sound of a different tuning-fork; on bringing both tuning-forks to the mouth of each bottle alternately, in each case that sound only will be heard which is reciprocated by the unisonant bottle.[3]

The phenomenon of a third sound, produced by the coincidences of the vibrations of two consonant sounds, is well known. From what has been premised, it is reasonable to suppose that if a column of air be caused to reciprocate this third sound, or grave harmonic as it is called, and not the two sounds which generate it, it might be heard, uncombined with the two other sounds. But on attempting this experiment, I was unable, on the following account, to succeed. The third sound is always unity when the ratio of the lowest sounds is reduced to its lowest terms; thus, with respect to a perfect fifth, 2: 3, the third sound 1, is an octave below the lower sound, and the grave harmonic 1, of a major third, 4: 5, is two octaves below the lower sound; we will suppose this fundamental sound, represented by unity, to be the C, corresponding with the sound of an open tube of four feet, or of a closed tube of two feet: in the first case, the column of air being capable, as explained in § 3, of vibrating in any number of aliquot parts, not only the grave harmonic = 1, but the sounds represented by 2: 3 and 4: 5, will also be reciprocated; and in the latter case, where the subdivisions are as the arithmetical progression 1, 3, 5, 7, 9, 11, &c., one sound of each consonance will be reciprocated, besides the grave sound. The expected result may probably still be obtained by employing columns or volumes of air, whose subdivisions are less regular.

§ 5. Among the Javanese musical instruments brought to England by the late Sir Stamford Raffles, there is one called the Gender, in which the resonances of unisonant columns of air are employed to augment, I may almost say to render audible, the sounds of vibrating metallic plates. Of these plates there are eleven; their sounds correspond with the notes of the diatonic scale, deprived of its fourth and seventh, and extend through two octaves. The mode of vibration of the plates is that with two transversal nodal lines; and they are suspended horizontally by two strings, one passed through two holes in the one nodal line, and the other through similar holes in the other nodal line of each plate. Under each plate is placed an upright bamboo, containing a column of air, of the proper length to reciprocate the lowest sound of the plate. If the aperture of the bamboo be covered with pasteboard, and its corresponding plate be struck, a number of acute sounds only (depending on the more numerous subdivisions of the plate) will be heard; but on removing the pasteboard, an additional deep, rich tone is produced by the resonance of the column of air within the tube.

THE GENDER. Musical Instrument of Java.

The instrument from which the annexed drawing was taken is at present in the museum of the Honourable East India Company; and there is another specimen in the possession of Lady Raffles.

The same principle appears to have been employed, in more rude forms, in the construction of several Asiatic and African musical instruments; but I am unaware of any instrument having yet been manufactured in Europe, in which the unisonant resonances of columns of air have been made available as a means of augmenting the intensity of sounds. I shall shortly publish an account of the several modes I have myself devised, and practically employed, for advantageously applying this principle.

§ 6. In the experiments I have hitherto detailed, the reciprocated vibrations have always been isochronous with those of the original sounding body; or, in other words, the resonance and the original sound have always been unisonant. The experiment I shall now bring forward will show that this is not universally the case, and that there are other phenomena of resonance which have never, hitherto, been investigated in their theory, or in their practical applications. I took a tube, closed at one end by a moveable piston, and placed before its open end the branch of a vibrating tuning-fork of the ordinary pitch, C; the length of the column of air was six inches; on diminishing the length of the column to three inches, the sound of the tuning-fork was no longer reciprocated, but its octave above (the sound of the column when it is directly excited) was produced. By employing a graver tuning-fork and tubes of very small diameter, and successively adjusting the lengths of the columns of air so as to be one half, one third, one fourth, one fifth, &c. of the column reciprocating the fundamental sound, the octave, twelfth, double octave, seventeenth, &c. to that sound will be produced. The relative numbers of the vibrations of these sounds, considering the vibrations of the fork as unity, are 1, 2, 3, 4, 5, &c. It therefore is evident, from experiment, that a column of air may vibrate by reciprocation, not only with another body whose vibrations are isochronous with its own, but also *when the number of its own vibrations are any multiple of those of the original sounding body*.

The converse of this law does not hold; for when the number of vibrations of a column of air are any sub-multiple of those of the original sounding body, there is no resonance. To prove this with regard to the octave, let the length of the column of air unison to the sound of a fork be doubled, and not the slightest trace of the octave below (i. e. the real sound of the column) will be perceptible: this negative experiment must be tried with a closed tube which is incapable of producing a harmonic octave; an open tube would resound unisonantly to the fork by its subdivision.

§ 7. On the law experimentally established in the preceding paragraph depends the explanation of the production of sounds by the guimbarde or Jew's harp. This simple instrument consists of an elastic steel tongue, riveted at one end to a frame of brass or iron, the form of which is represented in the annexed figure; the free extremity of the tongue is bent outwards to a right angle, so as to allow the finger easily to strike it when the instrument is placed to the mouth, and firmly supported by the pressure of the parallel extremities of the frame against the teeth.

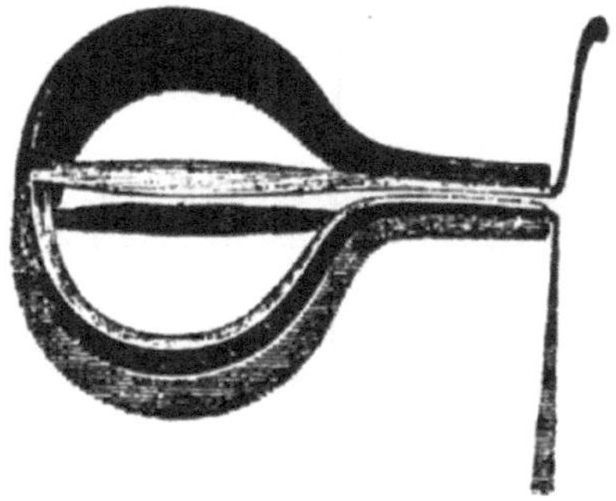

The vibrations of the tongue itself correspond with a very low sound; but being placed before the cavity of the mouth, the form and dimensions of which are capable of various alterations by the motions of the tongue and the lips, when the number of vibrations of the contained volume of air is any multiple of the original vibrations of the tongue, a sound is produced, corresponding to the modification of the oral cavity.

Supposing the primitive sound of the tongue to be C^1, the series of sounds which it can produce by multiple reciprocation will be as follows.

Multiples of the original vibrations of the tongue.

1...2...3...4...5...6...7...8...9...10...11...12...13...14...15...16 etc. 32.

Corresponding sounds.

C^1 C^2 G C^3 E G B♭ C^4 D E F$_+$ G A- B♭ B C^5 etc. C^6

If the original sound be any other note, another series will arise, but the intervals of the successive sounds will always preserve the above relations. By the usual jew's harps the three first sounds of the series cannot be produced, the dimensions of the cavity of the mouth not being sufficiently large to reciprocate them.

The scale above exhibited is evidently too incomplete and too defective, to allow even the most simple melodies to be performed on a single jew's harp; but the deficiencies may be supplied by employing two or more of these instruments. Mr. Eülenstein, by availing himself of the resources afforded him by the scales of sixteen jew's harps, is able to modulate through every key, and to produce effects truly original, and of extreme beauty. Those who have heard only the rude twanging to which the performance of this instrument in ordinary hands is confined, can have no idea of the melodious sounds which, under Mr. Eülenstein's management, it is capable of producing.

§ 8. The following experiment will prove the accuracy of the preceding explanation, and will establish the true theory of the production of sounds by the guimbarde, beyond the possibility of doubt.

I fixed a jew's harp firmly at the two points where ordinarily it rests against the teeth, allowing sufficient space between the two supports for the tongue to vibrate freely to its greatest extent; and I tuned the tongue by applying wax to its free extremity, until it sounded C, corresponding to the sound of a closed tube four feet in length. I then placed before the tongue the open end of a closed tube, containing a column of air two feet in length and one inch in diameter, and furnished with a moveable piston, by which the column could be shortened to any required length. On striking the tongue, the octave of the fundamental sound was produced, being the sound 2 of the scale in the preceding section; by shortening the column of air successively to one third, one fourth, one fifth, one sixth, one seventh, &c. &c. up to very numerous aliquot subdivisions, all the notes of the series were correctly obtained. By marking the different lengths of the piston rod for each sound of the series, I was able to produce the notes of the scale, ascending or descending regularly, or to fix any sound at pleasure.

§ 9. No other sounds can be produced by reciprocation from columns of air, but those perfectly identical with the multiples of the original vibrations of the tuning-fork, or the tongue of the jew's harp. On inquiring what takes place when the length of the column is intermediate between the lengths appropriate to reciprocate two succeeding multiples, – it will be found, that though each sound of the series is heard most audibly when the column is accurately adapted to it, yet it may also be heard, unaltered in tune, though diminished in intensity, when the column is lengthened or shortened within a certain extent, which is greater for the lower sounds of the series, and less for the higher, on account of the wider intervals between the sounds in the former case.

It may now be understood in what manner a column of air is capable of reciprocating simultaneously two or three sounds of a chord. In Mr. Eülenstein's performance, this effect is thus produced: suppose the perfect major chord of C is required; three jew's harps, incapable of producing lower sounds than the fourth of the series, are selected; and the mouth being made to correspond with the C, the other two sounds, E and G, are likewise reciprocated, though faintly, because these sounds are nearer that to which the mouth is adapted, than any other multiples of the original vibrations of the tongues are.

§ 10. When two imperfect unisons are sounded together, the interferences of their undulations give rise to periodical alternations in the intensity of the sound; this effect is called beating, and the greater the deviation from unison is, the more rapid are the beats. From what has been above stated, the same column of air may reciprocate sounds differing in pitch; if, therefore, two vibrating tuning-forks, imperfect unisons to each other, be brought together over the embouchure of a flute, or the open end of a tube containing an appropriate column of air, the periodical recurrence of the beats will be rendered strikingly evident, slowly or rapidly succeeding each other, accordingly as the forks are more or less in tune with each other.

NOTES

1 Dr. Savart has observed a similar effect by sounding a bell before a large tube inclosing an unisonant column of air. *Recherches sur les Vibrations de l'Air*. Annales de Chimie, tom. 24.

2 Lord Bacon may be said to have anticipated this experiment in the following passage in his Sylva Sylvarum: – "The experiments of sympathy may perhaps be transferred from stringed instruments to others; as, if there were two bells in unison in one steeple, to try whether striking the one would move the other, more than if it were a different chord; and so in pipes, of equal bore and sound, to try whether a light straw or feather would move in one pipe, when the other is blown in unison with it." *Art.* Phonics, *Sect.* xxi. *On the Sympathy or Antipathy of Sounds with one another.*

3 Had Sir Isaac Newton been acquainted with these, or with any similar facts he might have illustrated his theory of the reflection of colours by an experimental, instead of a suppositious analogy with the reciprocation of sounds. As the passage in which this comparison is made, is remarkable, I will quote it. "If light be considered without respect to any hypothesis, I can as easily conceive that the several parts of a shining body may emit rays of different colours, and other qualities, of all which light is constituted, as that the several pipes of an organ inspired all at once, or all the variety of sounding bodies in the world together, should produce sounds of several tones, and propagate them through

the air, confusedly intermixed. And, if there were any natural bodies that would reflect sounds of one tone, and stifle or transmit those of another, then, as the echo of a confused aggregate of all tones would be that particular tone, which the echoing body is disposed to reflect; so, since there are bodies apt to reflect rays of one colour, and stifle or transmit those of another, I can as easily conceive that those bodies, when illuminated by a mixture of all colours, must appear of that colour only which they reflect." *Philosophical Transactions, No.* 88.

Editorial Headnote

9. Charles Wheatstone, "On the transmission of musical sounds through solid linear conductors, and on their subsequent reciprocation", *Journal of the Royal Institution of Great Britain*, no. 5 (London: John Murray, Dec. 1831), pp. 223–38.

The movement of Wheatstone's acoustic marvels from his shop on The Strand to the Royal Institution's lecture theatre constituted an escalation from popular entertainment to a subject of considered scientific discussion. None of Wheatstone's contrivances would achieve this transition with greater efficiency than his Enchanted Lyre mechanism, a staple of the West End's popular entertainment since 1821. In 1830, Wheatstone worked with Faraday to set up the Enchanted Lyre in the Royal Institution's lecture theatre as the centrepiece for a lecture on sonorous communications through solid linear conductors, which Faraday would deliver on 5 March. Wheatstone subsequently published an account of this in 1831 in the *Journal of the Royal Institution of Great Britain*. Wheatstone's lecture combined the principles of the Enchanted Lyre with a description of Chladni's acoustic figures, bringing together the phenomenon of sonorous vibrations with that of resonance. The key, Wheatstone explained, to successfully transmitting waves through a solid material was that the sounding body was capable of resonating at the same frequency as the original transmission, such as in the movement of vibrations from a violin's strings to its wooden soundboard via its bridge. The Enchanted Lyre demonstrated this perfectly. Wheatstone and Faraday had bored a hole through the Royal Institution's lecture theatre's floor and inserted a forty-foot rod, connecting an external tuning fork to a sounding board in the theatre itself. When vibrated, the tuning fork caused the sounding board to emit the same notes throughout the theatre. Wheatstone then connected the rod's exterior end to a piano, which could then be used to make a lyre or guitar play music for the Royal Institution audience.[1] Importantly, Wheatstone was quick to emphasize the potential of this arrangement for long-distance sonorous communication. Before long, he speculated, tubes carrying rods would be capable of conveying music across London and to domestic homes, whose inhabitants would be able to "hear at pleasure the performance in any of these rooms, by merely attaching a reciprocating instrument to the conductors".[2] Given that his rods could communicate sound at a velocity of 18,000 feet per second, Wheatstone boasted that music might be transmitted 200 miles in a minute, potentially linking performances in London to listeners as far away as Aberdeen. Nor were such claims lacking in credibility.

The *Philosophical Magazine*, for one, put great store into Wheatstone's promise, as did the *Literary Gazette*, which felt his research showed

> the possibility of telephonic communication. It was stated, that as sound has proved to travel through several conducting substances at the extraordinary velocity of 18,000 feet in a second, were it possible to transmit audible sound so far, a phonic communication might be made between London and Edinburgh – a distance of nearly 400 miles – in less than two minutes.[3]

Wheatstone's work on solid linear conductors of sound was more than a spectacle or a subject of philosophical inquiry but promised to usher in a utopian age of rapid communication.

NOTES

1 On Wheatstone's Royal Institution lectures on sound, see Edward J. Gillin, *Sound Authorities: scientific and musical knowledge in nineteenth-century Britain* (Chicago: University of Chicago Press, 2022), pp. 42–61.
2 Wheatstone, *The Scientific Papers of Sir Charles Wheatstone*, pp. 57–60.
3 (Anon.), "Royal Institution", *The Literary Gazette* (1830, London, England), pp. 223–24, 224.

9

ON THE TRANSMISSION OF MUSICAL SOUNDS THROUGH SOLID LINEAR CONDUCTORS, AND ON THEIR SUBSEQUENT RECIPROCATION

Charles Wheatstone

Source: Charles Wheatstone, "On the Transmission of Musical Sounds Through Solid Linear Conductors, and on their Subsequent Reciprocation", *Journal of the Royal Institution of Great Britain*, no. 5, (London: John Murray, Dec. 1831), pp. 223–38.

§ 1.

THE fact of the transmission of sound through solid bodies, as when a stick or a metal rod is placed at one extremity to the ear, and is struck or scratched at the other end, did not escape the observation of the ancient philosophers: but it was for a long time erroneously supposed, that an aëriform medium was alone capable of receiving sonorous impressions; and in conformity with this opinion, Lord Bacon, when noticing this experiment, assumes that the sound is propagated by spirits contained within the pores of the body.[1] The first correct observations on this subject appear to have been made by Dr. Hooke in 1667; who made an experiment with a distended wire of sufficient length to observe that the same sound was propagated far swifter through the wire than through the air.[2] Professor Wunsch, of Berlin, made, in 1788, a similar experiment, substituting 1728 feet of connected wooden laths for the wire, and confirmed Dr. Hooke's results.[3] Other experiments of a similar nature were subsequently made by Herhold and Rafn,[4] Hassenfratz and Gay Lussac,[5] &c.; but the first direct observations of the actual velocity of sound through solid conductors were made by Biot, assisted at different times by Bouvard and Martin. These experiments were made on the sides of the iron conduit-pipes of Paris, through the length of 951m. 25; and the mean result of two observations made in different ways gave 3459 metres, or 11,090 feet per second, for the velocity of sound in cast iron.[6]

Previously to these last-mentioned experiments, Chladni had, in an ingenious manner, inferred the velocity of sound in different solid substances; and his results are fully confirmed by calculations from other grounds. His method was founded

DOI: 10.4324/b23406-9

on Newton's demonstration, that sound travels through a space of a given length, filled with air, in the same time that a column of air of the same length, contained in a tube open at both ends, makes a single vibration.[7] His own discovery of the longitudinal vibrations of solid bodies, which are exactly analogous to the ordinary vibrations of columns of air, enabled him to apply this proposition to solid bodies, and to establish the general law, that sound is propagated through any elastic substance in the same time in which this substance makes one longitudinal vibration. In this manner he ascertained the velocities of sound in the following substances, among others: tin 7,800, silver 9,300, copper 12,500, glass and iron 17,500, and various woods from 11,000 to 18,000 feet in a second.

From the experiments of M. Perolle,[8] it would appear that the intensity with which sound is communicated through solid matters is nearly in proportion to the velocity of its transmission.

§ 2.

In all the experiments above alluded to, the sounds transmitted were either mere noises, such as the blow of a hammer, or, as in Herhold and Rafn's experiment, a single musical sound, produced by striking a silver spoon attached to one end of the conducting wire; and in no case were any means employed for the subsequent augmentation of the transmitted sound. I believe that, previous to the experiments which I commenced in 1820, none had been made on the transmission of the modulated sounds of musical instruments; nor had it been shown that sonorous undulations, propagated through solid linear conductors of considerable length, were capable of exciting, in surfaces with which they were in connexion, a quantity of vibratory motion, sufficient to be powerfully audible when communicated through the air. The first experiments of this kind which I made were publicly exhibited in 1821, and notices of them are to be found in the Literary Gazette, Ackerman's Repository, and other periodicals of that year. On June 30, 1823, a paper of mine was read by M. Arago, at the Academy of Sciences in Paris, in which I mentioned these experiments, and a variety of others relating to the passage of sound through rectilinear and bent conductors.[9] I propose, in the present instance, to give a more complete detail of these experiments than I have yet published; and at the same time to add what additional facts my subsequent experience has furnished me with on the same subject.

§ 3.

Before proceeding any further, it will be necessary to make a few observations on the augmentation of sound which results from the connexion of a vibrating body with other bodies capable of entering into simultaneous vibration with it. This participation of the vibrations of an original sounding body is called *resonance*, or reciprocation of sound.

Sonorous bodies are audible (the extent of their excursions being supposed equal) in proportion to the quantity of their vibrating surfaces. Thus, a plate of glass or metal is capable of producing powerful sounds without accessory means; but the sound of vibrating bodies of smaller dimensions, such as insulated strings, or tuning-forks, are scarcely audible at a moderate distance from the ear; but the sounds of the latter are capable of considerable augmentation when communicated to surfaces, as when they are placed to a table, or to the sounding-board of a musical instrument.

There are several circumstances which influence the intensity of the resonance of a sounding-board. The principal of these is the plane in which the vibrations of the sounding body are made with respect to the reciprocating surface. Thus, its vibrations may be so communicated as to be perpendicular, or normal to the surface, in which case the sound is the most greatly augmented; or they may be tangential to, or in the same plane with the surface, when the sound is the most feeble. The first of these cases may be illustrated by placing a vibrating tuning-fork perpendicularly to the surface of a flat board; and the second, by placing it perpendicularly to one of the edges of the board. In intermediate positions – viz. when the vibrations are communicated obliquely to the surface – the sound will be found to have intermediate degrees of intensity.

These facts, which the extensive investigations of Savart place in full evidence, being understood, the peculiarities of the sounding-boards of various musical instruments admit of easy explanation.

The sounding-board of the piano-forte is better disposed than that of any other stringed instrument, as the planes of the vibrations of the strings are, on account of the direction in which they are struck by the hammers, always perpendicular to its surface. The difference of intensity when a string vibrates in this way, and when it vibrates parallel to the surface, is very obvious, and may be easily tried by striking it with the finger in these two directions.[10] There is no other instrument now in use, in which the strings make their vibrations perpendicular to the sounding-board.

In the guitar, lute, &c., the strings are also parallel to the sounding-board, but the vibrations must, for the convenience of performance, be made obliquely to it. If the sides of the instrument be of inconsiderable depth, and the back be flat, the difference of intensity between the perpendicular and oblique vibrations will be very sensible. But if the sides be deep, very little difference will be perceived, as the vibrations which are tangential to the front sounding-board are perpendicular to the sides, which thus enter readily into normal vibrations; this fact may be proved by placing the ear to the side of a guitar while a string is made to sound with its plane of vibration successively parallel and perpendicular to it. In some instruments, as the lute, mandoline, &c. the back is polygonal or curved; by this construction a considerable portion of the resonant surface enters into normal or nearly normal vibrations when the strings are struck obliquely to the principal sounding-board.

These laws are not so immediately applicable to the violin, and other instruments of the same class; an extensive series of experiments will yet be necessary to enable us to account for the peculiarities of their forms, their various curvatures, and the functions of that irregular conductor, resting on the sounding-board at two points only, which in these instruments is called the bridge. The investigations of Savart still leave much to be desired on this head.

In no instrument are the strings perpendicular to the sounding-board; for in such case, however a string were made to vibrate, its communicated vibrations would be tangential. But they are sometimes placed obliquely, as in the harp, and then the same changes of intensity may be observed as when the strings are parallel to the board; for if the plane of their vibrations coincide with that of the inclination of the board, the communicated vibrations of the board will be oblique to its surface, and the intensity will be at its maximum; but if they be perpendicular to this plane, the communicated vibrations must be tangential to this surface, and consequently the intensity will be at its minimum.

Besides the proper adaptation of sounding-boards, there are other circumstances on which the tones of a stringed instrument materially depend; one of the most important of these is, the proper dimensions of the volume of air contained within the sides; the laws of these resonant cavities have occupied the attention of Savart, but the obvious use of the bars placed within these cavities to divide the mass of air, and thus to enable it to vibrate more readily in separate portions, seems to have escaped his notice.

§ 4.

In the piano-forte, the guitar, &c., the ends of the strings are not in immediate contact with the sounding-board, but they rest on bars of wood, which are called bridges, through which the vibrations are communicated to the board. In these instruments the bridge is usually about half an inch in height, and in the violoncello does not exceed three inches. To ascertain how far the distance might be extended between the string and the sounding-board of a piano-forte without injury to the tone, I substituted a glass rod five feet in length for the bridge, and by placing at its end a string stretched on a steel bow, I found that the sound of the string was as distinctly audible as when it was immediately in contact with the board; a tuning-fork placed at the end of the rod gave the same result. These experiments, which were the first I made on the subject, and which suggested all the subsequent ones, have been repeated in the theatre of the Royal Institution on a larger scale. A series of connected deal rods, forty feet in length, was suspended so as to extend, in a straight line, obliquely from an open window of the cupola, to within a short distance of the floor of the room; on the upper end of this conductor, an assistant placed the stem of a vibrating tuning fork; when no sounding-board was placed at the lower extremity of the conductor, no sound was heard, but it became powerfully audible the instant the communication was made: this experiment was repeated with different acute and grave toned tuning-forks, employed both in combination and in succession.

Tuning-forks are the most convenient instruments for making experiments on the transmission of sound, because their vibrations are almost inaudible by themselves, and only become strongly audible when augmented by resonant surfaces.

§ 5.

The vibrations of the sounding-board of any stringed instrument may be communicated in the same manner as those of a string, or of a tuning fork, to a distant sounding-board by means of a metallic, glass, or wooden conductor; but in this case it is necessary to prevent the original sounds from being heard through the air, otherwise the communicated sounds will not be distinguishable from them. This may be effected by placing the originally vibrating, and the reciprocating instruments in different rooms, and allowing the conductors to pass through the floor or wall separating the two rooms.

In the passage of the conducting-rod or wire through these partitions, care must be taken to prevent its touching their sides; for this purpose, a tin tube, covered at its two ends with leather, or India rubber, may be inserted in the partition, and the conductor be made to pass through holes in these coverings, so as not to touch the side of the tube.

A square piano-forte is a very convenient instrument to employ in these experiments. If the sound is to be transmitted upwards, nothing more is requisite than to open or remove the lid of the instrument, and to allow the conductor to rest upon the sounding-board. A metallic wire is not sufficiently rigid to support itself thus without bending; a rod of some straight-fibred wood, as lancewood or deal, is therefore better adapted for this form of the experiment; the lower end of the rod must be reduced in thickness, so as to allow it to pass between two adjacent strings; and the best place to make the contact will be found to be about a quarter of an inch from the bridge, among the middle notes, and on the side occupied by the unvibrating portions of the strings. The reciprocating instrument in the room above, may be a guitar or any other similar instrument, or a harp; in which latter case, the rod may be brought in contact with the inner surface of the belly of the instrument, through one of the apertures of the swell. These were the forms under which the experiments have been repeated at the Royal Institution.

If the sounds of the piano-forte are to be transmitted downwards, a brass wire, about the thickness of a goose-quill, will suffice for the communication, as the weight of a reciprocating instrument suspended from it below will keep it sufficiently straight. To bring the conducting-wire into contact with the under surface of the sounding-board of the piano-forte, an aperture must be made in the bottom of the instrument immediately below the intended point of contact; and to ensure a perfect connexion with the sounding-board, it is advisable to furnish the wire with a shoulder just below its entrance into the aperture, and to occasion an upward pressure on this shoulder, by a piece of leather stretched on a ring (as in the insulating-tube above described) and placed at the end of a strong steel spring; the other end of which is screwed firmly to the bottom of the instrument. To assist in supporting the wire, another shoulder may be made on it, so as to rest upon the upper covering of the insulating-tube which passes through the floor; and the reciprocating instrument may be suspended by inserting the end of the wire into the sounding-board, and then securing it by a nut and screw on the opposite side. The form of the resounding instrument is a matter of choice; but, in order to obtain the freest and loudest tones, it is requisite to have the principal vibrating surface perpendicular to the conducting-wire. It is instructive to observe the gradual changes in the intensity and the quality of the transmitted sounds, when the sounding-board is made to pass through the various degrees of obliquity from a perpendicular direction to the conductor, until it is in the same plane with it; or, to employ Savart's language, as the communicated vibrations change from normal to tangential; in the latter case, the sounds have a subdued, and what is ordinarily called a metallic quality.

In the first public experiments I made in 1821, the reciprocating instrument, which was the representation of an ancient lyre, was so constructed as to produce tangential vibration; the tones were consequently far inferior to what I have since been able to produce. The transmitted sounds are not sensibly impaired when the wire is separated at several places, and the disunited parts fastened together by mechanical contact; the annexed wood-cut represents the divisions of the conducting-wire, which I found it convenient to make in the original form of the experiment, for the sake of portability and facility of removal; but, if the apparatus be intended as a fixture, it will be easier and better to employ but one length of wire. The wire consisted of four portions: the first part touched the sounding-board of the piano-forte, and reached half-way to the floor; the second passed through the insulating-tube in the floor, and terminated when it reached the ceiling of the

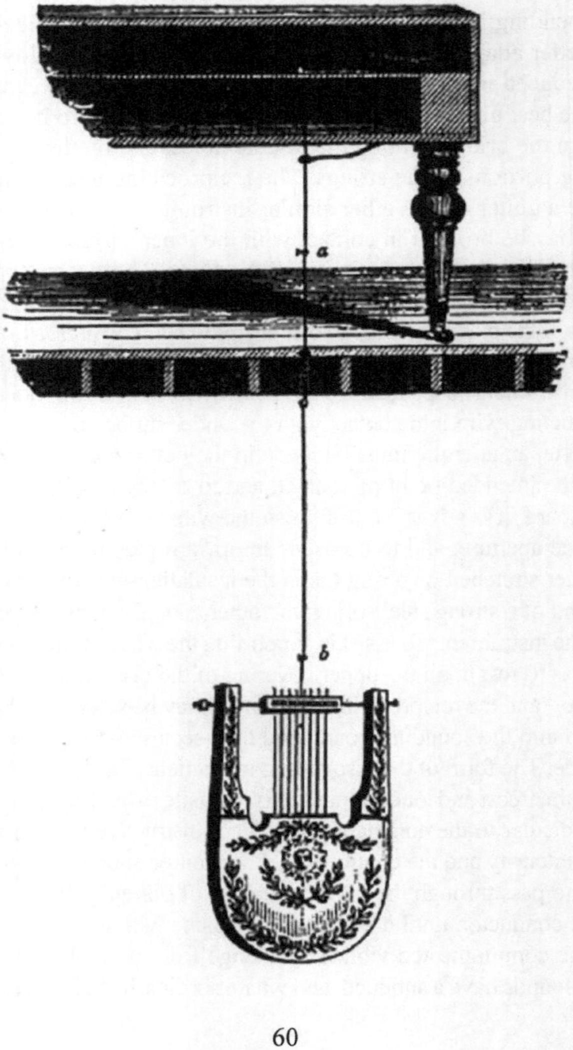

room below in a hook; a third part was suspended from this hook by a loop; and the fourth, after identifying itself with one of the apparent wires of the lyre, passed within the instrument, and was ultimately fixed, at its lower end, to the point marked at the end of the dotted line on the sounding-board; each of the disunited parts were allowed to overlap each other at *a* and *b*, and were fastened together by means of a clamp with a screw-nut. The whole apparatus thus prepared may be easily removed; the clamps being unscrewed and the resounding instrument removed, the lower wire must be unhooked from the ceiling, the hook unscrewed, and the middle wire withdrawn from the insulating-tube: the time for fixing or removing the apparatus need not exceed a few minutes.

From what has preceded, it will be obvious in what manner two square piano-fortes or two harps may be so connected as mutually to reciprocate each other's sounds; by such an arrangement, two performers in different rooms may play a duet together to two distinct audiences, or one may echo the performance of the other. If the transmission is required to be horizontal, *i.e.*, between two rooms on the same floor, cabinet piano-fortes must be employed.

The sounds of an instrument may be at the same time transmitted to more than one place; for instance, communications may be made from a square piano-forte to a resounding instrument above, and to another below; and the communication may be even continued through a series of reciprocating instruments. If the instruments be not in adjacent rooms, but be further removed from each other, a person in the intermediate room, through which the conductor passes, will hear no sound but what is communicated by the ordinary means. Hence it would be possible to extend a horizontal conductor through a series of rooms belonging to different houses, and (provided the instrument connected with one of its extremities be constantly played upon) to hear at pleasure the performance in any of these rooms, by merely attaching a reciprocating instrument to the conductor; on removing this instrument, the sonorous undulations would pass inaudibly to the next apartment. These observations will equally apply to the transmission of other musical sounds, which will be hereafter noticed (§ 6, 7).

§ 6.

The transmission of the sounds of those stringed instruments which produce sustained sounds, as the violin, violoncello, &c., is equally effective. The conducting-rod may be applied either to the back or the front of the instrument; no precise directions can be given with respect to the points at which the contact should be made; but, in general, the effect has appeared to me better when the end of the conductor has not been too far removed from the situation of the sound-post.

§ 7.

I have been able to effect the transmission of the sounds of reed wind-instruments through solid conductors as perfectly as that of instruments dependent on the vibrations of sounding-boards. In the clarionet, or any other reed instrument, the column of air and the vibrating tongue (or reed) mutually influence each other in such a manner, that whether the sounds be communicated to the atmosphere from the column of air, or to a solid conductor from the vibrating tongue, the quality (timbre) of the sound undergoes no change.

To connect the conducting wire, which may be of brass, and about a tenth of an inch in diameter, with the tongue of the clarionet, the end of the wire must be bent for about a quarter of an inch, and then filed flat on both sides. This flattened end must be fastened to the fixed end of the tongue by the silk wrapping which usually fastens the tongue only, and the angle of the bend be adjusted so as to suit the position of the performer. If the sound is to be transmitted downwards, the embouchure of the clarionet must be placed in the performer's mouth in the usual way, viz. the tongue of the reed resting on the under lip; but if the sound is to be transmitted upwards, the performer must play, as some eminent masters of this instrument do, with the tongue applied to the upper lip. For the bassoon or the hautbois, it is equally convenient to the performer, whether the wire be applied to the reed above or below.

The resounding instrument may, as in the experiments above detailed, be either a harp, a piano-forte, or a guitar. It is a singular effect to hear the sounds of a wind-instrument thus reproduced by a sounding-board.

§ 8.

The experiments I have made with respect to other classes of wind-instruments have not been equally successful. It is not possible to communicate the vibrations of the air to a solid conductor without an enormous loss of intensity: if, however, the intermediation of other bodies which enter readily into vibration, from the agitations of the air, be employed, the transmission may in some measure be effected. Thus, if the end of the conducting-wire be placed in the most strongly vibrating part of the column of air in a flute, there is but little perceptible transmission of sound; but if the wire touch the side of the instrument, it will more readily transmit the sounds, as the side is susceptible of entering into vibration. Even in this latter case, the sounds are scarcely audible, unless the ear be held close to the resounding instrument.

In a similar manner, the sounds of an entire orchestra may be transmitted, viz. by connecting the end of the conductor with a properly constructed sounding-board, so placed as to resound to all the instruments. The effect of an experiment of this kind is very pleasing; the sounds, indeed, have so little intensity as scarcely to be heard at a distance from the reciprocating instrument; but on placing the ear close to it, a diminutive band is heard, in which all the instruments preserve their distinctive qualities; and the pianos and fortes, the crescendos and diminuendos, their relative contrasts. Compared with an ordinary band, heard at a distance through the air, the effect is as a landscape seen in miniature beauty through a concave lens, as compared with the same scene viewed by the ordinary vision through a murky atmosphere.

§ 9.

In the preceding experiments on the transmission of sound through solid bodies, the conductors have been represented as straight; but, though sound is transmitted the more readily through straight conductors, it will yet pass, though with diminished intensity, through rods with angular and curved bendings. If a vibrating tuning-fork be placed at one end of a straight brass rod, the other end of which rests perpendicularly upon a sounding-board, the vibrations will, in accordance with what has been above stated, be powerfully transmitted; on gradually bending the rod at any part of its length, while the vibrations of the tuning-fork are kept in the same plane with the angle of the bent rod, the transmitted sound will progressively decrease in intensity, and will be very feeble when the angle has become a right one: as the bending is continued so as to make the angle between the two parts of the rod more acute, the intensity of the sound will increase in the same order in which it had before diminished; and when the two parts of the rod are nearly parallel, the sound will be nearly as loud as when the transmission was rectilineal. If, during the gradual bending of the rod, the plane of the vibrations of the tuning-fork be perpendicular to the plane of the angle made by the two parts of the rod, the same changes will be observed, but in a more obvious manner, than in the former case; and when the angle becomes a right one, the sound will be scarcely perceptible. At intermediate inclinations of the two planes, the gradations of intensity, occasioned by the bending of the rod, will be found to be intermediate.

The changes of intensity dependent on the variation of the angle of the two planes may be instructively shewn by bending the rod permanently to a right angle, and placing, as before, the stem of a tuning-fork so as to form the prolongation of one of the parts of the rod, the other part of the rod resting on the sounding-board. On gradually turning the tuning-fork round the axis of its stem, without inclining it to the rod, the plane of the vibrations will assume every angle with respect to the plane in which the two parts of the rod is bent. During the revolution it will be observed, that when the planes coincide the intensity will be at its maximum, and when they are perpendicular to each other, at its minimum; thus, supposing the sound to commence when the two planes are parallel, it will gradually diminish until they make an angle of 90°; it will then increase through the same changes of intensity, in an inverted order, until it acquires its maximum at 180°; it will again decrease between this and 270°, and increase until it arrives at its first position at 0°. If the stem of the tuning-fork be placed perpendicularly on the side of a conducting-rod resting on a sounding-board, the same phenomena may be observed; the stem of the tuning-fork is, in fact, a short conductor, forming a right angle with the rod.

Were it necessary for the transmission of sound that the undulations should propagate themselves only rectilinearly, it is obvious that they would not pass through a bent rod; and, on the other hand, had they the property of diffusing themselves equally in all directions, we should not observe any differences of intensity in the experiments above noticed. These experiments lead us to conclude, that sound diffuses itself in all directions, though unequally; that it is communicated more readily in the plane in which the original vibrations are made, and with the greatest degree of intensity in the direction of these vibrations.

§ 10.

In most of the experiments relating to the transmission of the sounds of musical instruments, which I have in the preceding paragraphs detailed, the conductor has been represented as receiving its impulses from a surface vibrating normally, to which it was perpendicularly attached; the communication was consequently effected by longitudinal undulations in the conducting wire. But if, while the conductor retains its position, the surface were to vibrate tangentially or obliquely, the communication would be effected by transversal or oblique undulations.

In practice it is preferable to employ the longitudinal undulations for the purpose of transmitting musical sounds to a distance; for, firstly, the transmission is more efficacious; and, secondly, the transverse undulations have a great tendency to communicate themselves laterally from the conductor to the surrounding medium, and thereby to become audible without the assistance of a reciprocating instrument. This lateral dispersion is scarcely observable with small conductors but is very obvious when a rod of considerable diameter is employed.

I had an opportunity of observing this fact while repeating some of the preceding experiments at the Royal Institution. A square piano-forte was placed in the apartment beneath the lecture-room; and a conductor, placed perpendicularly to its sounding-board, passed through the floor separating the two rooms, but no reciprocating sounding-board was placed at its upper end. By this arrangement, longitudinal undulations were communicated to the conductor; and, whether this was a brass wire one-tenth of an inch in diameter, or a square deal rod half an inch thick, the insulation appeared to be equally perfect. But it was not so when the conductor, instead of being placed on the sounding-board of a piano-forte, was made to rest on the top of the bridge of a violin, and the strings, put into vibration by drawing a bow across them, communicated transverse vibrations to the conductor; it was now observed, that the metal wire insulated the sound tolerably well, but that when the wooden rod was employed, the sound communicated to the air from the entire surface of the portion of the rod above the floor, was nearly as loud as if a sounding-board were placed at its extremity.

§ 11.

I have in this paper given the general results of a variety of experiments made at different and distant periods during the last ten years; but they are far from forming so complete a course as I have been desirous of making. To extend these experiments much farther would be attended with some difficulties: but as the velocity of sound is much greater in solid substances than in air, it is not improbable that the transmission of sound through solid conductors, and its subsequent reciprocation, may hereafter be applied to many useful purposes. Sound travels through the air at the rate of 1142 feet in a second of time; but it is communicated through iron wire, glass, cane, or deal-wood rods, with the velocity of about 18,000 feet per second, so that it would travel the distance of 200 miles in less than a minute.

When sound is allowed to diffuse itself in all directions as from a centre, its intensity, according to theory, decreases as the square of the distance increases; but if it be confined to one rectilinear direction, no diminution of intensity ought to take place. But this is on the supposition that the conducting body possesses perfect homogeneity, and is uniform in its structure, – conditions which never obtain in our actual experiments. Could any conducting substance be rendered perfectly equal in density and elasticity, so as to allow the undulations to proceed with a uniform velocity without any reflections or interferences, it would be as easy to transmit sounds through such conductors from Aberdeen to London, as it is now to establish a communication from one chamber to another. Whether any substance can be rendered thus homogeneous and uniform remains for future philosophers to determine.

The transmission to distant places, and the multiplication of musical performances, are objects of far less importance than the conveyance of the articulations of speech. I have found by experiment that all these articulations, as well as the musical inflexions of the voice, may be perfectly, though feebly, transmitted to any of the previously described reciprocating instruments by connecting the conductor, either immediately with some part of the neck or head contiguous to the larynx, or with a sounding-board, to which the mouth of the speaker or singer is closely applied. The almost hopeless difficulty of communicating sounds produced in air with sufficient intensity to solid bodies, might induce us to despair of further success; but could articulations similar to those enounced by the human organs of speech be produced immediately in solid bodies, their transmission might be effected with any required degree of intensity. Some recent investigations lead us to hope that we are not far from effecting these desiderata; and if all the articulations were once thus obtained, the construction of a machine for the arrangement of them into syllables, words, and sentences, would demand no knowledge beyond that we already possess.

NOTES

1 'If a rod of iron or brass be held with one end to the ear, and the other be struck upon, it makes a much greater sound than the same stroke upon the rod, when not so contiguous to the ear. By which, and other instances, it should seem that sounds do not only slide upon the surface of a smooth body, but also communicate with the spirits in the pores of the body.' – *Sylva Sylvarum,* Phonics, § 3.

'The pneumatical part, which is in all tangible bodies, and has some affinity with air, performs, after a sort, the office of the air. Thus the sound of an empty barrel is in part created by the air on the outside, and in part by that in the inside; for the sound will be less or greater, as the barrel is more or less empty; though it communicates also with the spirit in the wood, through which it passes from the outside to the inside.' – *Sylva Sylvarum,* Phonics, § 2.

2 'And though some famous authors have affirmed it impossible to hear through the thinnest plate of Moscovy glass, yet I know a way by which 'tis easy to hear one speak through a wall a yard thick. It has not yet been thoroughly examined, how far otacoustics may be improved, nor what other ways there may be of *quickening* our hearing, or *conveying* sounds through *other bodies* than the air; for that that is not the only medium I can assure the reader, that I have, by the help of a *distended wire,* propagated the sound to a very considerable distance in an instant, or with as seemingly quick a motion as that of light; at least, incomparably swifter than that which at the same time was propagated through the air; and this not only in a straight line, or direct, but in one bended in many angles.' – Preface to Hooke's 'Micrographia.'

3 Acad. Berl. Deutsch, abh. 1788, 87.

4 Reil's Archiv für die Physiologie, vol. iii., No. iii., p. 178.

5 Annales de Chimie, tome liii., p. 64.

6 Mémoires de la Société d'Arcueil, tome ii., p. 403.

7 A single vibration is here considered as the motion of the vibrating body between the two opposite limits of its excursion, and with this signification the expression is adopted by Chladni. Other authors, however, regard this, with Newton and Sauveur, as a semi-vibration, and call an entire vibration the motion of the vibrating body from one limit of its excursion until it again arrives at the same limit. This difference of meaning attached to the same term has given rise to several mistakes.

8 Journal de Physique, tome xlix., p. 382.

9 An abridgment appeared in the Ann. de Chimie, July, 1823, and the entire paper in the Annals of Philosophy, August, 1823.

10 It sometimes happens, when the impulse is oblique to the direction in which the string presses on the bridge, that its plane of vibration assumes a rotatory motion; the periodical changes of intensity thus occasioned, produce an effect similar to that of the beating of imperfect unisons. This phenomenon is generally erroneously attributed by tuners to a faulty string.

Editorial Headnote

10. Robert Willis, "On the Vowel Sounds, and on Reed Organ-pipes", *Transactions of the Cambridge Philosophical Society* (Cambridge: Cambridge University Press, 1829), pp. 1–38.

In 1820, Robert Willis (1800–1875) attended a demonstration of Hungarian inventor Wolfgang von Kempelen's (1734–1804) "Turk" automaton, which had bedazzled audiences across Europe with its ability to play chess with considerable skill. Kempelen's Turk was extremely alarming in its implication that human intelligence could be emulated through mechanical ingenuity, but, as it would eventually be discovered, the automaton was a hoax: beneath the chessboard, inside the Turk's table, a human operated the machine and directed its movements. Nevertheless, the potential of the mechanical to imitate human functions was not lost on young Willis, who would become a leading authority on questions of engineering, architectural construction, and mechanics. After studying at Gonville and Caius College Cambridge between 1822 and 1826, he subsequently became a college fellow, before securing the University of Cambridge's Jacksonian Chair of Natural Philosophy in 1837.[1] In this role, which he held until 1875, Willis worked to place mechanics, or rather the "science of mechanism", on the university's curriculum as a subject of mathematical analysis. His 1841 *Principles of Mechanism* delivered a mathematical treatment of motions, both in human-built machines and in natural specimens, such as the joints of a crab's claw.[2] Willis would also contribute a considerable amount of scholarship to the growing study of architectural history. Based on his 1832–33 honeymoon travels, Willis published his *Remarks on the Architecture of the Middle Ages, especially of Italy* in 1835. In particular, Willis was keen to subject Gothic buildings to mechanical analysis in terms of their means of weight suspension. Reducing constructions to their "mechanical" and "decorative" elements, Willis conceived of columns, ribs, vaults, buttresses, pinnacles, and flying buttresses as forming unifying systems that were both structurally sound and visually satisfying. In this analysis, he contributed to William Whewell's (1794–1866) and Thomas Rickman's (1776–1841) earlier efforts to categorize medieval Gothic architecture and reduce the elements of its construction to a series of fixed laws, which accounted for the historical development of the style.[3] Some of Willis's earliest work, however, concerned phonetics, especially towards the aim of replicating human speech through careful mechanical analysis and construction. In 1829, Willis published "On the Vowel Sounds, and on Reed Organ-pipes" in the *Transactions of the Cambridge Philosophical Society*, this being the content of two papers he had delivered to the society in 1828 and 1829.

Willis built on Kempelen's own enterprise, back in 1784, with the construction of a speaking machine modelled on the human form. Willis built a machine of his own, employing a pipe that could be adjusted in length by a plunger, blowing air onto a reed, to produce different vowel sounds. Willis would go on to publish his *On the Mechanism of the Larynx* in 1833, examining the larynx through mechanical analysis.[4]

NOTES

1 Alexandrina Buchanan, *Robert Willis (1800–1875): and the foundation of architectural history* (Woodbridge, UK: Boydell Press, 2013), pp.46, 48 and 50; James Q. Davies, "Instruments of Empire", in Davies and Lockhart (eds.), *Sound Knowledge* (Chicago: University of Chicago Press), pp.145–74, 156–9; Ben Marsden, "Willis, Robert (1800–1875)" *Oxford Dictionary of National Biography* (Oxford: Oxford University Press, 2004); online edition, Oct. 2009, www.oxforddnb.com/view/article/29584, accessed 1 March 2023; Robert Willis, *An attempt to analyse the automaton chess player, of Mr. De Kempelen* (London: J. Booth, 1821).

2 Ben Marsden, ""The progeny of these two "Fellows"": Robert Willis, William Whewell and the sciences of mechanism, mechanics and machinery in early Victorian Britain", *British Journal for the History of Science*37, no. 4 (Dec. 2004), pp.401–34; Robert Willis, *Principles of Mechanism, designed for the use of students in the universities, and for engineering students generally* (London: John W. Parker, 1841).

3 Robert Willis, *Remarks on the architecture of the Middle Ages, especially of Italy* (Cambridge: J. & J. J. Deighton, 1835).

4 Robin Maconie, "Willis, Speech, Sound, and Music", in Alexandrina Buchanan, James W. P. Campbell, Javier Giron, and Santiago Huerta (eds.), *Robert Willis: science, technology and architecture in the nineteenth century. Proceedings of the international symposium held in Gonville and Caius College, Cambridge UK, 16th–17th September 2016*, (Madrid: Instituto Juan De Herrera, 2016), pp.75–84.

10

ON THE VOWEL SOUNDS, AND ON REED ORGAN-PIPES

Robert Willis, M.A.
Fellow of Caius College, and of The Philosophical Society.

Source: Robert Willis, "On the Vowel Sounds, and on Reed Organ-pipes", *Transactions of the Cambridge Philosophical Society* (Cambridge: Cambridge University Press, 1829), pp.1–38.

[Read Nov. 24, 1828, and March 16, 1829.]

THE generality of writers who have treated on the vowel sounds appear never to have looked beyond the vocal organs for their origin. Apparently assuming the actual forms of these organs to be essential to their production, they have contented themselves with describing with minute precision the relative positions of the tongue, palate and teeth, peculiar to each vowel, or with giving accurate measurements of the corresponding separation of the lips, and of the tongue and uvula, considering vowels in fact more in the light of physiological functions of the human body than as a branch of acoustics.

Some attempts, it is true, have been made at various times to imitate by mechanical means the sounds of the human voice. Friar Bacon, Albertus Magnus, and others, are said to have constructed machines of this kind, but they were probably mere deceptions, like some contrivances which may be found in the works of Kircher and other writers of the same description,[1] The abbè Mical (according to Rivarol)[2] made two colossal heads which were capable of pronouncing entire sentences, but the artist having destroyed them in a fit of disappointment at not receiving his expected reward from the government, and having left no trace of their construction, we are left completely in the dark, as to the means employed by him to produce the different sounds. He died about the year 1786. The only attempts which have a claim to a scientific character, are those of Kratzenstein and Kempelen; these gentlemen were both occupied about the year 1770, in the mechanical imitation of the voice, and have both in the most candid manner disclosed the means employed by them, and the results of their experiments, the first in a prize Essay presented to the Academy of Petersburgh in 1780,[3] the second in a separate treatise.[4]

Kratzenstein's attempts were limited to the production of the vowels *a, e, o, u, i*, by means of a reed of a novel and ingenious construction attached to certain pipes,

DOI: 10.4324/b23406-10

some of them of most grotesque and complicated figure, for which no reason is offered, save that experience had shewn these forms to be the best adapted to the production of the sounds in question.

Kempelen's treatise abounds with original and happy illustrations, and the author is no less remarkable for his ingenuity and success, than for the very lively and amusing way in which he has treated his subject. None of these writers, however, have succeeded in deducing any general principles.

Kempelen's mistake, like that of every other writer on this subject, appears to lie in the tacit assumption, that every illustration is to be sought for in the form and action of the organs of speech themselves,[5] which, however paradoxical the assertion may appear, can never, I contend, lead to any accurate knowledge of the subject. It is admitted by these writers,[6] that the mouth and its apparatus, was constructed for other purposes besides the production of vowels, which appear to be merely an incidental use of it, every part of its structure being adapted to further the first great want of the creature, his nourishment. Besides, the vowels are mere affections of sound, which are not at all beyond the reach of human imitation in many ways, and not inseparably connected with the human organs, although they are most perfectly produced by them: just so, musical notes are formed in the larynx in the highest possible purity and perfection, and our best musical instruments offer mere humble imitations of them; but who ever dreamed of seeking from the larynx, an explanation of the laws by which musical notes are governed. These considerations soon induced me, upon entering on this investigation, to lay down a different plan of operations; namely, neglecting entirely the organs of speech, to determine, if possible, by experiments upon the usual acoustic instruments, what forms of cavities or other conditions, are essential to the production of these sounds, after which, by comparing these with the various positions of the human organs, it might be possible, not only to deduce the explanation and reason of their various positions, but to separate those parts and motions which are destined for the performance of their other functions, from those which are immediately peculiar to speech (if such exist.)

In repeating experiments of this kind, it must always be kept in mind, that the difference between the vowels, depends entirely upon contrast,[7] and that they are therefore best distinguished by quick transitions from one to the other, and by not dwelling for any length of time upon any one of them. A simple trial will convince any person, that even in the human voice, if any given vowel be prolonged by singing, it soon becomes impossible to distinguish what vowel it is.

Vowels are quite a different affection of sound from both pitch and quality, and must be carefully distinguished from them. By quality, I mean that property of sound, by which we know the tone of a violin from that of a flute or of a trumpet. Thus we say, a man has a clear voice, a nasal voice, a thick voice, and yet his vowels are quite distinct from each other. Even a parrot, or Mr. Punch, in speaking, will produce A's, and O's, and E's, which are quite different in their *quality* from human vowels, but which are nevertheless distinctly A's, and O's, and E's. Again, as to pitch, all the vowels may be sung upon many notes of the scale, but of this more hereafter.

Euler[8] has discriminated these affections of sound, and distributed them among the different properties of the aerial pulsations as follows. The pitch depends on the number of vibrations in a given time. Loudness on the greater or less extent of the excursion of the particles. Quality and the vowel sounds, he thinks must depend on the form of the curve by which the law of density, and velocity in the pulse is defined, or upon the latitude of the pulse, but this he offers as a mere opinion, unsupported by experiment, save that[9] to account for the peculiar property of sound, by which we know a flute from a trumpet, &c. he remarks, that as the vibrations of each instrument are excited in a manner peculiar to itself, its pulsations must also follow peculiar laws of condensation and motion, by which he thinks the sound will be characterized.

It is agreed on all hands, that the construction of the organs of speech so far resemble a reed organ-pipe, that the sound is generated by a vibratory apparatus in the larynx, answering to the reed, by which the pitch or number of vibrations in a given time is determined; and that this sound is afterwards modified and altered in its quality, by the cavities of the mouth and nose, which answer to the pipe that organ builders attach to the reed for a similar purpose. Accordingly, the whole of the phenomena I am about to describe, will be found to result from the application of reeds to pipes and cavities of different and varying magnitude.

Now it is important to the success of these experiments, that the tone produced by the reed should be as smooth and pure as possible. The coarse tone of the common organ reed completely unfits it for the purpose, and hence we find both Kempelen and Kratzenstein endeavouring to ameliorate it. Kempelen made the tongue of ivory, and covered its under side as well as the portion of the reed against which it beat, with leather. Kratzenstein succeeded better, by introducing a most important improvement in its construction. Instead of allowing the tongue to beat upon the edge of the reed, he made it exactly to fit the opening, leaving it just freedom enough to pass in and out during its vibrations.[10] By this construction, when carefully executed, the tone of the reed acquires altogether a new character, becoming more like the human voice than any other instrument we are acquainted with, besides possessing within certain limits the useful quality of increasing its loudness, with an increased pressure of air without altering its pitch.

All the reeds I have made use of are constructed on this principle, in one or other of the forms represented in Figures 10.1 and 10.2. They are all attached to blocks *fgh* furnished with a circular tenon *fg*, by which they can be fitted into the different pieces of apparatus represented in the plate; their place in all the figures being distinguished by the letter *R*. In Figure 10.1, *ab* is a tongue of thin brass fixed firmly at its upper extremity *a*, but capable of vibrating freely in and out of a rectangular aperture in the side of the small brass tube or reed *ced*[11] which it very nearly fits. This is the original construction of Kratzenstein. In Figure 10.2, *ab* is the tongue, attached at *a*, and capable of vibrating through an aperture, which it very nearly fits, in a brass plate screwed to the upper surface of the block.[12] This is similar to the construction of the *Mundharmonicon*, or *Eolina*, lately introduced into this country from the continent, but it appears to have been originally suggested by Dr. Robison.[13]

My first object being to verify Kempelen's account of the vowels, I fitted one of my reeds *R* to the bottom of a funnel shaped circular cavity open at top, of which *Z*, Figure 10.5, is a section (the pipe *TV* standing on the wind-chest in the usual way),[14] and by imitating his directions for the positions of the hand within the funnel, I obtained the vowels very distinctly.[15] I soon found that on using a shallower cavity than his, these positions became unnecessary, as a flat board *LM* sliding on the top of the funnel, and gradually enlarging the opening *KL*, would give the series U, O, A, just as well, and by making the funnel much shallower, as in Figure 10.4,[16] I succeeded in getting the whole series in the following order: U, O, A, E, I.[17] Cylindrical, cubical and other shaped cavities answered as well, under certain restrictions, but I forbear to dwell on this form of the experiment because subsequent ones have rendered this more intelligible, and I merely mention it to shew the steps by which I was led from Kempelen's original experiment.

The success of this attempt induced me to try the effect of cylindrical tubes of different lengths, and for the more complete investigation of this case, I constructed the apparatus represented in Figure 10.5 and 10.7. *TV* is a tube or *portevent* bent at right angles, and connected with the wind-chest at the extremity *T*, this tube terminates in a wooden piston *PQ*, provided with a socket for the reception of the reed *R* which is represented in its place. A piece of drawn telescope-tube *ABCD* is fitted to the piston which is leathered so as to be air-tight, but allows the tube to be drawn backwards or forwards, so as to alter the length of the portion *PB* beyond the reed at pleasure; the horizontal position is given to the tube to make it more manageable. There are other tubes *EFGH* of the same diameter as *ABCD* furnished with sockets at *EG* which fit on to *BD*, and their lengths are different multiples of *AB*.[18]

If therefore any reed be fitted to *PQ*, and the tube *AB* be gradually drawn out, it will shew the effect of applying to this reed a cylindrical tube of any length from nothing to *AB*, then if the tube be pushed back, and a joint *EF* equal to *AB* be fitted on, a fresh drawing out of the tube will shew the effect of any length from *AB* to double *AB*, and in this way with different joints we may go on to any length we please. The results of experiments with this apparatus, I will describe in general terms,

No. 1.

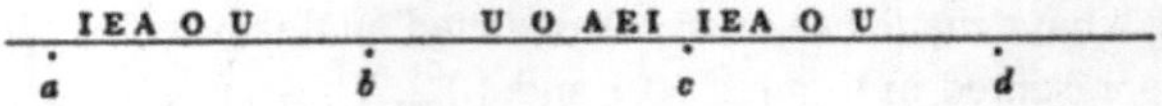

Let the line *abcd* represent the length *PB* of the pipe measured from *a*, and take *ab, bc, cd*, &c., respectively equal to the length of the stopped pipe in unison with the reed employed, that is, equal to half the length of the sonorous wave of the reed.

The lines in these diagrams must, in fact, be considered as measuring rods placed by the side of the tube *ABCD* with their extremity *a* opposite to the piston *P*, the letters and other indications upon them shewing the effect produced when the extremity *B* of the pipe reaches the points so marked, and the distance therefore from these points to *a* respectively being the length of the pipe producing the effects in question.

Now if the pipe be drawn out gradually, the tone of the reed, retaining its pitch, first puts on in succession the vowel qualities I E A O U; on approaching *c* the same series makes its appearance in inverse order, as represented in the diagram, then in direct order again, and so on in cycles, each cycle being merely the repetition of *bd*, but the vowels becoming less distinct in each successive cycle. The distance of any given vowel from its respective center points *a, c*, &c. being always the same in all.

No. 2.

If another reed be tried whose wave = *a.c.* (No. 2.) the centers of the cycles *a.*, *c.*, *e. e.*, &c. will be at the distance of the sonorous wave of the new reed from each other, but the vowel distances exactly the same as before, so that generally, if the reed wave $ac = 2a$, and the length of the pipe producing any given vowel measured from $a = v$, the same vowel will always be produced by a pipe whose length $= 2na \pm v$, *n* being any whole number.

When the pitch of the reed is high, some of the vowels become impossible. For instance, let the wave of the reed = *ac* (No. 3.) where ½ *ac* is less than the length producing U.

No. 3.

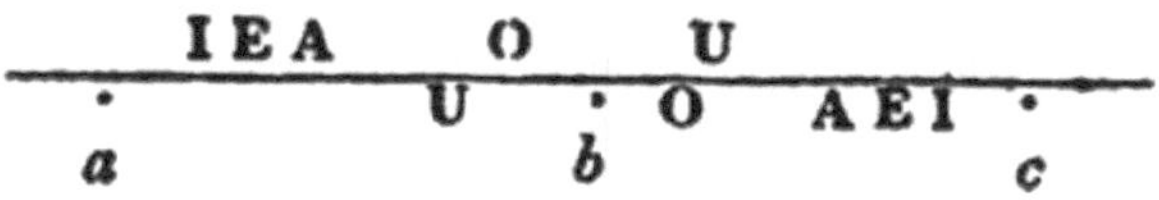

In this case it would be found that the series would never reach higher than O; that on passing *b*, instead of coming to U, we should begin with O again, and go through the inverse series. In like manner, if still higher notes be taken for the reed, more vowels will be cut off. This is exactly the case in the human voice, female singers are unable to pronounce U and O on the higher notes of their voice. For example, the proper length of pipe for O, is that which corresponds to the note *c''*, and beyond this note in singing, it will be found impossible to pronounce a distinct O.

The short and long U however are indefinite in their lengths, the short U (as in *but*) seems to be the natural vowel of the reed, and as this is but little affected by the pipe, except in loudness, between O and *b* (No. 1.) this vowel will be found to prevail through a long space, and upon approaching *b*, to change gradually into the long U, (as in *boot*), which always appears more perfect the longer the distance *ab* can be made. When reeds of high pitch are used (as in No. 3.) the vowels always become indistinct in the neighbourhood of *b*, and with bass reeds there appears to be a series of vowels like the former, on both sides the points *b, d*, &c. (No. 1.) but differing from the other, both by being much less distinct, and also by each vowel occurring at twice the distance from these points that it does in the other series from *a, c*, &c. after the manner of No. 4, where this new series is marked with an accent.

No. 4.

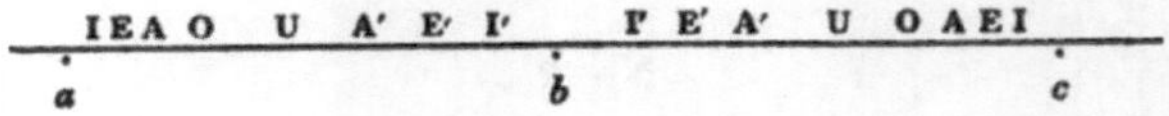

Cylinders of the same length give the same vowel, whatever be their diameter and figure. This may be conveniently shewn, by attaching a reed *R*, Fig. 11, to a portevent *T*, terminating in a horizontal flat plate *WX* covered with soft leather[19].[20] Tin or wooden tubes of any figure, open at both ends, and made flat at the bottom so as to fit air-tight to the leather plane, may then be applied, as in the figure, and their effect upon the reed tried. Another such leather plane, Fig. 12, may be provided, but furnished with an embouchure at *W*, like that of a common organ-pipe. By means of this, we can ascertain the note proper to any cavity, such as the cone *Z* placed above it in the figure. If this be transferred to the former plane its vowel will be ascertained.

As far as I have tried this, I have always found that any two cavities yielding the identical note when applied to Fig. 12, will impart the same vowel quality to a given reed at Fig. 11, or indeed to any reed, provided the note of the reed be flatter than that of the cavity, according to the principle explained in No. 3.[21]

The vowel distances in the first series, that is, those measured from *a* (No. 1.) are always rather less than those measured from the center points *c, e*, &c. This diminution varies with different reeds, and appears to be due to some disturbing effect of the reed itself, or the short pipe annexed to it which I have not been enabled as yet to examine so satisfactorily as I could wish. For this reason I have preferred in the following table obtaining the vowel lengths from the second and third series, by bisecting their respective distances from each other measured across *c*, which appear liable to no such alterations. These lengths, in inches, occupy the third column. For want of a definite notation, I have given in the second column the English word containing the vowel in question. The fourth contains the actual note of the musical scale corresponding to a stopped pipe of the vowel length, supposing O to yield c'' which it does as nearly as possible. In effect its length = 4.7 inches which with Bernouilli's correction[22] gives 4 inches for the length of the pitch pipe, and this will be found to give c''.

I have found this table as correct a general standard as I could well expect; for vowels, it must be considered, are not definite sounds, like the different harmonics of a note, but on the contrary glide into each other by almost imperceptible gradations, so that it becomes extremely difficult to find the exact length of pipe belonging to each, confused as we are by the difference of quality between the artificial and natural vowels. Future experiments, in more able hands than mine will,

Table I.

I	See	.38?	g^{v}
E	Pet	.6	c^{v}
	Pay	1	d^{iv}
A	Paa	1.8	f'''
	Part	2.2	$d''b$
A^{0}	Paw	3.05	g''
	Nought	3.8	$e''b$
O	No	4.7	c''
U	But	Indefinite	
	Boot		

I trust, determine this matter with greater accuracy, and I should not even despair of their eventually furnishing philologists with a correct measure for the shades of difference in the pronunciation of the vowels by different nations.

A few theoretical considerations will shew that some such effects as we have seen, might perhaps have been expected. According to Euler,[23] if a single pulsation be excited at the bottom of a tube closed at one end, it will travel to the mouth of this tube with the velocity of sound. Here an echo of the pulsation will be formed which will run back again, be reflected from the bottom of the tube, and again present itself at the mouth where a new echo will be produced, and so on in succession till the motion is destroyed by friction and imperfect reflexion. If it be a condensed pulsation that is echoed from the open end of a tube, the echo will be a rarefied one and *vice versâ*, but the direction of the velocity of translation of its particles will be the same. On the other hand when the reflexion takes place from the stopped end, the pulsation retains its density, but changes its velocity of translation. The effect therefore will be the propagation from the mouth of the tube of a succession of equidistant pulsations alternately condensed and rarefied, at intervals corresponding to the time required for the pulse to travel down the tube and back again; that is to say, a short burst of the musical note corresponding to a stopped pipe of the length in question, will be produced.

Let us now endeavour to apply this result of Euler's to the case before us, of a vibrating reed, applied to a pipe of any length, and examine the nature of the series of pulsations that ought to be produced by such a system upon this theory.

The vibrating tongue of the reed will generate a series of pulsations of equal force, at equal intervals of time, but alternately condensed and rarefied, which we may call the primary pulsations; on the other hand each of these will be followed by a series of secondary pulsations of decreasing strength, but also at equal intervals from their respective primaries, the interval between them being, as we have seen, regulated by the length of the attached pipe. Take an indefinite line Af_2 (No. 5) to represent the time, and let *ABC* ... &c. be the primary pulsations, $\underset{1}{a}\underset{2}{a}\ldots\underset{1}{b}\underset{2}{b}\ldots$ their respective secondaries, and for simplicity we will suppose that after the third they become insensible, also we will denote a condensed pulse by a stroke above the line, and a rarefied one by a stroke below. Suppose in the first place that the intervals of the secondary pulses are less than those of the primaries, and take AB, BC, &c. ($= a$) to represent the primary interval, and $\underset{1}{Aa}\ \underset{12}{aa}$ &c. ($= s$) the secondary. The order of the pulses in this case will be evidently that represented in the diagram.

No. 5.

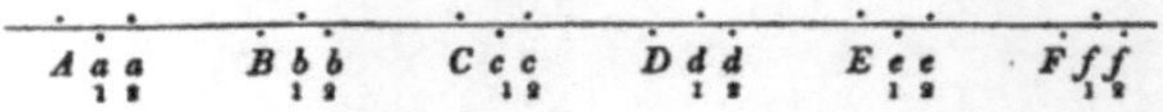

Suppose now that the secondary interval instead of being $= s$, should $= 2a + s$,

No. 6.

then taking ABC ... at equal intervals ($= a$) as before, to represent the primary pulses, we must place $\underset{1}{a}$ at the interval $2a + s$ from A, that is, at the distance s from C, $\underset{2}{a}$ at the distance $2s$ from E, and so on, and similarly for $\underset{12}{bb}$ &c. but in this way we plainly get, after the four first, a series of pulses precisely similar to those we obtained before, both in order of intervals, succession of intensity, and alternations of condensed and rarefied pulses. If we take the secondary interval $= 4a + s$, we shall find the same series occurring after the eighth primary, and in like manner when the secondary interval $= 2na + s$, we always obtain the same succession of pulses after the $4n$ first which may be neglected, and as we may take the distances AB, $\underset{1}{Aa}$ &c. to represent the length of stopped pipes, which give musical notes, the interval of whose pulses are in that ratio respectively; we may say that whatever effect be produced on the ear by applying a pipe, length $= s$ to a vibrating reed whose note is in unison with a stopped pipe, length $= a$, will also be produced by applying to the same, a pipe, length $= 2na + s$, n being any whole number.

Let now the secondary interval $= 2a - s$, a similar process will give the following series.

No. 7.

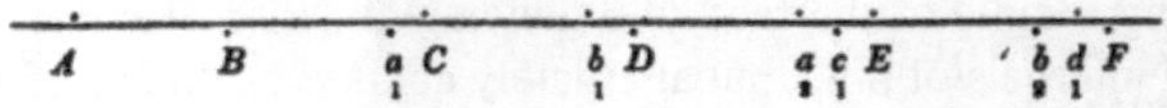

This after the four first is exactly similar to the former in the order of the intervals of the pulses, and the alternation of condensation and rarefaction, and only differs from it by the little groups of pulses increasing in intensity in this case, and diminishing in the other, a difference scarcely worth noticing, so that we may now say, that whatever effect be produced by applying a pipe ($= s$) to reed ($= a$), the same will be produced by a pipe $2na \pm s$.

No. 8.

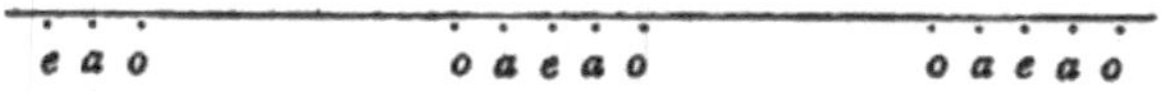

Hence if a given effect *a* be produced by a pipe whose length is *ea* (No. 8.) the same will be produced by a pipe whose length is *ee* – *ea, ee* + *ea*, &c. (taking *ee* = 2*a*). Again, if another effect *o* be produced by a pipe = *eo*, the same will be produced by a pipe = *ee* – *eo, ee* + *eo*, &c. and so on, and in this way it appears that if upon gradually lengthening the pipe we find a certain series of effects produced at the beginning, we shall have the same series in inverse and direct order alternately, the centers of each set of effects being separated by an interval (= *ee*) the length of the open pipe in unison with the reed. So far we find a perfect agreement with experiment, and it is plainly impossible to extend our reasoning much further with respect to the effect which a series of pulsations of this kind might be expected to produce on our organs of hearing.

If we examine the nature of our series, we shall find it merely to consist of the repetition of one musical note in such rapid succession as to produce another. It has been long established, however, that any *noise* whatever, repeated in such rapid succession at equidistant intervals, as to make its individual impulses insensible, will produce a musical note. For instance, let the musical note of the pipe be *g''*, and that of the reed *c'*, which is 512 beats in a second, then their combined effect is *g''* ... *g''* ... *g''* ... *g''* ... (512 in a second) in such rapid equidistant succession as to produce *c'*, *g''* in this case producing the same effect as any other noise, so that we might expect *à priori*, that one idea suggested by this compound sound would be the musical note *c'*.

Experiment shows us that the series of effects produced are characterized and distinguished from each other by that quality we call the vowel, and it shews us more, it shews us not only that the pitch of the sound produced is always that of the reed or primary pulse, but that the vowel produced is always identical for the same value of *s*. Thus, in the example just adduced, *g''* is peculiar to the vowel A^0 (Tab. I.): when this is repeated 512 times in a second, the pitch of the sound is *c'*, and the vowel is A^0: if by means of another reed applied to the same pipe it were repeated 340 times in a second, the pitch would be *f*, but the vowel still A^0. Hence it would appear that the ear in losing the consciousness of the pitch of *s*, is yet able to identify it by this *vowel* quality. But this vowel quality may be detected to a certain degree in simple musical sounds; the high squeaking notes of the organ or violin, speak plainly I, the deep bass notes U, and in running rapidly backwards and forwards through the intermediate notes, we seem to hear the series U, O, A, E, I, I, E, A, O, U, &c. so that it would appear as if in simple sounds, that each vowel was inseparable from a peculiar pitch,[24] and that in the compound system of pulses, although its pitch be lost, its vowel quality is strengthened. To confirm this, we ought to be able to shew that the note peculiar to each vowel in simple sounds, is identical with that of the secondary pulse in the compound sound, that is, with the note produced by a stopped pipe of the same length as that in Tab. I. corresponding to the given vowel, and as far as I have tried, it appears to be the case, but there is so much room for the exercise of fancy in this point, from the difficulty of fixing on the exact vowel belonging to a simple sound, that I do not mean to insist upon it.

No. 9.

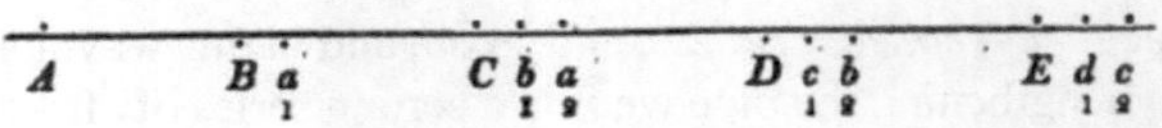

No. 10.

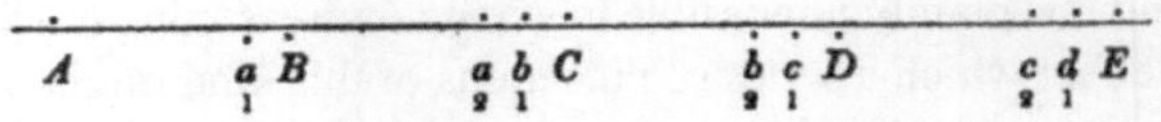

If we were to take the secondary interval = $a + s$ or $a - s$, we should obtain series of pulses like those in Nos. 9 and 10. These being after the two first precisely similar to No. 5, in the order of intervals and succession of intensity, would seem, at first sight, to prove that we ought to have a direct and inverse series of vowels like the others, on both sides of *b*, No. 1, and similarly, if the secondary interval were taken = $a \pm 2n+1$, we ought to have the same on both sides of *d, f*, &c. The pulsations in each of these groups, however, are either all condensed or all rarefied. Near *b*, therefore, they will tend to coalesce, and will scarcely be sufficiently distinguished from each other to impart the vowel quality in the same way that groups of alternate pulsations do. When they suggest any musical note at all, it will plainly be an octave higher than if the pulsations were alternate, because in the latter case, the interval between one *condensed* pulse and another, is twice that in the former. As the vowels have been shewn to be identified by the musical note of the secondary groups, it follows, therefore, that each vowel will, in this new series, be at twice the distance from *b, d*, &c. that it is in the original series from *a, c, e*, &c. I have already mentioned, that vowels of this kind are to be found on both sides of these points, *b, d*, &c. (See No. 4.)

Having shewn the probability that a given vowel is merely the rapid repetition of its peculiar note, it should follow that if we can produce this rapid repetition in any other way, we may expect to hear vowels. Robison and others had shewn that a quill held against a revolving toothed wheel, would produce a musical note by the rapid equidistant repetition of the snaps of the quill upon the teeth. For the quill I substituted a piece of watch-spring pressed lightly against the teeth of the wheel, so that each snap became the musical note of the spring. The spring being at the same time grasped in a pair of pincers, so as to admit of any alteration in length of the vibrating portion. This system evidently produces a compound sound similar to that of the pipe and reed, and an alteration in the length of the spring ought therefore to produce the same effect as that of the pipe. In effect the sound produced retains the same pitch as long as the wheel revolves uniformly, but puts on in succession all the vowel qualities, as the effective length of the spring is altered, and that with considerable distinctness, when due allowance is made for the harsh and disagreeable quality of the sound itself.

But there is another remarkable phenomenon to be observed in the experiment with Figure 10.6, which, as it is wholly unconnected with the vowels, I have hitherto, to prevent confusion, omitted to notice. I stated that during the experiment the reed retained its pitch, in fact however there is a considerable change in it which I proceed to describe.

In Fig. 18, the indefinite line *abcb'* ... is taken, as before, to represent the measuring rod; the ordinates of the curve, traced below this line as an axis, are the lengths of the pitch-pipe in unison with the note produced by the apparatus, when the extremity of the pipe reaches their respective feet.[25] Let *ab, bc, cb'* ... be respectively equal to the half-wave of the reed, therefore *c* is the center of the second series of vowels.

Let the pipe be gradually drawn out; the pitch will remain constant till its extremity approaches *f*, about midway between *a* and *b*. Here the note begins to flatten (which is represented by the increase of the ordinates), proceeding still farther, it continues to flatten till the pipe reaches a point *n* beyond *b*, where the note suddenly leaps back to one about a quarter tone sharper than the original, which however it soon after drops gradually to, and preserves till the pipe reaches *c*; proceeding from *c* through the successive cycles, the same phenomena will be found at the corresponding points of each, but not in so great a degree. The whole amount by which the note is flattened varies, but is generally about a tone, it is accurately stated in the Tables.[26]

The point *n* is by no means fixed, a jerk of the apparatus, a too hasty motion of the pipe, will make the reed leap back to the original note much sooner than it would do with a very gentle and gradual motion; it varies too with different reeds, although their pitch be the same, and it varies with the pressure of the bellows, the note changing sooner with a greater pressure.

Now if we begin from any point beyond *n*, and shorten the pipe gradually, its extremity may be brought considerably behind *n*, (say to *m*) before it leaps back to the other note, so that in fact at any point *p* between *m* and *n* there are two notes *pq, pr*, which it may be made to produce, neither of them exactly its proper one, but one a little *flatter* and the other a little *sharper*. Some reeds upon fixing the extremity of the pipe between *m* and *n*, may be made to change these notes from one to the other by a dexterous jerk of the bellows upwards or downwards, and to retain either of them for any length of time, at pleasure.

With others again there is a point between *m* and *n*, where the two notes appear actually to be heard at once, producing a most singular effect. The real fact in this case seems to be that the reed is producing its two notes alternately, but changing quickly and periodically from one to the other, so that they seem to be both going on at once.

To produce these effects completely, a considerable pressure of wind is sometimes required.[27] Should the reed be stiff, or the pressure on it not sufficiently great, it becomes quite silent for a considerable extent on both sides of *p*.

The note always increases in loudness on approaching *f*, and diminishes when it begins to flatten. After the leap it is commonly husky and bad for a little distance, and at every point *c*, that is, whenever the pipe is a multiple of the whole wave of the reed, the tone of the reed appears to be perfectly unaffected by the pipe, and is, if any thing, rather less loud than it would be without it.

Similar phenomena have been imperfectly observed by the old organ builders, and have been described by Robison, who set a reed in a glass foot, and adapted a sliding telescope tube to it. Again by Biot, who also used a glass foot, but made the length of the reed to vary instead of that of the pipe.[28] All these experiments were made however with the old-fashioned reed, whose oscillations were disturbed by the tongue beating on the edge of the tube, so that the phenomena could not be defined with so much precision as they are, when the free reeds are made use of. To discover how the motion of my reed was affected, I therefore adapted a telescope tube *ABC*, Fig. 17, to a glass *portevent DdEeT*. Behind the tongue *R*, but at such a distance as not to disturb its action, I placed a micrometer scale *mn*, supported by a wire *o*, by which I could measure the extent of its excursions towards *m*, which are very well defined: of course in the other direction they are concealed by the brass reed. I found that the excursions were constant as long as the pitch remained constant: when the flattening or sharpening took place, the excursions were diminished. (See Note B.). The ordinates of the dotted curve above the axis, in Fig. 18. represent the semi-excursions of the reed in this experiment. It was remarked that when the reed was managed so as to produce the double note, that its excursions were no longer well defined, but it seemed thrown into strange convulsions.

Let us now consider how to account for these effects. We have seen that an aerial wave travelling along a pipe will be regularly deflected backwards from the extremity, changing the sign of its density, and preserving that of its velocity when the pipe is open; but if closed, then retaining the sign of its density, and changing that of its velocity. Let the curve (Fig. 19.) represent in the usual manner, the densities of the series of waves generated from a reed *R*, in a pipe *AB*, and proceeding in the direction of its axis, the ordinates above the axis indicating condensation, and those below, rarefaction.

To take the simplest case, let the pipe be stopped at *A* and *B*. The series therefore will have been reflected from *B* to *A*, and back again continually, diminishing in force each time, till the effect of the waves becomes insensible. The whole effect upon any given particle within the pipe at any time, will, upon the principle of the superposition of small motions, be the sum of the effects produced by all these reflected waves, upon that particle at that moment. If therefore *BA'*, *A'B'*, &c. be taken equal to *AB*, the actual density of every point of the pipe, at the instant when any given portion *R* of a wave is issuing from *A*, will be represented by the curve formed by combining all the alternate portions *AB*, *A'B'* taken directly, with all the portions *BA*, *B'A''*, ... &c. taken reversely, and allowing for the gradual diminution of force.

If *B* be open, the curve must at each point *B*, *B'*, &c. be reversed with respect to the axis, as is shewn by the dotted lines, since the density is reversed at the open end of the pipe.

Suppose now that the tube is exactly the length of a wave, as *AB*, (Fig. 20,) and first let *A* and *B* be both closed. In this case the direct portions A_1B_1, A_2B_2, &c. will be exactly similar to *AB*, also all the retrograde portions A_1B, A_2B_1 will be exactly similar whatever portion of the wave is issuing from *A*, therefore all the direct waves will unite and assist each other, and in like manner all the retrograde waves, so that the column of air in the pipe will vibrate in the well known manner of a common organ-pipe. We should obtain a similar result if the pipe had been taken equal to any multiple of the wave.

But if *B* be open, the alternate portions BB_1, B_2B_3, &c. will be inverted: in this case A_1B_1 being exactly similar to *AB*, but on the opposite side of the axis, will tend to counteract it; again A_2B_2 destroys A_1B_1 and so on, and in like manner the retrograde waves alternately tend to neutralize each other, so that, in fact, were it not for the gradual decay of the secondary pulsations, there ought to be no sound at all in this case.

Again, if we consider a tube *ab* equal in length to half a wave, we shall find, in a similar way, that when *b* is stopped, the secondary pulses mutually *interfere* and destroy each other, and when open that they strengthen each other, and the same results would be found if the length of the pipe were taken equal to any odd multiple of the half wave. In the same way we may examine the case of pipes open at both ends, and so generally we find, that in pipes stopped or open at both ends, the pulsations strengthen each other when the pipe equals a wave or its multiple, and destroy each other when the pipe equals a half wave or its odd multiple, and that in pipes stopped at one end, the reverse is the case.

This mutual destruction of the pulsations, when the pipe is a multiple of the wave is confirmed by experiment. It has been shewn that such a pipe rather weakens than strengthens the tone of the reed.

We may now be partly able to see how the reed is affected by different lengths of pipe. Its motion, taken alone, is well known to be considerably under the dominion of the current of air by which its pulsations are maintained,[29] and if a pipe be attached to it, it will plainly be also subjected to the periodic return of the secondary pulsations, by which we may expect to find the time and mode of its oscillations affected. We have seen, however, that when the pipe is about the length of some odd multiple of the half wave, the secondary pulses are united. Here it is then that we may look for the greatest disturbance of the motion, and accordingly our experiments have shewn us, that the pitch of the reed, and consequently its oscillations, are only affected on approaching such lengths of the attached pipe. At other lengths it would appear that their impulses being separated, and falling in succession on the reed, are incapable of producing any sensible effect on its motion, although they are able to qualify the tone by imparting to it the vowel qualities.

Now as the reed moves outwards, a rarefied pulse is generated, which travels to the extremity of the pipe, and returns condensed; and supposing the pipe equal half a wave, the beginning of this secondary pulse will just coincide with the extremity of the first, so that as the reed returns, it is met by this reflected pulse, which tends to check the extent of its excursion, and, as it seems, to increase the time. The condensed pulse now generated will, in like manner, present a new secondary rarefied one, which will retard the reed during its outward motion, and so on.

But as the length of the pipe was taken equal to half the wave of the reed in its *free* oscillations, it appears that the wave now produced, since the note is flatter, will be of greater extent, so that the pipe may be lengthened to correspond to the new value of the half wave. But this being done, the same reasoning will shew that the excursion of the reed should be still further diminished in extent, and augmented in time; and in this way we may go on increasing the length of the pipe, diminishing the extent of the excursions, and flattening the note, till we reach some point where it will be easier for the reed to oscillate in the original mode, and there it will of course return to that mode of vibration. But the length of pipe at which this change takes place, will manifestly depend upon the elasticity of the reed, and the pressure of the current, and may therefore vary with different reeds and pressures as we have found it to do.

But we have seen that when the interval of the secondary pulses is somewhat greater than that of an oscillation of the reed, there is a slight tendency to accelerate the motion in diminishing its extent. If this once takes place, a similar line of reasoning will shew, that upon shortening the pipe to accommodate it to the new wave, it must go on rising in pitch till it reaches some point where it will be easier for it to vibrate in the other way.

That the lengths of pipe producing all these alterations of pitch are unfavorable to the reed's motion, is proved by the fact, that with them a less weight of wind reduces the reed to silence, although such less weight is quite sufficient to make the reed speak freely with favorable lengths. It also appears by the same test, that the lengths with which the sharpening takes place are considerably more unfavorable to the reed's motion, than those which flatten it.

The fact of certain lengths of pipe being unfavorable to the motion of the reed, naturally recalls the experiments of M. Greniè.[30] This gentleman found, that with some notes of the scale, he was obliged to alter the length of the *portevent*, or tube conveying the wind from the bellows to the reed, as there appeared to be some given length of *portevent* for [Illegible Text] reed, which completely prevented the reed from speaking, but it did not appear that these lengths followed any law. As this phenomena appeared to be of the same nature, as that I have just described, I set about to investigate it with the following apparatus.

By way of *portevent* I took two brass telescope tubes *AB*, *CD*, Fig. 16, sliding tightly over each other, and each a foot in length, the internal diameter of *CD* being .5 inches, at *A* was a socket for the reception of the reed *R*. Two additional tubes (as *EF*), with sockets, served to increase the length, so that by combining the use of these with the slide *BC*, I could try the effect of a *portevent* of any length from one foot to four.

For shorter lengths than this, I made use of a pipe *AB*, Fig. 13, to the end of which was attached a piston *BV*, carrying a reed *R*, and fitting air-tight in the *portevent TVC*, which is represented as standing on the wind-chest of the organ.

The note of the first reed I tried was = 4.7 inches by the pitch-pipe, and upon drawing the piston *BV* up by means of its attached pipe *AB*, the reed refused to speak, when *TV* = 3.4 in. and sounded again when *TV* = 5.5

Now according to the former experiments, this should have begun to happen when *TV* was nearly equal to the half wave; but the note of the reed I employed = 4.7 inches by the pitch-pipe, which, corrected by Bernouilli's Table, gives about 5.3 for the half wave, so that the *portevent* was much too short for my theory.[31] The cause of this anomaly appeared to lie in the short passage *efg* already described, which conveys the wind from the wind-chest to the *portevent*. To get rid of this, I fairly set the end of the *portevent* into the lid of the bellows, as represented in Fig. 14, so that the lower extremity *T'* of the pipe, or vibrating column, should be perfectly well defined. Now indeed the reed became silent, when *T'V'* = 5.9, and spoke again at about 8, leaving a difference in length from the former experiment of 2.5 inches, so that there can be little doubt, but that the passage *efg* is to be reckoned as part of the *portevent*, and perhaps may serve to account for the anomalous results obtained by M. Greniè. Pursuing this enquiry, by means of the longer tubes, Figs. 15, and 16, I found as I expected, that these intervals of silence, flattening, &c. occurred regularly, whenever the *portevent* was made nearly equal to some odd multiple of the half-wave of the reed.[32] There are faint indications of vowels in this case. (See Note C).

I shall conclude with a few experiments, which appear to confirm the views already laid down. Take a piston *MN*, Figure 10.8, very nearly of the diameter of the tube *ABCD*, Figure 10.6.[33] By means of the slender handle *o*, it may be made to slide up and down the tube, as represented in Figure 10.8 If this is done, the sound of the reed goes on without interruption, putting on, but rather less distinctly than before, the vowel qualities, in direct and inverse order alternately, corresponding to the length of tube *BN*, between the end of the piston and the mouth, whatever be the length of *PB*: while, on the other hand, whenever the length *PM* is made nearly equal to the whole wave of the reed, or some multiple of it, the flattening and other phenomena take place, which in the former experiments were found to occur only at odd multiples of the half length. In this case, one portion of the pulses generated by the reed, is propelled past the sides of the piston to the mouth of the pipe, and thence echoed backwards and forwards between *B* and *N*, producing the vowels, and the remaining portion echoed between *M* and *P*, as if *PM* were a pipe stopped at both ends, in which case we have seen that the secondary pulses unite, and therefore disturb the motion of the reed, when the pipe equals the whole wave, or a multiple of it.

If the reed *R*, instead of being inserted into the piston as in Figure 10.6, be fixed to the end of a slender tube *WR*, (Figure 10.9), which slides air-tight through a collar of leather attached to the piston, so that the length *PS* admits of alteration at pleasure, the vowels and other phenomena are still found to depend upon the length *PB*, (all those of course that would have been produced from *P* to *S* being lost.)

If the reed be attached to the end of a *portevent TVR*, (Figure 10.10), and presented to the tube *ABCD*, which is provided with a solid piston *PQ*, the vowels and other phenomena are still produced as before, and still depend upon the actual length *PB* of the greater tube.

There is one curious circumstance about the two last experiments, that although the flattening takes place only when *PB*, (Figure 10.9, and 10.10) is equal to the half wave of the reed, or some odd multiple of it, the amount of it varies with the distance of the reed from the mouth of the tube; being greatest, when *BS* is the half wave, or an odd multiple of it, and almost imperceptible, when *BS* is the whole wave, or a multiple, or when *BS* vanishes, that is, when the reed is exactly at the mouth of the pipe.[34] This appears to depend upon the different variation of density, at different parts of the pipe. We have seen, that when the pipe was equal to the half wave, or its odd multiples, that the variation of density took place as in a common organ-pipe, being nothing at the mouth, or at a multiple of the wave from it, where our flattening nearly disappears, and greatest at the distance of the half wave, or its odd multiples, where our flattening is greatest. Now as it has been shewn, that this alteration of pitch is occasioned by the reaction of the secondary pulses which disturb the motion of the reed by a periodic variation of density, it is clear, that if in the stratum of air *S*, (Figure 10.9, and 10.10,) where the reed is placed, this variation is destroyed or diminished by the pulses returning from *BD*, that the effect on the reed's motion will be proportionally diminished.

Lastly, if instead of presenting a stopped pipe to a reed, as in Figure 10.10, we substitute one open at both ends, such as a telescope without the glasses, or for shorter lengths, tin tubes, the flattening takes place when the tube equals the wave, or a multiple of it, and the vowels at double the length they did with stopped pipes. That this ought to happen is plain from what has been already said, and from the explanation of Nos. 9, and 10, which it will be easily seen, are the vowel diagrams for this case.

Some useful hints may now be deduced, for improving the construction of the reed-pipes of organs. Organ builders have been in the habit of attaching pipes to reeds, for two purposes. First, as in the ordinary reed stops, by applying to the reed a conical pipe, giving the same note, they hoped that the vibrations of reed and pipe would assist each other, and the whole effect be improved, and this view of the matter has always been taken by the theorists: the builders however have been perplexed by finding the motion of the reed obstructed, and its note flattened by such a pipe, and have therefore always made the pipe a little sharper than the reed, and its principal use appears to be, that by its conical shape, it augments the loudness of the note, on the principle of the speaking trumpet. But we have seen, that although when the reed and pipe give the same note, the undulations of the pipe disturb the motions of the reed, whether the pipe be open or stopped at both ends, or stopped at one end only; (always recollecting that the organ builder's reed-pipe must be considered as stopped at the reed end), yet that the amount of this disturbance varies with the place of the reed in such a pipe, being greatest at the bottom of the pipe, and least at its mouth,[35] so that it seems that the place hitherto chosen for the reed is the worst possible, and that it ought rather to be at the mouth of the pipe, in short that the best arrangement would be that of Figure 10.10, supposing the reed to be at *BD*, and the pipe placed vertically. This too would give the means of conveniently tuning the pipe to the reed by the plug *PQ*. With this arrangement the note of the reed is greatly augmented, and its motions being undisturbed, it speaks more readily.

The other application of pipes to reeds has been in the voxhumana stop, where very short pipes, not in unison with the reed, are employed: the principles of this, however, have never been thoroughly understood, and the whole has always been a puzzle to the builders.[36]

If by a voxhumana stop, we understand a stop possessing the same vowel quality in every note, the thing is easily to be obtained on the principles I have laid down, and I think would be a source of very pleasing variety; but it must be remembered, that there will always be a natural limit to the extent of such a stop upwards, from the impossibility of imparting each vowel quality to the notes of the scale beyond its own peculiar one. Thus we might have an O stop, but it could not extend above c'', an A stop might reach to f''', and an E to d^{iv},[37] and so on.

I shall here, for the present, conclude; reserving the application to the action of the human organs, of the facts and principles I have endeavoured to bring forward, to a future opportunity.

ROBERT WILLIS
Caius College,
March 16, 1829.

NOTES

1 Kircher. *Musurgia,* p.303. Bp. Wilkins. *Dædalus,* p.104. Schottus. *Mechanica. Hyd. Pneum.* p.240, and *Magia Univ.* II. 155. B. Porta. *Magia Nat.* p.287. The *Invisible Girl* was a contrivance of this kind. See *Nich. Journ.* 1802, p.56, 1807, p.69.

2 Rivarol. *Discours sur l'universalitè de la langue françoise.* Borgnis. *Traitè des machines imitatives,* p.160.

3 The abstract of this Essay will be found in the *Act. Acad. Petrop.* for 1780, and the whole Essay in the *Journal de Physique,* Vol. XXI. See also Young's *Nat. Phil.* I. p.783.

4 *Le Mecanisme de la parole suivi de la description d'une Machine parlante, par M. de Kempelen. Vienne* 1791. Dr. Darwin must also be reckoned among the mechanical imitators of speech. See *Darwin's Temple of Nature* 1803, Note XI.

5 Kempelen's definition of a vowel, for instance, is deduced entirely from the organs of speech, "Une voyelle est donc un son de la voix qui est conduit par la langue aux lèvres, qui le laissent sortir par leur ouverture. La différence d'une voyelle à l'autre n'est

produite que par le passage plus ou moins large que la langue ou les lèvres, ou bien ces deux parties ensemble accordent à la voix," §. 106.

6 Kempelen, §. 98.

7 Kempelen has remarked this with his usual acuteness. Describing one of his early experiments, with a machine something like fig. 5, from which he obtained some of the vowels by covering its mouth with his left hand, he says, "J'obtins d'abord diverses voyelles, suivant que j'ouvrois plus ou moins la main gauche. Mais cela n'arrivoit que lorsque je faisois rapidement de suite divers mouvemens avec la main et les doigts. Lorsqu'au contraire je conservois pendant quelque tems la même position quelconque de la main, il me paroissoit que je n'entendois qu'un A. Je tirai bientôt de eeci la conséquence, que les sons de la parole ne deviennent bien distincts que par la proportion qui existe entr'eux et qu'ils n'obtiennent leur parfaite clarté que dans la liaison des mots entiers et des phrases." p.400.

8 *De motu aëris in tubis*. Schol. 11. and 111. Prop. 73. *Nov. Comm. Petrop*. XVI. and *Mem. Acad. Berlin* 1767, p.354. See also Bacon *Hist. Nat.* §. 290.

9 Schol. II. Prob. 80.

10 Both these expedients are mentioned by M. Biot (*Physique* II.) but he has ascribed the first to M. Hamel (p.170) the second to M. Greniè (p.171) being apparently not at all aware of the existence of these Memoirs of Kratzenstein and Kempelen. There can be very little doubt but that Kratzenstein is to be regarded as the true inventor of this *anche libre*. His paper was published in 1780. See Young's *Nat. Ph.* Vol. I. p.783.

11 This brass tube is the *reed,* I believe, in the modern language of organ builders, and *ab* the *tongue;* but the term *reed* is often applied to the whole machine, and as such I have used it in this paper. In fact, in its primitive form, I suspect the whole was cut out of a single piece of reed or cane near a joint as in Fig. 3, which is copied from Barrington's drawing of the reed of the Anglesey *Pibcorn* (*Archael.* III. 33.)

12 A section of Fig. 2, is seen in its place at *R,* Fig. 11. In all the other figures *R* is a section of Fig. 1.

13 Art. Musical Trumpet. *Enc. Britt.* Supplement to 3d ed. 1801. Works IV. 538. Wheatstone in Harmonicon, Feb. and Mar. 1829.

14 A reference to Fig. 13, in which a pipe is represented as standing on the wind-chest, will serve to illustrate the subject to persons unacquainted with the structure of the organ, and at the same time afford me an opportunity of defining certain technical terms which I shall be compelled to make use of.

A large pair of bellows kept in motion by the foot, and so constructed as to afford a constant pressure is connected with a long horizontal trunk, or *wind-chest,* of which *klno* is a cross section, and which is therefore constantly filled with condensed air. The upper side of this trunk consists of a very thick board *pkql,* which is pierced with a number of passages similar to *efg,* every one of which is furnished with a valve or *pallet kl* moving on a joint at *k,* and kept closed by a spring *m;* a wire terminating in a knob or key *r,* passes through a hole in the upper board, and rests on the pallet, so that on pressing down this knob the pallet opens and a current of air immediately rushes through the corresponding passage *efg,* and passes into any pipe *AT* which may be placed over the aperture. In this instance *R* is the vibrating reed, to be set in motion by the current. The term *portevent* is always used for that part of the tube *TR* which lies between the reed and the wind-chest, and the term *pipe* for the portion *BA,* which is between the reed and the open air, and it is to be understood that in all the figures the lower end of the tube *TV* is supposed to be placed upon one of the holes of the wind-chest. The pressure of the air is always measured by the number of inches of water it will support in a common bent tube manometer; this is generally about 3 inches in organs.

15 I think it necessary to mention that the whole of the experiments described in this paper were performed before the Philosophical Society on the same evenings that the paper was read.
16 Kempelen's funnel is 2 inches diameter at the mouth, and 3 inches deep from the mouth to the reed. Fig. [10.]4, is only ½ inch deep and same diameter.
17 I use these letters throughout with the continental pronunciation.
18 The inside diameter of *ABCD* = 1.3 in., its length = 1 ft. 6 in., and the whole length = 12 feet when all the joints are combined.
19 In speaking of musical notes, I shall denote their place in the scale by the German tablature, the octave from tenor *c* to *b* on the third line of the treble is marked once thus *c'*, the next above, twice, *c''* and so on. As a standard for the pitch, I use a pitch-pipe which is made to sound by a small pair of attached bellows yielding a constant pressure of 2.5 in. The internal dimensions are .85 by .9 in., and 1 foot long. *Lumiere* .16 in. An attached scale is graduated to shew the actual length of the pipe (that is, the distance from the bottom of the pipe to the bottom of the piston) in English inches and decimals.
20 The diameter of my plate = 3 inches.
21 We can now connect the results of Figs. 4 and 5 with those of Fig. 6. If the cavity *Z*, Fig. 4, be placed on Fig. 12, and the flat board *EM* slid over its mouth, a scale of notes will be heard. If now any position of *LM* be taken by which *Z* is made to yield the same note as a given cylinder, then both will yield the identical vowel upon Fig. 11.
22 *Mem. Ac. Par.* 1762, p.460. Biot, *Phys*. II. p.134.
23 Prob. 77, and Cors. Schol. to Prop. 76. and Schol. 2 and 3, Prop. 78. in *Nov. Comm. Petrop*. XVI. *Mem. Acad. Berl.* 1767. See also *Enc. Metrop*. Art. Sound, p.776.
24 Kempelen has a curious remark about this, §. 110 .... "Il me semble, que lorsque je prononce des voyelles differentes sur le même ton, elles ont pourtant quelque chose qui donne le change à mon oreille, et me fait penser qu'il y a une certaine melodie, qui cependant, comme je le sais très-bien ne peut-être produite que par la variation des tons en aigus et graves .... "
25 To save room on the plate, but at the same time to shew the real nature of this change of pitch, these ordinates were constructed for the particular case of a reed (pitch = 6.6) and drawn to the same scale as the rod (¼ of the original) but taken equal, not to the length of the pitch-pipe, but to the excess of its length above six inches. If the line *ab* were drawn an inch and a half higher up, the ordinates would be exactly proportional to the real ones. (See Note A.)
26 Vide Note A.
27 As much as six inches.
28 *L'Art du Facteur d'Orgues,* par D. B. de Celles, pp.439, 440. (Vide Note E.) Robison, *Works* IV. 508, and *Enc. Brit.* Biot, *Physique* II. 169.
29 Biot, *Physique* II. 167.
30 Biot, *Physique,* II. 173.
31 Especially as the waves in these experiments always come out rather *longer* than the corrected length of the pitch-pipe, (See Note A.)
32 As therefore there are certain lengths of *portevent,* which prevent the reed from speaking, it is necessary to have the power of adjusting *TVR,* Fig. 6, to the reed made use of: this is done by a sliding joint at *V,* which may be fixed to the required length by a clamp, all this is omitted in the figure to avoid confusion.
33 Diameter of tube = 1.3 in., of piston = 1.25 in.
34 See Note (D).
35 See Note (D).

36 According to Biot, the pipes of the voxhumana are nearly of the same size, (*Phys*. II. 171.) but in the diapason given by B. de Celles, they differ considerably, although very short (*Facteur d'Orgues,* pp.84, 366). Their form is a cylinder set upon a truncated inverted cone, and half closed at the mouth. The length of *cc* = 9.1 in. diameter 1.4 in. length of c''' = 3.3 in. diameter 1 in.

37 The pipes attached to the reeds might be made to give E, and shades be brought over all their mouths simultaneously, by a pedal, so as to enable them to give any vowel at pleasure, imitating Fig. 4, where however the cavity *Z* may be made of any convenient form, and the shade *LM* need not rest on its edges, but be made to approach the mouth in any way most suitable to the mechanism employed. The common reeds *may* be used, but their vowels are by no means so distinct as those produced by the free reeds, which should always be employed in experimenting.

APPENDIX

NOTE A

The following Table will serve to illustrate the remarks on the change of pitch, and, in fact, is the one from which Fig. 18. was constructed. The pitch of the reed = 6.6 inches or g' nearly. I have in all the Tables denoted the pitch by the actual length of the pitch-pipe, as the most accurate method; all the measures are in inches and decimals.

Table II.

Length of tube PB, Figure 6		*Pitch*		
	4	6.63		
	5	6.71		
	6	6.76		
	7	6.85		
	8	7.1		
1st duplication	8.74=*am*			6.43
	8.8=*ap*	7.5	or	6.45
	8.89=*an*	7.55		
	9			6.53
	10			6.6
center of 2nd Cycle	17.1=*ac*			
2nd duplication	26=*am'*	7.03	or	6.5
	26.45=*an'*	7.11	–	6.53
center of 3rd Cycle	34.7			
4th duplication	43.3	6.96	–	6.53
	44.1	7.01	–	6.58
5th duplication	60.7	6.87	–	6.55
	62	6.96	–	6.6

Table III.

Pitch of Reed = 2.5.							
Length of tube at successive doubling points as the pipe was lengthened. }	4.75	13.8	23.1	32.15	41.45	50.85	59.9

Table IV.

Pitch of reed, an octave above 4.2. In this case the pitch of the reed being high, the vowels were all lost. The reed, instead of altering its pitch, and doubling at successive periods, became silent for a little space and then resumed its note, but always gradually increasing in loudness to a painful degree, just before its silent places.

Length at silent points.	2.55	8.9	15.3	21.7	28.05	34.45	40.85
– when note resumed.	3.65	9.9	16.1	22.55	28.9	35.2	41.6

I have stated that the cycles of phenomena recur at distances equal to the sonorous wave of the reed. I must remark, however, that this distance is always somewhat greater than it ought to be, if calculated from the length of the pitch-pipe corrected by Bernouillis's Table. (Biot. *Phys*. II. 135. *Mem. Acad. Par*. 1762.) The difference is about an eighth or tenth of the wave. Thus,

Reed, Tab. II. by a mean of the different distances gives the half wave }		= 8.8
But pitch = 6.6, and this corrected		= 7.2
	Diff.	= 1.6
Reed, Tab. III. gives half wave		= 4.65
Pitch = 2.5, and corrected		= 3.4
	Diff.	= 1.25
Reed, Tab. IV	half wave	= 3.2
Pitch = octave above 4.2		
Corrected = 4.95, ∴ true pitch		= 2.475
	Diff.	= .725

NOTE B

Table V contains a set of experiments with Fig. 17, and may serve to shew the effect which the length of the pipe, and the force of the wind has on the excursions of the reed. The reed employed, was one of those which produced the double sound.

Table V.

Pressure 6 inches			Pressure 4 inches			Pressure 2 inches		
Length	*Pitch*	*½ Excursion*	*Length*	*Pitch*	*½ Excursion*	*Length*	*Pitch*	*½ Excursion*
12	11.05	.076	12	10.93	.07	12	11	.046
15	11.52	.06	14	11.13	.066	14	11.1	.03
16	11.6	.05	15	11.3	.056	15	11.3	.03
17		.046	16	11.6	.046		Silence	
18		.036	17	11.8	.034	17	9.6	
18.75		.03	18	Silence		18	10.2	
19	Double	.024?	19	10.65?	.02	20	10.4	.02
19.5		.03	19.4	10.7	.04	21	10.75	.03
20	10.8	.06	20		.052	24 to 33 }		
21	10.87	.07	21	10.75	.064		10.87	.054
24 to 33 }	10.9	.08	22 to 33 }	10.8	.074	40.8	11.2	.030
35	11.05	.076	34	10.83	.074			
39	11.15	.07	35	10.95	.072			
40	11.24	.068	39	11.1	.066			
40.8	11.4	.06	40.8	11.3	.05			

NOTE C

Table VI. is a set of results from Fig. 16, applied to the wind-chest, as in Fig. 13. Table VII. is a set from Fig. 13, with the same reed. It will be seen, that in this case the length of *AB* affects the results as might have been expected.

Table VI. 6 Inches pressure.

Length of DA from Wind-chest	*Pitch*
15.25	5.2
19	5.25
20.75	5.5
22.5	5.7
23.25	5.9
23.4	Leap.
23.9	Fluttering Tone.
24.1	5.1 } Husky Tone.
24.5	
25	Good.
27.4	5.15
29.8	5.2
32	5.25
34	5.35
35.5	5.4
36.75	5.65
37.6	5.7 and Leap
38.1	5.
38.75	5.1
39.8	5.15
43.5	5.2
49.2	5.35
51.25	5.45
53.5	5.65

Table VII.

Length of TV	*Pitch*		
	AB = 12	*AB = 20*	*AB = 22*
3	5.23		
4	5.25		5.45
5	5.3		5.5
6	5.6	5.63	5.65
6.4	5.65 Silence at 6.5	4.75	5.05
8	4.75 Tone bad.	5.2	5.33
9.2	5.	5.3	5.4
10.7	5.2	5.3	5.45

NOTE D

Table VIII. contains a set of experiments with Figure 10.10 The tube *ABCD* was fixed with respect to *TVR* in such a manner that *R* was protruded into the tube at distances expressed in the first column. The piston was then moved till the flattening took place, and the note jumped back to the original one, or rather as before-mentioned to one a little sharper. The distance of the piston from the end of the tube, when this took place is seen in the second column. The note to which the pitch had sunk in the third column, and the sharp note to which it returned in the fourth.

Table VIII. PITCH OF REED, 4.87.

Length of BS	*Length of BP*	*Pitch*	
		Flat	*Sharp*
First Leap			
0	6.45	4.95	
.5	6.38	5.05	
1	6.28	5.1	4.83
2	6.03	5.15	4.65
4	6.06	5.17	4.5
5	6.19	5.23	4.4
Second Leap			
0	20.78	4.95	
1	20.7	5.05	
2	20.53	5.01	4.85
4	20.58	5.07	4.83
5	20.65	5.16	4.72
6	20.85	5.17	4.75
8	21.	5.13	4.74
10	21.22	5.06	4.73
12	21.2	5.03	4.85
14	20.73	4.94	
16	20.35	5.03	4.83
18	20.3	5.13	4.70

Here the flattening is almost insensible at the mouth of the tube, and when BS = about 14, and is greatest midway. As the first leap takes place when BP = 6.45, and the second when BP = 20.78, we may infer the wave of the reed to be 13.4, nearly, so that here the least effect is produced, when the pipe is nothing or the whole wave, and the greatest when it equals the half wave, as I had stated. The value of BP is very nearly constant, and its variations appear to follow a regular law, its mean value being nearly at the half waves.

NOTE E

As this book is scarce, I have given the passage, which is curious, (*Facteur d'Orgues*, par D. Bedos de Celles 1766–70, p.439.)

"On doit remarquer que lorsqu'on veut mettre un Tuyau d'Anche au ton qui lui est propre selon la longueur où il se trouve, on le fait monter en baissant la rasette (je suppose que celle-ci touchoit le coin,) le son devient mâle, harmonieux. Si l'on baisse un peu plus la rasette, le son devient plus doux, plus tendre, mais moins mâle et moins éclatant. Si l'on baisse encore la rasette, le son diminué, il s'éteint et devient sourd; si l'on baisse encore la rasette, le son double, c'est à dire, qu'il monte tout à coup d'un ton ou d'un tierce et quelquefois davantage; il change d'harmonie, et ce son ne vaut rien. On le fait redescendre en rehaussant la rasette, jusqu'à ce qu'il revienne à son vrai ton, qui doit-être mâle, èclatant et harmonieux, jusqu'à faire sentir un Bourdon qui parleroit ensemble avec le Tuyau d'Anche,"

p.439.

.... "Pour les éprouver, on mettra la main dessus un instant tandis qu'ils parlent; comme si on vouloit les boucher, alors le Tuyau commencera à doubler; mais il se remettra de lui même au ton aussitôt qu'on aura oté la main. S'il ne se remet pas de lui même, ce sera une marque qu'il sera un peu trop long, &c." ......

p.440.

Erratum.
Page 11, line 11, *for* with bass reeds, *read* with bass and tenor reeds.

Editorial Headnote

11. Charles Wheatstone, "On the Vowel Sounds, and on Organ-Pipes", in *The Scientific Papers of Sir Charles Wheatstone, D.C.L., F.R.S* (London: Taylor and Francis, 1879), pp. 348–67.

Be it by sonorous impulses through linear conductors, such as in the Enchanted Lyre, or by electrical signals through telegraph cables, the realization of long-distance communication remained central to Charles Wheatstone's research ambitions throughout the 1820s and 1830s. Beginning in 1834, Wheatstone began working on the transmission of electricity through copper wire, the same year in which he became Professor of Experimental Philosophy at the newly established King's College London (KCL). Labouring in the vaults beneath KCL's buildings, Wheatstone calculated the velocity of electricity to be 250,000 miles per second. His ultimate ambition, however, was to fashion a working electric telegraph system that was both practical and commercially viable. Wheatstone's work caught the attention of William Fothergill Cooke (1806–1879), an English inventor and entrepreneur, who met with the KCL professor in February 1837. Together they developed and patented a "five needle" electric telegraph in May, securing a contract for its use along the Great Western Railway. In 1844, they won a twenty-year contract with the Admiralty for an annual income of £1,500 to connect London and Portsmouth.[1] Yet, while Wheatstone had abandoned sound as a medium of long-distance communication, the transmission of speech remained a utopian dream: the impulse might have been conveyed electrically, but Wheatstone remained optimistic that it might result in the replication of spoken words over great geographical space. Little wonder then that Robert Willis's "On the Vowel Sounds, and on Reed Organ-pipes" article for the *Transactions of the Cambridge Philosophical Society* was of such interest to Wheatstone. The transmission of speech by electrical telegraphy would, clearly, require some sort of device for producing speech at the receiver's end of the communication link and this, Wheatstone believed, would require both mechanical ingenuity and an understanding of how to imitate the human vocal organs. As Wheatstone himself put it, while the communication of music over great distance would provide amusement, such a feat was of much "less importance than the conveyance of the articulations of speech".[2] Willis's work on vowel sounds was of extreme value because, as Wheatstone explained, "if all the articulations were once thus obtained, the construction of a machine for the arrangement of these syllables, or words, and sentences, would demand no knowledge beyond what we already possess".[3] The publication of Wheatstone's own review article, "On the Vowel Sounds, and on Reed Organ-pipes", originally in October 1837 in the *Westminster Review*, was very much part of this broader work towards rapid spoken communication over vast physical distances. Nor was Wheatstone alone in this ambition. In 1846, the Austrian inventor Joseph Faber (1786–1866) exhibited his Euphonia in London, rousing the mirth of the satirical journal *Punch*. This talking machine, consisting of a piano, bellows, and mechanical replication of the vocal organs, prompted *Punch* to speculate that such devices would be ideal for Parliament. "Combined with Mr Babbage's Calculating Machine, the Euphonia might replace, with perfect propriety, a Chancellor of the Exchequer", proposed *Punch*, while "by the aid of a simple grinder at the bellows, long speeches might be uttered with all the best benefits of emphasis and oratory, without a Scotch accent like Lord Brougham's, or a high key like Mr. Shiel's, or a conventicle twang like Sir Robert's".[4] More seriously, at the fifty-second-anniversary festival of the Charitable Printers' Pension Corporation, Thomas Huxley (1825–1895) prophesized on the future of journalism. Huxley praised Britain's journalists and printers but warned that

> as writing had superseded speaking and printing had superseded writing, so, perhaps, before many generations were over, speaking would resume something of its ancient sway. Such instruments as the microphone and the telephone were not now more imperfect than the printing press of 400 years ago was compared with the machines of the present day . . . the Englishmen of the future, instead of dividing his attention in the morning between his *Times* and his bacon, might be able to give his entire attention to his bacon by having the speeches made on the previous night in the House of Commons, and the general intelligence of the day, ground off by a small barrel organ on the breakfast-table.[5]

Huxley was convinced that, before long, a technological revolution in speech communication would completely reorganize the sharing of news and intelligence, including the proceedings of Parliament. Away from such speculations, Wheatstone's work proved influential. Having had a replica of Kempelen's talking machine built, on which he delivered a report to the British Association for the Advancement of Science in 1835, Wheatstone subsequently displayed this machine at KCL. In 1860, a thirteen-year-old Alexander Graham Bell (1847–1922) would see the device, years before developing his own telephonic apparatus. In 1878, in his presidential address to the Society of Telegraphic Engineers, Carl Wilhelm Siemens (1823–1883) would identify Wheatstone as the initiator of the development of the telephone.[6]

NOTES

1 Ben Marsden and Crosbie Smith, *Engineering Empires: a cultural history of technology in nineteenth-century Britain* (Basingstoke: Palgrave Macmillan, 2005), pp. 189–97.
2 Wheatstone, *The Scientific Papers of Sir Charles Wheatstone*, p. 63.
3 Ibid., p. 63.
4 (Anon.), "The Speaking Machine", *Punch* XI (1846, London, England), p. 83.
5 (Anon.), "Printers' Pension Corporation", *The Times*, no. 29567 (14 May 1879, London, England), p. 7.
6 (Anon.), "Telegraphy", *The Times*, no. 29160 (24 Jan. 1878, London, England), p. 11.

11

ON THE VOWEL SOUNDS, AND ON ORGAN-PIPES

Charles Wheatstone

Source: Charles Wheatstone, "On the Vowel Sounds, and a Reed Organ-pipes", in *The Scientific Papers of Sir Charles Wheatstone, D.C.L., F.R.S* (London: Taylor and Francis, 1879), pp. 348–67.

(1.) *On the Vowel Sounds, and on Reed Organ-Pipes. By* ROBERT WILLIS, *M.A., Fellow of Caius College, and of the Cambridge Philosophical Society*. (From the Transactions of the Cambridge Philosophical Society. Cambridge, 1829.)
(2.) *Le Mécanisme de la Parole, suivi de la Description d'une Machine Parlante. Par* M. DE KEMPELEN, Conseiller Aulique Actuel de sa Majesté l'Empereur Roi. Vienna, 1791.
(3.) C. G. KRATZENSTEIN. *Tentamen Coronatum de Voce*. Petrop., 1780.

[From the *London and Westminster Review*, no. xi. & liv., Oct. 1837.]

(The article is signed C[HARLES] W[HEATSTONE].)

WE propose in this article to give an account of the various attempts which have been made to imitate the articulations of speech by mechanical means, and to show how far the united labours of the philosopher and the mechanician have advanced the inquiry respecting the physical causes upon which these articulations depend.

Before we proceed, it may not be uninteresting to take a glance at the various accounts recorded of the speaking statues of the Ancients, and of those pretended speaking machines which at different times were exhibited prior to the present century.

The voice, as is well known, is easily and audibly transmitted to distant places by means of communicating tubes; this simple means of conveying words to the lips of a statue did not escape the observation of the priests of antiquity, and we have evidence that some of their oracles spoke in this manner. It is true that in general the priests considered it more safe to deliver the answers themselves, or to cause them to be delivered by women instructed for the purpose. But we read frequently that their idols spoke; and the same thing is stated with regard to the images of saints.

Oracular responses were delivered by a speaking head at Lesbos; it predicted, though in equivocal terms, the violent death of Cyrus the Great, which terminated his expedition against the Scythians. This was the head of Orpheus.

DOI: 10.4324/b23406-11

The celebrated head of Memnon, though in general it emitted only a *musical* sound when the rising sun fell upon its lips, yet it is proved, by inscriptions engraven on the colossus, that the priests, proportioning the miracle to the credulity of the votary, caused the statue *sometimes* to speak.

The speaking heads of the Middle Ages had more of a philosophic character. They were not brought forward for the purpose of imposing on a superstitious multitude the belief that they were miracles emanating from supernatural intelligence; they professed merely to be the products of human ingenuity – the inventions of men distinguished from the rest of mankind only by their superior knowledge.

Gerbert, who, under the name of Sylvester II., occupied the Papal chair from 999 to 1003, constructed a speaking head of brass. For this exertion of his ingenuity he was, of course, accused of magic, the common charge against mechanical inventors in those times.

In the thirteenth century Albertus Magnus, the extent of whose knowledge was astonishing for the age in which he lived, made a head of earthenware. It is said that his disciple, Thomas Aquinas, was so terrified when he heard it speak that he broke it to pieces, upon which the mechanist exclaimed, "There goes the labour of forty years." In the same century lived the celebrated Roger Bacon, who is reported to have made a similar automaton.

It is both amusing and instructive to recur sometimes to those crude hypotheses respecting physical phenomena which were advanced in the early dawn of science. Walchius thought it possible so to contrive a trunk or hollow pipe that it should preserve the voice entirely for certain hours or days, so that a man might send his words to a friend instead of his writing. There being always a certain space of intermission, for the passage of the voice, betwixt its going into these cavities and its coming out, he conceives that if both ends were seasonably stopped, whilst the sound was in the midst, it would continue there till it had some vent. "Huic tubo verba nostra insusurremus, et cum probe munitur tabellario committamus," &c. When the friend to whom it is sent shall receive and open it, the words shall come out distinctly, and in the same order in which they were spoken. From such a contrivance as this (saith the author) did Albertus Magnus make his image, and Friar Bacon his brazen head, to utter certain words. This scientific hypothesis powerfully and worthily supports the credit of that Popish tradition concerning Joseph's "hah!" or the noise that he made (as carpenters usually do) in giving a blow – which is said still to be preserved in a glass bottle among other ancient relics.

The celebrated Jesuit, Athanasius Kircher, who lived in the seventeenth century, and was, if not the inventor, the first describer of the Æolian harp, the speaking trumpet, and many other important acoustical inventions, assures us, in the ninth book of his 'Musurgia Universalis,' &c., that a statue might be made, perfectly isolated, the eyes, lips, and tongue of which shall have a motion at will, which shall pronounce articulate sounds, and which shall appear to be alive. He had an intention of making one of this kind for Queen Christina, but he was prevented, it is said, either by want of time or by the expense.

After this time speaking machines of this kind were frequently exhibited in different parts of Europe, and the means of communication were sometimes so ingeniously concealed as to excite the astonishment, and impose upon the credulity, of even men of learning. A Thomas Irson exhibited a speaking head, which excited the wonder of Charles II. and his court. It answered in several languages to questions whispered in its ear. When the astonishment was at its height, one of the pages discovered in the adjoining room a Popish priest, who answered through a pipe.

In the last century a figure of Bacchus seated on a barrel was exhibited at Versailles. It pronounced, in a loud and intelligible voice, all the days of the week, and wished the company good day. Many persons were deceived by it, because the owner of the machine allowed them to inspect the inside of the figure and the barrel, where nothing was perceived but organ-pipes, bellows, wind-chests, wheels, cylinders, &c. But the deception did not last long. A person more inquisitive than the rest discovered a false wind-chest, in which a dwarf was concealed, who articulated the words, the sound of which was conveyed by means of a tube to the mouth of the figure.

Professor Beckmann describes a similar pretended automaton which he saw. The figure was that of a Turk placed on a box, which was filled, as in the former case, with pipes, bellows, &c; but the voice was communicated from a person in the adjoining room.

In consequence of the prevalence of these deceptions, suspicion was so much excited whenever a speaking machine was announced, that when De Kempelen first made public his invention in 1783, a pamphlet was published, entitled 'The Speaking Machine and Chess Player Exposed and Detected.' De Kempelen, we may remark, was the inventor of the latter curious figure; the sole merit of which, as he himself has acknowledged, consisted in the ingenuity of the deception.

As a specimen of the speculations respecting the mechanism of speech which were prevalent about a century and a half ago, and which then supplied the place of observation and experiment, we may mention a small work on this subject, written by Van Helmont, the celebrated alchemist. Its translated title is, 'A short Explanation of the true natural Hebrew Alphabet, which at the same time supplies a method by which those who are born deaf can instruct themselves, not only how to understand others when they speak, but likewise how to speak themselves.' Van Helmont in this tract endeavours to establish that all the letters of the Hebrew alphabet are written in exact imitation of the position which the tongue assumes in pronouncing them. He then goes further, and proves to his own satisfaction that the letters of the alphabet ought to follow each other in the order they do, and in no other; because the tongue, whilst finishing the pronunciation of one letter, is already in the position for commencing the following. His imagination has represented the tongue in the most extravagant contortions. His work is illustrated by thirty-four plates.

When absurd speculations like this supplied the place of accurate knowledge, it cannot be supposed that any progress was made towards constructing a speaking machine properly deserving the name. Such an invention requires, above all, for its accomplishment, a full and exact knowledge of the mechanism of speech.

The first experimental investigation of the articulations of speech dates from 1779, when the Imperial Academy of St. Petersburg proposed for their annual prize the following two questions: –

First. What is the nature and character of the sounds of the vowels A E I O U, so different from each other?

Secondly. Can an instrument be constructed like the *vox humana* pipes of the organ, which shall accurately express the sounds of the vowels?

These questions were answered and the prize gained by Professor C. G. Kratzenstein. After examining the positions of the various organs, and measuring the apertures of the lips, teeth, &c. for the different sounds, he constructed a series of tubes, which, when applied to an organ-bellows, imitated with tolerable accuracy the five vowel sounds required. The forms of the tubes which he employed cannot readily be rendered intelligible without the aid of figures; four were rendered vocal by the application of freely vibrating reeds, the other in the manner of an ordinary organ-pipe.

Though Kratzenstein has the undoubted merit of having been the first who imitated any of the articulations of speech by mechanical means, yet we cannot give him the credit of having thrown any philosophic light on the subject. It has since been ascertained, as we shall presently see, that the forms of the tubes have not that relation to the different vowel sounds which he supposed to exist; but that tubes exactly the same in form will articulate different vowels when their dimensions are altered.

About the same time that Kratzenstein was engaged in this inquiry, M. de Kempelen of Vienna was occupied in a more extensive investigation of the same subject. He attempted not only to imitate the vowel sounds, but the consonants also: at present we will confine ourselves to his experiments with the vowels. De Kempelen found that by placing a conical tube, like the bell of a clarionet, before the reed he employed, different vowels were obtained according as the bell was more or less covered. On such a tube, U is heard when the tube is nearly closed, O when it is about half closed, and A when it is entirely open. It is only when a rapid transition is made from one of these sounds to another that each can be distinctly recognized; when the cover is kept upon the tube for some time in the same position, the sound appears always to be A. De Kempelen infers from this that the sounds of speech become distinct only from the contrast which exists between them, and that they obtain their perfect clearness only when connected in entire words and phrases.

Before proceeding to the researches of Mr. Willis on the vowel sounds, it will be necessary to explain, first, in what the vowels differ from the other articulations of speech, and, secondly, in what they differ from each other.

The vowels are formed by the voice, modified, but not interrupted, by the varied positions of the tongue and lips. Their differences depend on the proportions between the aperture of the lips and the internal cavity of the mouth, which is altered by the different elevations of the tongue. The vowel sound *aw*, as pronounced long in "fall" and short in "folly," is formed by augmenting the internal cavity by the greatest possible depression of the tongue, and at the same time enlarging the separation of the lips. Departing from this sound there are two series, the one represented in the vertical row of the following Table, the other in the horizontal row. In the first, the external aperture remains open, and the internal cavity gradually diminishes by the successive alterations of the tongue. In the second, the tongue remains depressed, but the aperture of the lips is gradually diminished. There is also an intermediate series of vowel sounds, obtained by different elevations of the tongue when the lips are partially closed; these, though abounding in many foreign languages, are not used in our own; they are shown in the second vertical column.

			4·7 C^3	Indefinite.
3·05	G^2			
3·8	$E\flat^2$	Aw	O	Oo
2·2	$D\flat^3$	Ah	Ou	
1·8	F^3			
1·0	D^4	Ae	Eù	
0·6	C^5	A	Eú	
0·38	G^5	E	Ü	

This fable indicates all the most usually pronounced vowel sounds, but practised ears might distinguish others intermediate in each series; for each vowel may pass to the next in order, either above or below it, by imperceptible gradations. Each of these vowels may be long or short, according to the duration of its sound in a syllable.

When these vowels are sounded, the soft palate is raised so as to prevent the sound from issuing through the nasal channels; when, on the contrary, the soft palate is depressed, the partial escape of the breath through the nostrils modifies all the preceding sounds in a very evident manner. These modifications, however, are never employed in the English language.

The extreme inadequacy of our written language in its representation of the articulations of speech is very obvious. We have six characters which are called vowels, each of which represents a *variety* of sounds quite distinct from each other, and while each encroaches on the powers of the rest, many simple vowels are represented by combinations of two letters. On the other hand, some simple vowel letters represent true diphthongs, consisting of two distinct simple vowels pronounced in rapid succession; thus, *a* consists of *a* and *e, i* of *ah* and *e*, and *u* of *e* and *u*. Again, the literal diphthongs are mostly simple vowel sounds, as *ea* in "bleak," *ie* in "thief," &c.

This want of correspondence between the characters of our written and the sounds of our spoken language has been a great obstacle to the proper understanding of the real elements of speech. A child is taught that the letters W, H, Y, make the syllable "why"; now, if we examine the sound of this word, we shall find it to be formed by the rapid succession of the vowel sounds *U, ah, E*. In attempting, therefore, to imitate by artificial means the sound of this word, we should pay no regard to the letters of which it is formed; the elementary *sounds* alone are the objects of our attention. The same observation is generally applicable to the words of our language.

Turn we now to the recent and interesting researches of Mr. Willis on the vowel sounds. His first object was to verify De Kempelen's observations. Having obtained the vowel sounds U, O, A very distinctly, he found that on using a shallower cavity than De Kempelen had employed, and sliding a flat board on the top of the funnel, he could obtain the entire series in the order U, O, A, E, I. The success of this experiment induced Mr. Willis to try the effect of cylindrical tubes of different lengths, and, for the more complete investigation of this case, he constructed an apparatus consisting of a tube or *port-vent*, bent at right angles to connect it with the wind-chest of the bellows. At the extremity of this tube is a plug, in which is fitted a freely vibrating reed. Another tube of equal length is fitted to this, so that, on being gradually drawn out, it will show the effect of applying to this reed a cylindrical tube of any length, from nothing to about two feet; then if the tube be pushed back and another joint equal to it be fitted on, a fresh drawing out of the tube will show the effect of any length from that of the first tube to double the length, and in this way with different joints we may go on to any length we please. The result of experiments with this apparatus we will describe in general terms.

Let the line *a b c d* represent the length of the pipe measured from *a*, and take *a b, b c, c d*, &c. respectively equal to the length of the stopped pipe in unison with the reed employed; that is, equal to half the length of the sonorous wave occasioned by the reed.

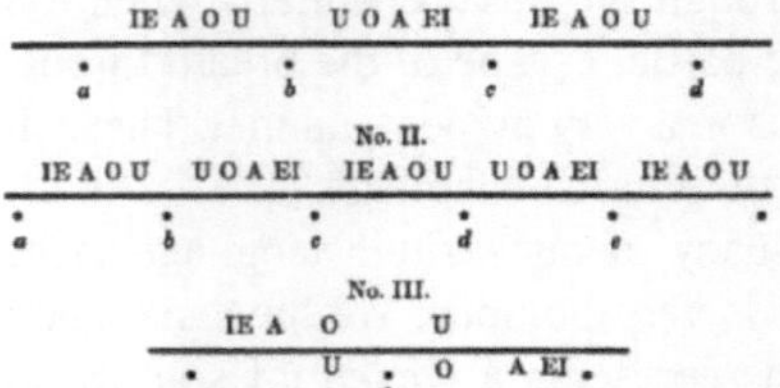

The lines in these diagrams must be considered as measuring-rods placed at the side of the tube, with their extremity *a* opposite to the point at which the reed is placed; the letters and other marks show the effects produced when the open extremity of the pipe reaches the points so marked, and the distance, therefore, from these points to *a* respectively is the length of the pipe producing the effects in question.

Now, if the pipe be drawn out gradually, the tone of the reed, retaining its pitch, first puts on in succession the vowel qualities IE A O U; on approaching *c* the same series makes its appearance in *inverse order*, as represented in the diagram; then in direct order again, and so on in cycles – each cycle being merely the repetition of *b d*, but the vowels becoming less distinct in each successive cycle. The distance of any given vowel from its respective centre points is invariably the same in all.

If another reed be tried which gives a higher sound, and whose wave, represented by the second figure, is consequently shorter, the centres of the cycles *a, c, e*, &c., will be at the distance of the sonorous wave of the *new* reed from each other, but the vowel distances (and this is an important point to be remarked) will be exactly the same as before. It may be stated as a general rule, that if a certain length added to the reed gives to its sound a determinate vowel character, the same vowel is heard when the same length is added to or subtracted from a length equal to any simple multiple of the original wave ($2\ n\ a + v$).

When the pitch of the reed is high, some of the vowels become impossible. For instance, let half the wave of the reed (No. III.) be less than the length producing U. In this case it is found that the series never ascends higher than O, and that on passing *b*, O of the reverse series commences. In like manner, if reeds of higher notes be taken, more vowels are cut off. This is exactly, Mr. Willis remarks, the case in the human voice; female singers are unable to pronounce U and O in the higher notes of their voice; for example, the proper length of the pipe for O is that which corresponds to the note C on the third space in the treble, and beyond this note in singing it becomes impossible to pronounce a distinct O.

We have added to the table of vowel sounds at page 354 the lengths of a stopped pipe which Mr. Willis has ascertained to correspond with the different vowel-sounds, and also the musical notes which these pipes would give were they sounded themselves.

Mr. Willis further states that cylinders of the same length give the same vowel, whatever be their diameter or figure; and that, so far as he has tried, he has always found that any two cavities yielding the identical note when applied to the embouchure of an organ-pipe, will impart the same vowel-quality to a given reed, or, indeed, to any reed provided its note be flatter than that of the cavity.

From these experiments it is evident that the forms stated by Kratzenstein, as producing the different vowels, are perfectly arbitrary. The entire series of vowels can be produced from tubes of either of his forms by merely changing its dimensions.

Mr. Willis finally concludes, from his experiments, that the vowel quality, added to any sound, is merely the coexistence of its peculiar note with that sound; this accompanying note being excited by the successive reflections of the original wave of the reed at the extremities of the added tube.

This view of the matter naturally associates the phenomena of vowel-sounds with those of multiple resonance, a subject first investigated by Professor Wheatstone.

The phenomena of simple or unisonant resonance are so well known that we need only call attention to one or two of the most striking facts. If a vibrating body be brought near a column or volume of air, which would be capable of producing the same sound were it immediately caused to sound as an organ-pipe or otherwise, then the sound of the vibrating body is greatly reinforced, as when an harmonica-glass is brought before an unisonant cavity, or when a tuning-fork is placed at the embouchure of a flute, the apertures of which are stopped, so that, if blown into, the flute would sound the same note; in the latter case the experiment is more remarkable, as the sound of the tuning-fork is scarcely itself audible. The same effect takes place when the cavity of the mouth is adjusted so as to be in unison with the tuning-fork.

We now come to the new facts of resonance. A column of air will not only enter into vibration when it is capable of producing the *same* sound as the vibrating body which causes the resonance, but also when the number of the vibrations which it is capable of making is any simple multiple of that of the original sounding body, or, in other words, if the sound to which the tube is fitted is any harmonic of the original sound.

For instance, if a tube closed at one end by a movable piston is taken, and its length adjusted to six inches, it will resound as an unison to a C tuning-fork; and if we shorten the length of the tube to three inches, the unison will no longer be reciprocated, but its octave will be heard. The same effect is produced by altering the cavity of the mouth.

By placing a vibrating lamina which produces a lower sound than can be obtained from a tuning-fork (the tongue of a Jew's harp, for instance), and successively adjusting the column of air so as to be one half, one fourth, one fifth, &c. of the column reciprocating the fundamental sound, the octave, twelfth, double octave, seventeenth, &c. will be produced. The relative numbers, considering the vibrations of the tongue as unity, are 1, 2, 3, 4, 5, &c. The mouth produces precisely the same effect as this changeable tube does, and all the beautiful sounds which Mr. Eulenstein manages with so much skill [on the Jew's harp] are produced by this means; they are multiple resonances of the column of air, and not the vibrations of the tongue itself, as was formerly supposed.

Similar results are obtained when the vibrating tongue of an Æolina is brought before a tube, and its length is altered; and this case resembles Mr. Willis's arrangement. The same multiple resonances are produced also when the cavity of the mouth is substituted for the tube. In these cases the fundamental sound is louder than when the Jew's harp is employed. The sound of the larynx itself may be substituted for that of the vibrating tongue, and similar harmonic sounds will be heard; this experiment may easily be made by placing a piston-tube before the mouth while the voice continues to sound rather a low note.

About two years ago a young man named Richmond exhibited a novel kind of musical performance with the voice. On examining the circumstances under which the sounds were produced, it was ascertained that the continued sound or drone was produced by the larynx, and that he had acquired the art of adjusting the cavity of the mouth so as to fit it for resounding to any multiple. In this way he was able to command these subordinate sounds in any succession, and even to dwell upon them; and he could thus perform a great number of airs.

Some kinds of sounds are better suited to produce these multiple resonances than others; and it is an universal fact that wherever these subordinate sounds can be distinguished, there also the vowel qualities are heard; and, reciprocally, when a sound puts on successively different vowel qualities, these multiple resonances are audible. The tongue of a Jew's harp, which so readily gives rise to these subordinate sounds, is obedient not only to the vowel sounds, but to almost all the articulations of speech. The free reed, or Æolina tongue, when it is such as can enter readily into vibration, is affected in a similar manner; but when it is too rigid, though it may produce as clear a musical sound as before, the multiple resonances and the vowel qualities are equally lost, not, perhaps, because they do not exist, but because they are overpowered by the original sound of the reed.

We do not mean to assert that each multiple resonance is a distinct vowel sound: but we infer that, when a tube is added to a reed or vibrating tongue, whatever may be its length, a quality is added to the original sound, which depends on the feeble vibrations of the air in the added tube; these increase in number in proportion to the shortness of the tube; and when the number of vibrations thus excited is any multiple of the original vibrations of the reed, the energy of the resonance is so greatly augmented as to produce the effect of a superadded musical sound.

Thus it is evident that the vowel qualities and multiple resonances are different forms of the same phenomena.

In considering the consonants to which we may now proceed, we may apply to them the same remark which has been already applied to the vowels. In attempting to class them, for the purpose of ascertaining in what respect these sounds agree with or differ from each other, we must pay no attention to the alphabetic characters which so inefficiently represent them, or to the names by which these characters are designated. For we find that a simple character sometimes represents a compound sound, as G and X; and a simple sound is as frequently denoted by two letters, as Sh, Th, Dh, Ng, in "song," &c. The same character often represents more than one sound; for instance, S has four different powers in *s*ea, hi*s*, vi*s*ion, and *s*ure. Many other redundancies and deficiencies might be noticed, but these will suffice to show the danger of classifying letters instead of sounds. The letters *f, v, r, l, p, t, k, d, m*, and *n* are, however, generally constant in their signification. In this Table all the elementary consonant sounds are arranged, indicated by those characters by which they may be most readily recognized. These sounds will be best compared by articulating them all uniformly, followed by the same vowel, as *Pe, Be*, &c.

It would occupy too much space to explain the particular positions of the different organs during the pronunciation of all these sounds; it must suffice at present to state that for the six sounds characterized in the above table by the term explosive, the breath or voice is entirely obstructed, while for all the others it is allowed to pass, but with partial interruptions.

Consonants.

Mutes		*Sonants*	*Narisonants*
Explosive.	P	B	M
	T	D	N
	K	G	Ng
Continuous.	F	V	
	•	Y	
	Sh	J	
	S	Z	
	Th	Dh	
	•	R	
	Ll	L	
	Kh	Gh	

But there is one comprehensive and important division of these articulations which we cannot pass over in silence. It is found that when the organs are in the proper position for uttering any one of the articulations of the first vertical column, without the slightest change being made in that position, a different quality may be given to it, according as it is accompanied or not by the sound of the larynx, or the voice properly so called. It is in this respect alone that F differs from V, &c.

Every sonant has thus its corresponding mute. This rule seems to be violated, however, with regard to the sounds L and R. But though the corresponding mutes of these sounds are not in the English language, they are to be found in other tongues. The sound denoted in Welsh by Ll is the mute corresponding to our L; and though it is extremely difficult to catch this sound from merely hearing it pronounced, attention to the rule will enable any one to articulate it properly in a few minutes. The organs must be in the proper position for articulating L, but the voice must not sound; or, in other words, the L must be whispered, and then followed by the vowel; as, for example, Llewellyn, Llangollen.

The importance of this rule is obvious; for if in the construction of a speaking machine the voiceless articulations indicated in the first vertical column of the preceding Table were perfectly obtained, all the sounds in the second column could be immediately produced by adding the vocal sound, or what is substituted for it, to those of the first column. A single motion or key would therefore be sufficient to add eleven new articulations to the utterance of the instrument.

The positions of the organs for the three sounds M, N, Ng, are exactly the same as for B, D, G, only, in the former, the soft palate is depressed so as to allow the voice to escape through the nostrils; by this the articulation is rendered continuous. If in a speaking machine, therefore, these three sounds were perfect, the addition of a single key would convert them into six elementary sounds. To imitate all the usual articulations of speech, a machine would not require such complicated motions as might at first be supposed: thirteen keys would be sufficient to command the twenty-six consonant sounds of the above table. The difficulty in the construction of such a machine is therefore reduced to the perfect imitation of these eleven sounds and the vowels, which we have already seen are not beyond our command. How far these difficulties have been overcome we shall now proceed to show.

De Kempelen's machine consisted externally of a square box, and a bellows, of the ordinary construction, placed on a board. The bellows were pressed down by the right arm, and it expanded itself, by means of a weight or spring, when the pressure was removed. There were two openings in the box, for the purpose of introducing the hands to act on the keys &c. within, and numerous small round apertures concealed by silk, to prevent stifling the sound. When this cover was removed, a small box or wind-chest was seen communicating with the pipe of the bellows at one end, and with an India-rubber bell or funnel, from which the sound ultimately issued, on the other; on this wind-chest were various keys to be touched by the fingers of the right hand for the sounds S, Sh, R, &c. The sound was produced by an ivory or brass reed, covered, on the side which vibrated against the edges of the aperture, with very thin leather; this reed, representing the larynx of the vocal organs, was placed within, between the wind-chest and the narrow part of the India-rubber funnel or mouth. The sound P was produced by suddenly removing the left hand from the front of the mouth, which it had previously completely stopped; the sound B by the same action, but, instead of closing the mouth completely, a very minute aperture was left, so that the sound of the reed might not be entirely stifled; M was heard on opening two small tubes, representing the nostrils, placed between the wind-chest and the mouth, while the front of the mouth was stopped, as for P. A few vowel modifications, and the sounds F and V, were produced by modifying the form and size of the aperture of the mouth by the left hand; the continuous consonants S and Sh were obtained by causing the wind to pass through small tubes of particular forms, the passage of the wind being governed by keys; and R was imitated by occasioning a vibration or trembling of the reed when an appropriate lever was depressed.[1]

"In the space of three weeks," says De Kempelen, "any one may acquire wonderful skilfulness in performing on the speaking machine, especially if he applies himself to the Latin, French, and Italian languages; for the German language is much more difficult, on account of the consonants which so frequently occur in it, the sibilant sounds, and the mute letters which so often terminate the words. I can pronounce immediately every French or Italian word I am asked; a long German word, on the contrary, costs me much trouble, and it is rare that I perfectly succeed. As to entire phrases I can produce but few; and these must be short, because the bellows is not sufficiently large to furnish the necessary quantity of wind. For example: *Vous êtes mon ami – Je vous aime de tout mon cœur;* or, in Latin: *Leopoldus secundus – Romanorum imperator – semper Augustus*, &c. I am perfectly convinced," says the author, "that without much skill the machine may be arranged with keys like the harpsichord, or organ, so that it would be much more easy to play it than in the present manner; but this is another step towards perfection which I leave to those who may be inclined to give some attention to this new invention in its infancy, and who, by their meditations and labours, are willing to carry it further."

De Kempelen's is not the only speaking machine which has been made; the Abbé Mical, a celebrated French mechanician, constructed two colossal brazen heads, which are said to have uttered not only words, but entire phrases; he submitted these master-pieces of his ingenuity to the Academy of Sciences in July 1783, in the same year which De Kempelen was at Paris with his machine. No details of the construction of these were ever published; and the inventor is said to have broken them to pieces on being disappointed of the reward which, on the recommendation of the Academy, he had expected from the Government. He died shortly after, in 1789. We can, therefore, only judge of the merits of these automata from the accounts of writers who saw them. From the description which Rivarol gives, it appears that Mical applied to one of the heads a set of keys which were acted upon by pins fixed on a cylinder, so that a determinate number of phrases were produced, as tunes are upon a barrel-organ; the intervals between the words, and the accentuation, Rivarol states to have been marked correctly. The other head had a key-board, like that of a pianoforte, the keys corresponding to the different sounds and tones of the French language, which were reduced to a small number by an ingenious method peculiar to the inventor. "With a little practice and skill," says Rivarol, "any person may talk with the fingers as with the tongue; and he may give to the language of the heads, the rapidity, the stops, and in fine all the characters which a language can possess which is not animated by the passions. Foreigners may take the 'Henriade,' or 'Telemachus,' and cause them to be recited from one end to the other, by placing the volumes on the vocal instruments, as we place the book of an opera on a pianoforte."

No doubt Rivarol's enthusiasm has led him greatly to overrate the performance of these automata. It would be a very easy matter to add either a key-board or a pinned cylinder to De Kempelen's instrument, so as to make the syllables which it utters follow, each with their proper accentuations and rests; but unless the articulations were themselves more perfect, it would not be worth the trouble and expense. According to a more sober writer, Mical's, in this respect, was not superior to De Kempelen's invention. Vicq d'Azyr, the celebrated anatomist, gives the following account of the Abbé Mical's heads, in his report upon them to the Academy: –

> "The heads," says this writer, "covered a hollow box, the different parts of which were connected together by hinges, and in the interior of which the inventor had disposed artificial glottises of different forms over stretched membranes. The air passing through these glottises, was directed on these membranes, which gave sounds of different pitches; and from their combination there resulted a very imperfect imitation of the human voice."

It is a curious circumstance that Kratzenstein, De Kempelen, and the Abbé Mical were all employed in their researches at the very same time, though in very different parts of Europe. Since that period no progress has been made in imitating the articulations of speech, and even the experiments of these mechanicians have been almost forgotten. The advantages that would result from the completion of a speaking machine render the subject worthy the attention of the philosopher and the mechanician. Dr. Darwin supposed that they might be made hereafter with voices so loud and sonorous as to command an army or instruct a crowd; but the most important uses to which they will perhaps in future times be applied, will be to fix and perpetuate the pronunciation of different languages.

There are not doubt a great many difficulties yet to be contended with before we can succeed in perfectly imitating the articulations of speech; but the partial success of the attempts of which we have laid an account before our readers ought to encourage further trials. It is not too much to say, in the words of Sir David Brewster, "We have no doubt that, before another century is completed, a talking and a singing machine will be numbered among the conquests of science."

NOTE

1 A speaking machine, made by Professor Wheatstone from De Kempelen's description, with some improvements, is at present in the collection of Philosophical Instruments of KCL.

12

ROYAL INSTITUTION

(Anon.)

Source: (Anon.), "Royal Institution", *The Dublin Literary Gazette, or Weekly Chronicle of Criticism, Belles Lettres, and Fine Arts* I (Dublin, Ireland, 1830), pp. 395–96.

Albemarle-street, London, June 11.

The Friday evening meetings of the season are brought to a close: the nineteenth and last of those intellectual treats was presented to us this evening in Mr. Faraday's concluding discourse on the subject of Phonics. What an extraordinary being is this Michael Faraday! Only give him a subject, no matter what, and he will descant upon it as delightfully as Cowper did upon his accidental theme of the sofa. Bid him discourse upon the music of the spheres, the hieroglyphics of Egypt, or the Entelecheia of Aristotle, and he will be found equally at home – equally delightful, impressive, and informing. In short, I verily believe, there is nothing in heaven or earth that is not dreamt of in Mr. Faraday's philosophy: he is as versatile and as comprehensive as the poet's eye in a fine frenzy rolling. He is the improvisatore of the scientific world – the organ of new inventors – the mouth-piece of dumb mechanic geniuses. Do you wish to make known your discoveries in any art or science? Talk to Mr. Faraday for ten minutes, and he will announce your labours to the public ear in the most captivating form imaginable. Besides *nihil tetigit quod non ornavit*, as Johnson wrote of poor Goldsmith – an idea is never transmitted through his mosaic brain without being adorned and gilded in the passage. On the present occasion, with much modesty and candour, he informed us that throughout his discourses on Phonics, he was entirely indebted to Mr. Wheatstone for his materials – he was himself but the instrument of their communication. Mr. Wheatstone was, it seems, the prompter or performer behind the scenes, while Mr. Faraday gesticulated before the audience. I have got a peep, however, at this Mr. Wheatstone, a young man of anxious, spectacle-on-nose appearance. He ventures out from his hiding-place after lecture, and offers further practical explanation of what Mr. Faraday has said. He keeps a music-shop, I am told, somewhere here in London.

With regard to the discourse of the evening – on the laws of the coexisting vibrations of strings and rods – it is curious to observe what beautiful illustration the most trivial phenomena may receive, when treated by a man of talent and taste. Things with which we have been familiar from our very childhood are brought before us as matter of philosophical investigation: our earliest amusements are

DOI: 10.4324/b23406-12

taken up as illustrative of some scientific novelty, and they are thus doubly delightful from the additional zest they derive from cherished associations. After proving the possibility of the actual coexistence of various vibrations in one and the same extended string, he turned to those of rods, which by the ingenious device of the Kaleidophone are made visible to the naked eye. The simplicity of the Kaleidophone is as remarkable as the appearances it presents, are beautiful and surprising. A slender rod of highly tempered steel, tipped at one end with a brilliant spherule, is fixed at the other extremity in a sounding board – a violin bow is then drawn across the rod, when a musical tone is uttered, and the spherule is seen to describe some of the most fanciful and extraordinary curves, circles in every variety of combination and concatenation, ellipses, spirals "in wandering mazes lost," – the bounding lines composed of minute and indefinable gyrations of fantastic tracery – description fails to convey any thing like an idea of their dazzling brilliancy and effect!

This introduced next the curious consideration of the time occupied by impressions on the retina. In the rapid whirl of a bright object – such as a polished ball suspended by a string, or the end of a cane on fire – a circle of light perfectly continuous is presented to the eye. The circle is formed by the revolution of the brilliant point with such rapidity that the impression remains upon the sight from the instant it quits a given position in the circumference till it reaches the same spot again. Another very beautiful method of exhibiting the phenomena in question, is by laying, over an illuminated or transparent surface, a revolving disk, opaque, but having a single radiating slit proceeding from its centre. – Through this slit, upon the rapid revolution of the disk, the whole of the illuminated surface is distinctly seen. Mr. Faraday mentioned a circumstance which lately occurred to him, as affording further illustration of the subject. On passing by the coach, between Greenwich and Woolwich, he observed that the arsenal was distinctly visible to him, and in its whole extent, through the apertures of the paling which skirts the road; whilst through any single aperture in it, no appreciable impression could have reached the eye. The whole picture was, in this instance, evidently formed by the combination of those separately almost imperceptible glimpses. We all see, by the bye, such things as that Mr. Faraday saw; but the same appearances call up very different reflections in the mind of the man of genius, from what they excite in those of unthinking heedless observers. Godwin makes some fine remarks on this topic in one of his works. But to bring my notes to a conclusion. Every single instantaneous impression, such as has been just now described, occupies a distinct portion of duration, though not appreciable by our ordinary methods of measuring time; yet an attempt to ascertain it, or to render it perceptible to the senses, seems not impossible by the application of the laws of vibrations in strings of given lengths and tones: for we can by an ingenious extension of the principles of the Kaleidophone make these vibrations apparent and capable of numeration.

The discourse was terminated by a suggestion of Mr. Faraday's, that it might be possible ere long, by further attention to the measurement of minute particles of time, so to procure a kind of microscopic method of magnifying them, as that we might ultimately be able to determine, by observations in the transmission of the electric fluid, whether there were two electricities or not.

There were some fair listeners in the gallery, attracted no doubt, by the hope of enjoying some more of those musical performances on strange instruments, which sent them home so delighted this night three weeks. If they failed, however, to be so much *amused* by the evening's entertainment, they must certainly have reaped greater advantage from these more serious illustrations of the philosophy of sound.

C.

Editorial Headnote

13. Michael Faraday, "On a Peculiar Class of Acoustical Figures, and on the Forms of Fluids Vibrating on Elastic Surfaces", *Philosophical Transactions of the Royal Society* 121 (London, England, 1831), pp. 299–340.

A close associate of Wheatstone, Michael Faraday (1791–1867) is best known for his contributions to the electrical sciences. Following Danish chemist Hans Christian Ørsted's (1777–1851) observation in 1820 that an electric current passing through copper wire could cause a magnetized needle to move, Faraday extended this connecting of electric and magnetic phenomena with his discovery of electromagnetic rotation in 1821. In 1831, he would go on to induce electricity by moving a magnet through a coiled helix of copper wire, establishing the principle of electromagnetic induction.[1] Faraday conducted his work in the underground laboratory of the Royal Institution where he had first secured appointment as Humphry Davy's (1778–1829) chemical assistant in 1813, rising to superintendent of the laboratory in 1821, and director in 1825. In this position, Faraday had access to some of London's leading scientific resources, including experimental apparatus and chemicals, as well as the laboratory itself, which was a rare commodity in early nineteenth-century Britain. So when Wheatstone persuaded Faraday to take an interest in acoustic science, this was a significant moment, given Faraday's skill as an experimentalist and the material resources he could bring to the subject.[2] The two experimentalists met sometime around 1825, which was unsurprising given Faraday's position at the Royal Institution and as Secretary of the Athenaeum Club at 107 Pall Mall, both just a short walk from Wheatstone's business premises in Pall Mall and The Strand. Faraday was himself a keen flautist and able vocalist during the 1810s, even playing music while working alongside Davy in the Royal Institution's laboratory. Before meeting Wheatstone, Faraday undertook little research on sound but, during the late 1820s grew increasingly interested in the subject. In early 1831, Faraday adapted Chladni and Wheatstone's techniques for observing the movement of sonorous waves through plates, using a range of media including silica and red lead. Throughout February and March, Faraday undertook a rigorous experimental programme, delivering his results to the Royal Institution on 21 March 1831 and the Royal Society in May. Faraday continued these experiments, with increasing complexity, until July. A key motivation in this work was Faraday's satisfaction with the order revealed through vibrating media, which was consistent with his religious, Sandemanian understanding of nature.[3] There was a real emphasis on creating aesthetic beauty running through Faraday's arrangements, such as his use of mercury and blue ink to produce silver "pearls" as the oscillations passed through the substances. Historians have often written this acoustic work off as essentially frivolous and something of a distraction to Faraday's wider philosophical interests, but it is significant that his work on measuring impulses through fluids, forming crispations, concluded in July 1831, just a month before he first successfully induced electricity from magnetism and movement.[4] After 1831, Faraday would focus his research on electrical and magnetic phenomena, but his earlier acoustic work provided an important context to these subsequent, more celebrated, experimental observations.

NOTES

1 On Faraday's career see, Geoffrey Cantor, David Gooding, and Frank A. J. L. James, *Michael Faraday* (New York: Humanity Books, 1991).

2 David Gooding, ""In Nature's School": Faraday as an Experimentalist", in David Gooding and Frank A. J. L. James (eds.), *Faraday Rediscovered: essays on the life and work of Michael Faraday, 1791–1867* (Basingstoke: Macmillan Press Ltd, 1985), pp. 105–135.

3 Geoffrey Cantor, *Michael Faraday: Sandemanian and Scientist. A study of science and religion in the nineteenth century* (Basingstoke: Macmillan, 1991); An account of Faraday's experiments, including his crispation observations can be found in Thomas Martin (ed.), *Faraday's Diary: being various philosophical notes of experimental investigation made by Michael Faraday*. Vol. I: Sep., 1820–11 Jun., 1832 (London: G. Bell and Sons, 1932).

4 For debates over the importance of these experiments see, James Hamilton, *Faraday: the life* (London: Harper Collins, 2002), pp. 233–44; Ryan D. Tweney, "Faraday's Discovery of Induction: a cognitive approach", in David Gooding and Frank A. J. L. James (eds.), *Faraday Rediscovered: essays on the life and work of Michael Faraday, 1791–1867* (Basingstoke: Macmillan Press Ltd, 1985), pp. 189–209, 205; Ryan D. Tweney, "Stopping time: Faraday and the scientific creation of perceptual order", *Physics*, 29 (1992), pp. 149–64, 151–3.

13

ON A PECULIAR CLASS OF ACOUSTICAL FIGURES, AND ON THE FORMS OF FLUIDS VIBRATING ON ELASTIC SURFACES

Michael Faraday

Source: Michael Faraday, "On a Peculiar Class of Acoustical Figures, and on the Forms of Fluids Vibrating on Elastic Surfaces", *Philosophical Transactions of the Royal Society* 121 (London, England, 1831), pp. 299–340.

Read May 12, 1831.

1. THE beautiful series of forms assumed by sand, filings, or other grains, when lying upon vibrating plates, discovered and developed by CHLADNI, are so striking as to be recalled to the minds of those who have seen them by the slightest reference. They indicate the quiescent parts of the plates, and visibly figure out what are called the nodal lines.

2. Afterwards M. CHLADNI observed that shavings from the hairs of the exciting violin bow did not proceed to the nodal lines, but were gathered together on those parts of the plate the most violently agitated, i. e. at the centres of oscillation. Thus when a square plate of glass held horizontally was nipped above and below at the centre, and made to vibrate by the application of a violin bow to the middle of one edge, so as to produce the lowest possible sound, sand sprinkled on the plate assumed the form of a diagonal cross; but the light shavings were gathered together at those parts towards the middle of the four portions where the vibrations were most powerful and the excursions of the plate greatest.

3. Many other substances exhibited the same appearance. Lycopodium, which was used as a light powder by OERSTED, produced the effect very well. These motions of lycopodium are entirely distinct from those of the same substance upon plates or rods in which longitudinal vibrations are excited.

4. In August 1827, M. SAVART read a paper to the Royal Academy of Sciences, in which he deduced certain important conclusions respecting the subdivision of vibrating sonorous bodies from the forms thus assumed by light

DOI: 10.4324/b23406-13

powders. The arrangement of the sand into lines in CHLADNI's experiments shows a division of the sounding plate into parts, all of which vibrate isochronously, and produce the same tone. This is the principal mode of division. The fine powder which can rest at the places where the sand rests, and also accumulate at other places, traces a more complicated figure than the sand alone, but which is so connected with the first, that, as M. SAVART states, "the first being given, the other may be anticipated with certainty; from which it results that every time a body emits sounds, not only is it the seat of many modes of division which are superposed, but amongst all these modes there are always two which are more distinctly established than all the rest. My object in this memoir is to put this fact beyond a doubt, and to study the laws to which they appear subject."

5. M. SAVART then proceeds to establish a secondary mode of division in circular, rectangular, triangular and other plates; and in rods, rings, and membranes. This secondary mode is pointed out by the figures delineated by the lycopodium or other light powder; and as far as I can perceive, its existence is assumed, or rather proved, exclusively from these forms. Hence much of the importance which I attach to the present paper. A secondary mode of division, so subordinate to the principal as to be always superposed by it, might have great influence in reasonings upon other points in the philosophy of vibrating plates; to prove its existence therefore is an important matter. But its existence being assumed and supported by such high authority as the name of SAVART, to prove its non-existence, supposing it without foundation, is of equal consequence.

6. The essential appearances, as far as I have observed them, are as follows. Let the plate before mentioned (2), which may be three or four inches square, be nipped and held in a horizontal position by a pair of pincers of the proper form, and terminated, at the part touching the glass, by two pieces of cork; let lycopodium powder be sprinkled over the plate, and a violin bow be drawn downwards against the middle of one edge so as to produce a clear full tone. Immediately the powder on those four parts of the plate towards the four edges will be agitated, whilst that towards the two diagonal cross lines will remain nearly or quite at rest. On repeating the application of the bow several times, a little of the loose powder, especially that in small masses, will collect upon the diagonal lines, and thus, showing one of the figures which CHLADNI discovered, will also show the principal mode of division of the plate. Most of the powder which remains upon the plate will, however, be collected in four parcels; one placed near to each edge of the plate, and evidently towards the place of greatest agitation. Whilst the plate is vibrating (and consequently sounding) strongly, these parcels will each form a rather diffuse cloud, moving rapidly within itself; but as the vibration diminishes, these clouds will first contract considerably in bulk, and then settle down into four groups, each consisting of one, two, or more hemispherical parcels (53), which are in an extraordinary condition; for the powder of each parcel continues to rise up at the centre and

flow down on every side to the bottom, where it enters the mass to ascend at the centre again, until the plate has nearly ceased to vibrate. If the plate be made to vibrate strongly, these parcels are immediately broken up, being thrown into the air, and form clouds, which settle down as before; but if the plate be made to vibrate in a smaller degree, by a more moderate application of the bow, the little hemispherical parcels are thrown into commotion without being sensibly separated from the plate, and often slowly travel towards the quiescent lines. When one or more of them have thus receded from the place over which the clouds are always formed, and a powerful application of the bow is made, sufficient to raise the clouds, it will be seen that these heaps rapidly diminish, the particles of which they are composed being swept away from them, and passing back in a current over the glass to the cloud under formation, which ultimately settles as before into the same four groups of heaps. These effects may be repeated any number of times, and it is evident that the four parts into which the plate may be considered as divided by the diagonal lines are repetitions of one effect.

7. The form of the little heaps, and the involved motion they acquire, are no part of the phenomena under consideration at present. They depend upon the adhesion of the particles to each other and to the plate, combined with the action of the air or surrounding medium, and will be resumed hereafter (53). The point in question is the manner in which fine particles do not merely remain at the centres of oscillation, or places of greatest agitation, but are actually driven towards them, and that with so much the more force as the vibrations are more powerful.

8. That the agitated substance should be in very fine powder, or very light, appears to be the only condition necessary for success; fine scrapings from a common quill, even when the eighth of an inch in length or more, will show the effect. Chemically pure and finely divided silica rivals lycopodium in the beauty of its arrangement at the vibrating parts of the plate, although the same substance in sand or heavy particles proceeds to the lines of rest. Peroxide of tin, red lead, vermilion, sulphate of baryta, and other heavy powders when highly attenuated, collect also at the vibrating parts. Hence it is evident that the nature of the powder has nothing to do with its collection at the centres of agitation, provided it be dry and fine.

9. The cause of these effects appeared to me, from the first, to exist in the medium within which the vibrating plate and powder were placed, and every experiment which I have made, together with all those in M. Savart's paper, either strongly confirm, or agree with this view. When a plate is made to vibrate (2), currents (24) are established in the air lying upon the surface of the plate, which pass from the quiescent lines towards the centres or lines of vibration, that is, towards those parts of the plates where the excursions are greatest, and then proceeding outwards from the plate to a greater or smaller distance, return towards the quiescent lines. The rapidity of these currents, the distance to which they rise from the plate at the centre of oscillation, or any other part, the blending of the progressing and returning

air, their power of carrying light or heavy particles, and with more or less rapidity or force, are dependent upon the intensity or force of the vibrations, the medium in which the vibrating plate is placed, the vicinity of the centre of vibration to the limit or edge of the plate, and other circumstances, which a simple experiment or two will immediately show must exert much influence on the phenomena.

10. So strong and powerful are these currents, that when the vibrations were energetic, the plate might be inclined 5°, 6°, or 8° to the horizon and yet the gathering clouds retain their places. As the vibrations diminished in force, the little heaps formed from the cloud descended the hill; but on strengthening the vibrations they melted away, the particles ascending the inclined plane on those sides proceeding upwards, and passing again to the cloud. This took place when neither sand nor filings could rest on the quiescent or nodal lines. Nothing could remain upon the plate except those particles which were so fine as to be governed by the currents, which (if they exist at all) it is evident would exist in whatever situation the plate was placed.

11. M. Savart seems to consider that the reason why the powder gathers together at the centres of oscillation is, "that the amplitude of the oscillations being very great, the middle of each of those centres (of vibration) is the only place where the plate remains nearly plane and horizontal, and where, consequently, the powder may reunite, whilst the surface being inclined to the right or left of this point, the parcels of powder cannot stop there." But the inclination thus purposely given to the plate, was very many times that which any part acquires by vibration in a horizontal position, and consequently proves that the horizontality of any part of the plate is not the cause of the powder collecting there, although it may be favourable to its remaining there when collected.

12. Guided by the idea of what ought to happen, supposing the cause now assigned were the true one, the following amongst many other experiments were made. A piece of card about an inch long and a quarter of an inch wide was fixed by a little soft cement on the face of the plate near one edge, the plate held as before at the middle, lycopodium or fine silica strewed upon it, and the bow applied at the middle of another edge; the powder immediately advanced close to the card, and the place of the cloud was much nearer to the edge than before. Figure 13.1 represents the arrangement; the diagonal lines being those which sand would have formed, the line at the top *a* representing the place of the card, and the × to the right the place where the bow was applied. On applying a second piece of card as at *b*, the powder seemed indifferent to it or nearly so, and ultimately collected as in the first figure: *c* represents the place of the cloud when no card is present.

13. Pieces of card were then fixed on the glass in the three angular forms represented in Figure 13.2; upon vibrating the plate the fine powder always went into the angle, notwithstanding its difference of position in the three experiments, but perfectly in accordance with the idea of currents intercepted more or less by the card. When two pieces of card were fixed on the plate as in Figure 13.3. *a*, the powder proceeded into the angle but not to the edge of the glass, remaining about

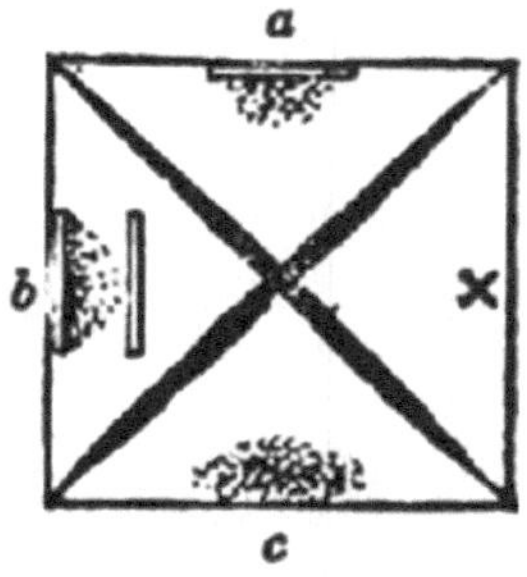

Figure 13.1

Figure 13.2

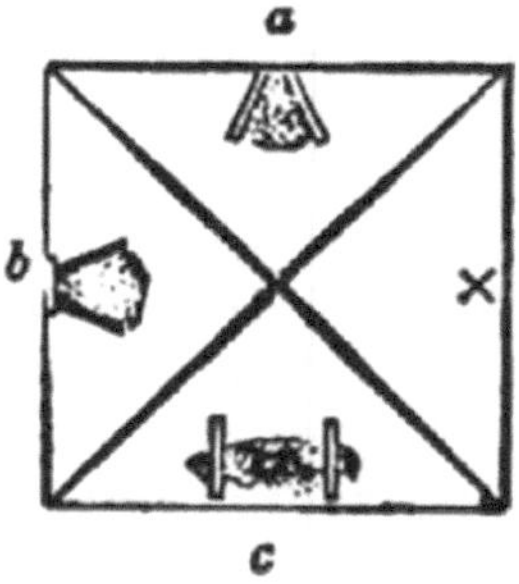

Figure 13.3

⅛th of an inch from it; but on closing up that opening, as at *b*, the powder went quite up into the corner.

14. Upon fixing two pieces of card on the plate as at *c* Figure 13.3, the powder between them collected in the middle very nearly as if no card had been present; but that on the outside of the cards gathered close up against them, being able to proceed so far in its way to the middle, but no further.

15. In all these experiments the sound was very little lowered, the form of the cross was not changed, and the light powders collected on the other three portions

of the plate, exactly as if no card walls had been applied on the fourth; so that no reason appears for supposing that the mode in which the plate vibrated was altered, but the powders seem to have been carried forward by currents which could be opposed or directed at pleasure by the card stops.

16. A piece of gold-leaf being laid upon the plate, so that it did not overlap the edge, Figure 13.4, the current of air towards the centre of vibration was beautifully shown; for, by its force, the air crept in under the gold-leaf on all sides, and raised it up into the form of a blister; that part of the gold-leaf corresponding to the centre of the locality of the cloud, when light powder was used, being frequently a sixteenth or twelfth of an inch from the glass. Lycopodium or other fine powder sprinkled round the edge of the gold-leaf, was carried in by the entering air, and accumulated underneath.

Figure 13.4

17. When silica was placed on the edge of another glass plate, or upon a book, or block of wood, and the edge of the vibrating plate brought as nearly as possible to the edge of the former, Figure 13.5, part of the silica was always driven on to the vibrating plate, and collected in the usual place; as if in the midst of all the agitation of the air in the neighbourhood of the two edges, there was still a current towards the centre of vibration, even from bodies not themselves vibrating.

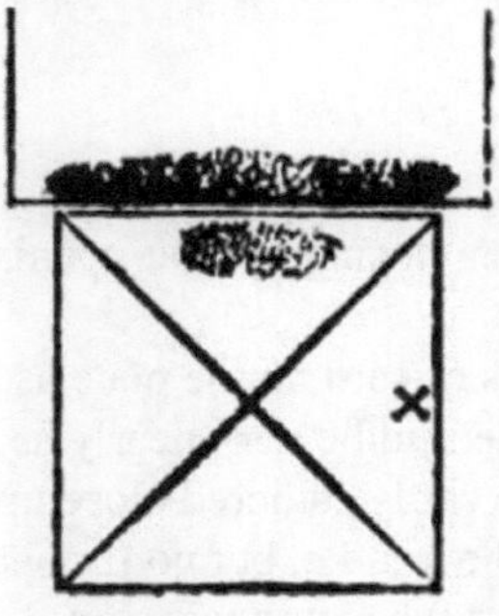

Figure 13.5

18. When a long glass plate is supported by bridges or strings at the two nodal lines represented in Figure 13.6, and made to vibrate, the lycopodium collects in three divisions; that between the nodal lines does not proceed at once into a line equidistant from the nodal lines and parallel to them, but advances from the edges of the plate towards the middle by paths, which are a little curved and oblique to the edges where they occur near the nodal lines, but are almost perpendicular to it elsewhere, and the powder gradually forms a line along the middle of the plate; it is only by continuing the experiment for some time that it gathers up into a heap or cloud equidistant from the nodal lines. But upon fixing card walls upon this plate, as in Figure 13.7, the course of the powder within the cards was directly parallel to them and to the edge, instead of being perpendicular, and also directly towards the centre of oscillation. To prove that it was not as a weight that the card acted, but as an obstacle to the currents of air formed, it was not moved from its place, but bent flat down outwards, and then the fine powder resumed the courses it took upon the plate when without the cards. Upon raising the cards the first effect was reproduced.

19. The lycopodium sprinkled over the extremities of such a plate proceeds towards places equidistant from the sides and near the ends, as at *a* Figure 13.8; but on cementing a piece of paper to the edge, so as to form a wall about one quarter or one third of an inch high, *b*, the powder immediately moved up to it, and retained this new place. In a longer narrow plate, similarly arranged, the powder could be made to pass to either edge, or to the middle, according as paper interceptors to the currents of air were applied.

20. Plates of tin, four or five inches long, and from an inch to two inches wide, fixed firmly at one end in a horizontal position, and vibrated by applying the fingers, show the progress of the air and the light powders well. The vibrations are of comparatively enormous extent, and the appearances are consequently more instructive.

21. If a tuning-fork be vibrated, then held horizontally with the broad surface of one leg uppermost, and a little lycopodium be sprinkled upon it, the collection

Figure 13.6

Figure 13.7

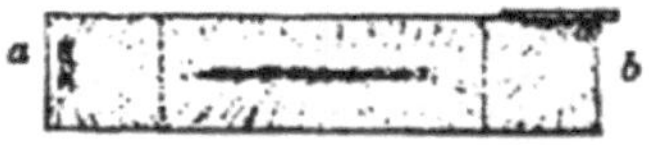

Figure 13.8

of the powder in a cloud along the middle, and the formation of the involving heaps also in a line along the middle of the vibrating steel bar, may be beautifully observed. But if a piece of paper be attached by wax to the side of the limb, so as to form a fence projecting above it, as in the former experiments (19), then the powder will take up its place close to the paper; and if pieces of paper be attached on different parts of the same leg, the powder will go to the different sides, in the different parts, at the same time.

22. The effects under consideration are exceedingly well shown and illustrated by membranes. A piece of parchment was stretched and tightly tied, whilst moist, over the aperture of a funnel five or six inches in diameter; a small hole was made in the middle, and a horse-hair passed through it, but with a knot at the extremity that it might thereby be retained. Upon fixing the funnel in an upright position, and after applying a little powdered resin to the thumbs and fore-fingers, drawing them upward over the horse-hair, the membrane was thrown into vibration with more or less force at pleasure. By supporting the funnel on a ring, passing the horse-hair in the opposite direction through the hole in the membrane, and drawing the fingers over it downwards, the direction in which the force was applied could be varied according to circumstances.

23. When lycopodium or light powders were sprinkled upon this surface, the rapidity with which they ran to the centre, the cloud formed there, the involving heaps, and many other circumstances, could be observed very advantageously.

24. The currents which I have considered as existing upon the surface of the plate, membranes, &c. from the quiescent parts towards the centres or lines of vibration (9), arise necessarily from the mechanical action of that surface upon the air. As any particular part of the surface moves upwards in the course of its vibration, it propels the air and communicates a certain degree of force to it, perpendicular or nearly so to the vibrating surface; as it returns, in the course of its vibration, it recedes from the air so projected, and the latter consequently tends to return into the partial vacuum thus formed. But as of two neighbouring portions of air, that over the part of the plate nearest to the centre of oscillation has had more projectile force communicated to it than the other, because the part of the plate urging it was moving with greater velocity, and through a greater space, so it is in a more unfavourable condition for its immediate return, and the other, i. e. the portion next to it towards the quiescent line, presses into its place. This effect is still further favoured, because the portion of air thus displaced is urged from similar causes at the same moment into the place left vacant by the air still nearer the centre of oscillation; so that each time the plate recedes from the air, an advance of the air immediately above it is made from the quiescent towards the vibrating parts of the plates.

25. It will be evident that this current is highly favourable for the transference of light powders towards the centre of vibration. Whilst the air is forced forward, the advance of the plate against the particles holds them tight; but when the plate recedes, and the current exists, the particles are at that moment left unsupported except by the air, and are free to move with it.

26. The air which is thus thrown forward at and towards the centre of oscillation, must tend by the forces concerned to return towards the quiescent lines, forming a current in the opposite direction to the first, and blending more or less with it. I endeavoured, in various ways, to make the extent of this system of currents visible. In the experiment already referred to, where gold-leaf was placed over the centre of oscillation (16), the upward current at the most powerful part was able to raise the leaf about one tenth of an inch from the plate. The higher the sounds with the same plate or membrane, i. e. the greater the number of vibrations, the less extensive must be the series of currents; the slower the vibrations, or the more extensive the excursion of the parts from increased force applied, the greater the extent of disturbance. With glass plates (2. 12) the cloud is higher and larger as the vibrations are stronger, but still not so extensive as they are upon the stretched membrane (22), where the cloud may frequently be seen rising up in the middle and flowing over towards the sides.

27. When the membrane stretched upon the funnel (22) was made to vibrate by the horse-hair proceeding downwards, and a large glass tube, as a cylindrical lamp-glass, was brought near to the centre of vibration, no evidence of a current entirely through the lamp-glass could be perceived; but still the most striking proofs were obtained of the existence of carrying currents by the effects upon the light powder, for it flew more rapidly under the edge, and tended to collect towards the axis of the tube; it could even be diverted somewhat from its course towards the centre of oscillation. A piece of upright paper, held with its edge equally near, did not produce the same effect; but immediately that it was rolled into a tube, it did. When the glass chimney was suspended very carefully, and at but a small distance from the membrane, the powder often collected at the edge, and revolved there; a complicated action between the currents and the space under the thickness of the glass taking place, but still tending to show the influence of the air in arranging and disposing the powders.

28. A sheet of drawing-paper was stretched tightly over a frame so as to form a tense elastic surface nearly three feet by two feet in extent. Upon placing this in a horizontal position, throwing a spoonful of lycopodium upon it, and striking it smartly below with the fingers, the phenomena of collection at the centre of vibration, and of moving heaps, could be obtained upon a magnificent scale. When the lycopodium was uniformly spread over the surface, and any part of the paper slightly tapped by the hand, the lycopodium at any place chosen could be drawn together merely by holding the lamp-glass over it. It will be unnecessary to enter into the detail of the various actions combining to produce these effects; it is sufficiently evident, from the mode in which they may be varied, that they depend upon currents of air.

29. A very interesting set of effects occurred when the stretched parchment upon the funnel (22) was vibrated under plates; the horse-hair was directed downwards, and the membrane, after being sprinkled over with light powder, was covered by a plate of glass resting upon the edge of the funnel; upon throwing the membrane into a vibratory state, the powder collected with much greater rapidity

than without the plate; and instead of forming the semi-globular moving heaps, it formed linear arrangements, all concentric to the centre of vibration. When the vibrations were strong, these assumed a revolving motion, rolling towards the centre at the part in contact with the membrane, and from it at the part nearest the glass; thus illustrating in the clearest manner the double currents caged up between the glass and the membrane. The effect was well shown by carbonate of magnesia.

30. Sometimes when the plate was held down very close and tight, and the vibrations were few and large, the powder was all blown out at the edge; for then the whole arrangement acted as a bellows; and as the entering air travelled with much less velocity than the expelled air, and as the forces of the currents are as the squares of the velocity, the issuing air carried the powder more forcibly than the air which passed in, and finally threw it out.

31. A thin plate of mica laid loosely upon the vibrating membrane showed the rotating concentric lines exceedingly well.

32. From these experiments on plates and surfaces vibrating in air, it appears that the forms assumed by the determination of light powders towards the places of most intense vibration, depend, not upon any secondary mode of division, or upon any immediate and peculiar action of the plate, but upon the currents of air necessarily formed over its surface, in consequence of the extra-mechanical action of one part beyond another. In this point of view the nature of the medium in which those currents were formed ought to have great influence over the phenomena; for the only reason why silica as sand should pass towards the quiescent lines, whilst the same silica as fine powder went from them, is, that in its first form the particles are thrown up so high by the vibrations as to be above the currents, and that if they were not thus thrown out of their reach they would be too heavy to be governed by them; whilst in the second form they are not thrown out of the lower current, except near the principal place of oscillation, and are so light as to be carried by it in whatever direction it may proceed.

33. In the exhausted receiver of the air-pump therefore the phenomena ought not to occur as in air; for as the force of the currents would be there excessively weakened, the light powders ought to assume the part of heavier grains in the air. Again, in denser media than air, as in water for instance, there was every reason to expect that the heavier powder, as sand and filings, would perform the part of light powders in air, and be carried from the quiescent to the vibrating parts.

34. The experiments in the air-pump receiver were made in two ways. A round plate of glass was supported on four narrow cork legs upon a table, and then a thin glass rod with a rounded end held perpendicularly upon the middle of the glass. By passing the moistened fingers longitudinally along this rod the plate was thrown into a vibratory state; the cork legs were then adjusted in the circular nodal line occurring with this mode of vibration; and when their places were thus found they were permanently fixed. The plate was then transferred into the receiver of an air-pump, and the glass rod by which it was to be thrown into vibration passed through collars in the upper part of the receiver, the entrance of air there being

prevented by abundance of pomatum. When fine silica was sprinkled upon the plate, and the plate vibrated by the wet fingers applied to the rod, the receiver not being exhausted, the fine powder travelled from the nodal line, part collecting at the centre, and other part in a circle, between the nodal line and the edge. Both these situations were places of vibration, and exhibited themselves as such by the agitation of the powder. Upon again sprinkling fine silica uniformly over the plate, exhausting the receiver to twenty-eight inches, and vibrating the plate, the silica went from the middle towards the nodal line or place of rest, performing exactly the part of sand in air. It did not move at the edges of the plate, and as the apparatus was inconvenient and broke during the experiment, the following arrangement was adopted in its place.

35. The mouth of a funnel was covered (22) with a well-stretched piece of fine parchment, and then fixed on a stand with the membrane horizontal; the horse-hair was passed loosely through a hole in a cork, fixed in a metallic tube on the top of the air-pump receiver; the tube above the cork was filled to the depth of half an inch with pomatum, and another perforated cork put over that; a cup was formed on the top of the second cork, which was filled with water. In this way the horse-hair passed first through pomatum and then water, and by giving a little pressure and rotatory motion to the upper cork during the time that the horse-hair was used to throw the membrane into vibration, it was easy to keep the pomatum below perfectly in contact with the hair, and even to make it exude upwards into the water above. Thus no possibility of the entrance of air by and along the horse-hair could exist, and the tightness of all the other and fixed parts of the apparatus was ascertained by the ordinary mode of examination. A little paper shelf was placed in the receiver under the cork to catch any portion of pomatum that might be forced through by the pressure, and prevent its falling on to the membrane.

36. This arrangement succeeded: when the receiver was full of air, the lycopodium gathered at the centre of the membrane with great facility and readiness, exhibiting the cloud, the currents, and the involving heaps. Upon exhausting the receiver until the barometrical gauge was at twenty-eight inches, the lycopodium, instead of collecting at the centre, passed across the membrane towards one side which was a little lower than the other. It passed by the middle just as it did over any other part; and when the force of the vibrations was much increased, although the powder was more agitated at the middle than elsewhere, it did not collect there, but went towards the edges or quiescent parts. Upon allowing air to enter until the barometer stood at twenty-six inches, and repeating the experiments, the effect was nearly the same. When the vibrations were very strong, there were faint appearances of a cloud, consisting of the very finest particles, collecting at the centre of vibration; but no sensible accumulation of the powder took place. At twenty-four inches of the barometer the accumulation at the centre began to appear, and there was a sensible, though very slight effect visible of the return of the powder from the edges. At twenty-two inches these effects were stronger; and when the barometer was at twenty inches, the currents of air within the receiver had force enough to cause the collection of the principal part of the lycopodium at the centre of

vibration. Upon again, however, restoring the exhaustion to twenty-eight inches, all the effects were reproduced as at first, and the lycopodium again proceeded to the lower or the quiescent parts of the membrane. These alternate effects were obtained several times in succession before the apparatus was dismounted.

37. In this form of experiment there were striking proofs of the existence of a current upwards from the middle of the membrane when vibrating in air, (24), and the extent of the system of currents (26) was partly indicated. The powder purposely collected at the middle by vibrations, when the receiver was full of air, was observed as to the height to which it was forced upwards by the vibrations; and then the receiver being exhausted, the height to which the powder was thrown by similar vibrations was again observed. In the latter cases it was nothing like so great as in the former, the height not being two-thirds, and barely one-half, the first height. Had the powder been thrown up by mere propulsion, it should have risen far higher in vacuo than in air: but the reverse took place; and the cause appears to be, that in air the current had force enough to carry the fine particles up to a height far beyond what the mere blow which they received from the vibrating membrane could effect.

38. For the experiments in a denser medium than air, water was chosen. A circular plate of glass was supported upon four feet in a horizontal position, surrounded by two or three inches of water, and thrown into vibration by applying a glass rod perpendicular to the middle, as in the first experiment in vacuo (34); the feet were shifted until the arrangement gave a clear sound, and the moistened brass filings sprinkled upon the plate formed regular lines or figures. These lines were not however lines of rest, as they would have been in the air, but were the places of greatest vibration; as was abundantly evident from their being distant from that nodal line determined and indicated by the contact of the feet, and also from the violent agitation of the filings. In fact, the filings proceeded from the quiescent to the moving parts, and there were gathered together; not only forming the cloud of particles over the places of intense vibration, but also settling down, when the vibrations were weaker, into the same involving groups, and in every respect imitating the action of light powders in air. Sand was affected exactly in the same manner; and even grains of platina could be in this way collected by the currents formed in so dense a medium as water.

39. The experiments were then made under water with the membranes stretched over funnels (22) and thrown into vibration by horse-hairs drawn between the fingers. The space beneath the membrane could be retained, filled with air, whilst the upper surface was covered two or three inches deep with water; or the space below could also be filled with water, or the force applied to the membrane by the horse-hair could be upwards or downwards at pleasure. In all these experiments the sand or filings could be made to pass with the utmost facility to the most powerfully vibrating part, that being either at the centre only, or in addition, in circular lines, according to the mode in which the membrane vibrated. The edge of the funnel was always a line of rest; but circular nodal lines were also formed, which were indicated, not by the accumulation of filings upon them, but by the tranquil state

of those filings which happened to be there, and also by being between those parts where the filings, by their accumulation and violent agitation, indicated the parts in the most powerful vibratory state.

40. Even when by the relaxation of the parchment from moisture, and the force upwards applied by the horse-hair, the central part of the membrane was raised the eighth of an inch or more above the edges, the circle not being four inches in diameter, still the filings would collect there.

41. When in place of parchment common linen was used, as becoming tighter rather than looser when wetted, the same effects were obtained.

42. Both the reasoning adopted and the effects described were such as to lead to the expectation that if the plate vibrating in air was covered with a layer of liquid instead of sand or lycopodium, that liquid ought to be determined from the quiescent to the vibrating parts and be accumulated there. A square plate was therefore covered with water, and vibrated as in the former experiments (2. 6.); but all endeavours to ascertain whether accumulation occurred at the centres of oscillation, either by direct observation, or the reflection from its surface of right-lined figures, or by looking through the parts, as through a lens, at small print and other objects, failed.

43. As however when the plate was strongly vibrated, the well-known and peculiar crispations which form on water at the centres of vibration, occurred and prevented any possible decision as to accumulation, it was only when these were absent and the vibration weak, and the accumulation therefore small, that any satisfactory result could be expected; but as even then no appearance was perceived, it was concluded that the force of gravity combined with the mobility of the fluid was sufficient to restore the uniform condition of the layer of water after the bow was withdrawn, and before the eye had time to observe the convexity expected.

44. To remove in part the effect of gravity, or rather to make it coincide with, instead of oppose the convexity, the under surface of the plate was moistened instead of the upper, and by inclining the plate a little, the water made to hang in drops at *a* or *b* or *c*, Figure 13.9, at pleasure. On applying the bow at ×, and causing the plate to vibrate, the drops instantly disappeared, the water being gathered up and expanded laterally over the parts of the plate from which it had flowed. On stopping the vibration, it again accumulated in hanging drops, which instantly

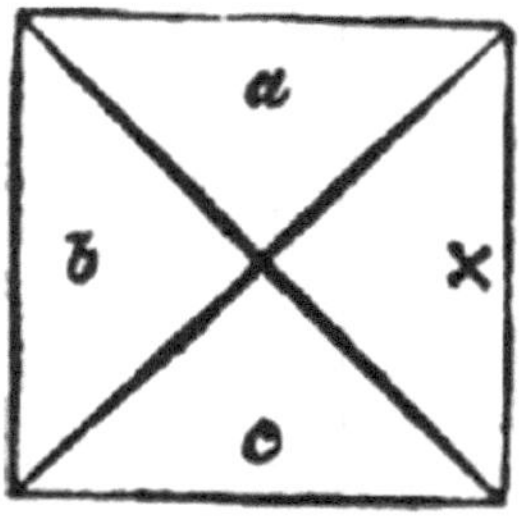

Figure 13.9

disappeared as before on causing the plate to vibrate, the force of gravity being entirely overpowered by the superior forces excited by the vibrating plate. Still, no visible evidence of convexity at the centres of vibration were obtained, and the water appeared rather to be urged from the vibrating parts than to them.

45. The tenacity of oil led to the expectation that better results would be obtained with it than with water. A round plate, held horizontally by the middle (6. 42), was covered with oil over the upper surface, so as to be flooded, except at ×, Figure 13.10, and the bow applied at × as before, to produce strong vibration. No crispation

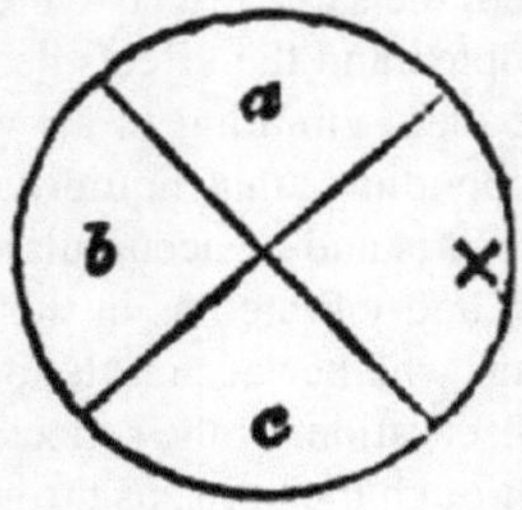

Figure 13.10

occurred in the oil, but it immediately accumulated at *a, b*, and *c*, forming fluid lenses there, rendered evident by their magnifying power when print was looked at through them. The accumulations were also visible on putting a sheet of white paper beneath, in consequence of the colour of the oil being deeper at the accumulations than elsewhere; and they were also rendered beautifully evident by making the experiment in sunshine, or by putting a candle beneath the plate, and placing a screen on the opposite side to receive the images formed at the focal distance.

46. When the vibration of the plate ceased, the oil gradually flowed back until of uniform depth. On renewing the vibration, the accumulations were re-formed, the phenomena of accumulation occurring with as much certainty and beauty as if lycopodium powder had been used.

47. To remove every doubt of the fluid passing from the quiescent to the agitated parts, centres of vibration were used, nearly surrounded by nodal lines. A square plate, Figure 13.11, being held at *c*, and the bow applied at ×, gave with sand, nodal

Figure 13.11

lines, resembling those in the figure. Then clearing off the sand, putting oil in its place, and producing the same mode of vibration as before, the oil accumulated at *a* and *b*, forming two heaps or lenses as in the former experiment (45).

48. The experiment made with water on the under surface (44) was now repeated with oil, the round plate being used (45). The hanging drop of oil rose up as the water did before, but the lateral diffusion was soon limited; for lenses were formed at the centres of vibration just as when the oil was upon the upper surface, and, as far as could be ascertained by general examination, of the same form and power. On stopping the vibration, the oil gathered again into hanging drops; and on renewing it, it was again disposed in the lens-like accumulations.

49. With white of egg the same observable accumulation at the centres of vibration could be produced.

50. Hence it is evident that when a surface vibrating normally, is covered with a layer of liquid, that liquid is determined from the quiescent to the vibrating parts, producing accumulation at the latter places; and that this accumulation is limited, so that if purposely rendered too great by gravity or other means, it will quickly be diminished by the vibrations until the depth of fluid at any one part has a certain and constant relation to the velocity there and to the depth elsewhere.

51. From the accumulated evidence which these experiments afford, I think there can remain no doubt of the cause of the collection of fine powders at the centres or lines of vibration of plates, membranes, &c. under common circumstances; and that no secondary mode of division need be assumed to account for them. I have been the more desirous of accumulating experimental evidence, because I have thought on the one hand that the authority of Savart should not be doubted on slight grounds, and on the other, that if by accident it be placed in the wrong scale, the weight of evidence against it should be such as fully to establish the truth and prevent a repetition of the error by others.

52. It must be evident that the phenomena of collection at the centres or lines of greatest vibration are exhibited in their purest form at those places which are surrounded by nodal lines; and that where the centre or place of vibration is at or near to an edge, the effects must be very much modified by the manner in which the air is there agitated. It is this influence, which, in the square plates (6. 12) and other arrangements, prevents the clouds being at the very edge of the glass. They may be well illustrated by vibrating tin plates under water over a white bottom, and sprinkling dark-coloured sand or filings upon various parts of the plates.

On the peculiar Arrangement and Motions of the heaps formed by particles lying on vibrating surfaces.

53. The peculiar manner in which the fine powder upon a vibrating surface is accumulated into little heaps, either hemispherical or merely rounded, and larger or smaller in size, has already been described (6. 28), as well also as the singular motion which they possess, as long as the plate continues in vibration. These heaps form on any part of the surface which is in a vibratory state, and not merely

under the clouds produced at the centres of vibration, although the particles of the clouds always settle into similar heaps. They have a tendency, as heaps, to proceed to the nodal or quiescent lines, but are often swept away in powder by the currents already described (6). When on a place of rest, they do not acquire the involving motion. When two or more are near together or touch, they will frequently coalesce and form but one heap, which quickly acquires a rounded outline. When in their most perfect and final form, they are always round.

54. The moving heaps formed by lycopodium on large stretched drawing-paper (28), are on so large a scale as to be very proper for critical examination. The phenomena can be exhibited also even by dry sand on such a membrane, the sand being in large quantity and the vibrations slow. When the surface is thickly covered by sand from a sieve, and the paper tapped with the finger, the manner in which the sand draws up into moving heaps is very beautiful.

55. When a single heap is examined, which is conveniently done by holding a vibrating tuning-fork in a horizontal position, and dropping some lycopodium upon it, it will be seen that the particles of the heap rise up at the centre, overflow, fall down upon all sides, and disappear at the bottom, apparently proceeding inwards; and this evolving and involving motion continues until the vibrations have become very weak.

56. That the medium in which the experiment is made has an important influence, is shown by the circumstance of heavy particles, such as filings, exhibiting all these peculiarities when they are placed upon surfaces vibrating in water (39); the heaps being even higher at the centre than a heap of equal diameter formed of light powder in the air. In water, too, they are formed indifferently upon any part of the plate or membrane which is in a vibratory state. They do not tend to the quiescent lines; but that is merely from the great force of the currents formed in water as already described (38), and the power with which they urge obstacles to the place of greatest vibration.

57. If a glass plate be supported and vibrated (6), its surface having been covered with sand enough to hide the plate, and water enough to moisten and flow over the sand, the sand will draw together in heaps, and these will exhibit the peculiar and characteristic motion of the particles in a very striking manner.

58. The aggregation and motion of these heaps, either in air or other fluids, is a very simple consequence of the mechanical impulse communicated to them by the joint action of the vibrating surface and the surrounding medium. Thus in air, when, in the course of a vibration, the part of a plate under a heap rises, it communicates a propelling force upwards to that heap, mingled as it is with air, greater than that communicated to the surrounding atmosphere, because of the superior specific gravity of the former; upon receding from the heap, therefore, in performing the other half of its vibration, it forms a partial vacuum, into which the air, round the heap, enters with more readiness than the heap itself; and as it enters, carries in the powder at the bottom edge of the heap with it. This action is repeated at every vibration, and as they occur in such rapid succession that the eye cannot distinguish them, the centre part of the heap is continually progressing

upwards; and as the powder thus accumulates above, whilst the base is continually lessened by what is swept in underneath, the particles necessarily fall over and roll down on every side.

59. Although this statement is made upon the relation of the heap, as a mass, to the air surrounding it, yet it will be seen at once that the same relation exists between any two parts of the heap at different distances from the centre; for the one nearest the centre will be propelled upward with the greatest force, and the other will be in the most favourable state for occupying the partial vacuum left by the receding plate.

60. This view of the effect will immediately account for all the appearances; the circular form, the fusion together of two or more heaps, their involving motion, and their existence upon any vibrating part of the plate. The manner in which the neighbouring particles would be absorbed by the heaps is also evident; and as to their first formation, the slightest irregularities in the powder or surface would determine a commencement, which would then instantly favour the increase.

61. It is quite true, that if the powder were coherent, that force alone would tend to produce the same effect, but only in a very feeble degree. This is sufficiently shown by the experiments made in the exhausted receiver (36). When the barometer of the air-pump was at twenty-eight inches, that in the air being about 29.2 inches, the heaps, or rather parcels, formed very beautifully over the whole surface of the membrane; but they were very flat and extensive compared with the heaps in air, and the involving motion was very weak. As the air was admitted, the vibration being continued, the heaps rose in height, contracted in diameter, and moved more rapidly. Again, in the experiments with filings and sand in water, no cohesive action could assist in producing the effect; it must have been entirely due to the manner in which the particles were mechanically urged in a medium of less density than themselves.

62. The conversion of these round heaps into linear concentric involving parcels, in the experiment already described (29. 31), when the membrane was covered by a plate of glass, is a necessary consequence of the arrangements there made, and tends to show how influential the action of the air or other including medium is in all the phenomena considered in this paper. No incompatible principles are assumed in the explication given of the arrangement of the forces producing the two classes of effects in question, and though by variation of the force of vibration and other circumstances, the one effect can be made, within certain limits, to pass into the other, no anomaly or contradiction is thus involved, nor any result produced, which, as it appears to me, cannot be immediately accounted for by reference to the principles laid down.

Royal Institution,
March 21, 1831.

APPENDIX

On the Forms and States assumed by Fluids in contact with vibrating elastic surfaces.

63. When the upper surface of a plate vibrating so as to produce sound (2. 6) is covered with a layer of water, the water usually presents a beautifully crispated appearance in the neighbourhood of the centres of vibration. This appearance has been observed by OERSTED,[1] WHEATSTONE,[2] WEBER,[3] and probably others. It, like the former phenomena which I have endeavoured to explain, has led to false theory, and being either not understood or misunderstood, has proved an obstacle to the progress of acoustical philosophy.

64. On completing the preceding investigation, I was led to believe that the principles assumed would, in conjunction with the cohesion of fluids, account for these phenomena. Experimental investigation fully confirmed this expectation, but the results were obtained at too late a period to be presented to the Royal Society before the close of the Session; and it is only because the philosophy and the subject itself is a part of that received into the Philosophical Transactions in the preceding paper, that I am allowed, by the President and Council, the privilege of attaching the present paper in the form of an Appendix.

65. The general phenomenon now to be considered is easily produced upon a square plate nipped in the middle, either by the fingers or the pincers (2. 6), held horizontally, covered with sufficient water on the upper surface to flow freely from side to side when inclined, and made to vibrate strongly by a bow applied to one edge, ×, Figure 13.12, in the usual way. Crispations appear on the surface

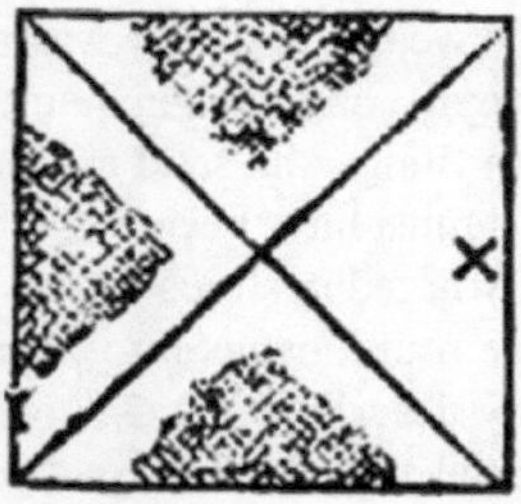

Figure 13.12

of the water, first at the centres of vibration, and extend more or less towards the nodal lines, as the vibrations are stronger or weaker. The crispation presents the appearance of small conoidal elevations of equal lateral extent, usually arranged rectangularly with extreme regularity; permanent[4] (in appearance), so long as a certain degree of vibration is sustained; increasing and diminishing in height, with increased or diminished vibration; but not affected in their lateral extent by such variations, though the whole crispated surface is enlarged or diminished at those times. If the plate be vibrated, so as to produce a different note, the crispations still appear at the centre of vibration, but are smaller for a high note, larger for a low one. The same note produced on different sized plates, by different modes of vibration, appears to produce crispations of the same dimension, other circumstances being the same.

66. These appearances are beautifully seen when ink diluted with its bulk of water is used on the plate.

67. It was necessary, for examination, both to prolong and enlarge the effect, and the following were found advantageous modes of producing it. Plates of crown-glass, from eighteen to twenty-two inches long, and three or four inches wide, were supported each by two triangular pieces of wood acting as bridges (18), and made to vibrate by a small glass rod or tube resting perpendicularly at the middle, over which the moist fingers were passed. By sprinkling dry sand on the plates, and shifting the bridges, the nodal lines were found (usually about one fifth of the whole length from each end), and their places marked by a file or diamond. Then clearing away the sand, putting water or ink upon the plate, and applying the rod or fingers, it was easy to produce the crispations and sustain them undisturbed, and with equal intensity for any length of time.

68. By making a broad mark, or raising a little ledge of bee's wax, or a mixture of bee's wax and turpentine, it was easy to confine the pool of water to the middle part of the plate, Figure 13.13, where, of course, the crispations were most

Figure 13.13

powerfully produced. Such a barrier is often useful to separate the wet and dry parts of the glass, especially when a violin bow is used as the exciter.

69. In other experiments, deal laths, two, three, or four feet long, one inch and a half wide, and three eighths or more of an inch in thickness, were used instead of the glass plates. These could be made to vibrate by the fingers and wet rod (67), and by either shifting the bridges or changing the lath an almost unlimited change of isochronous vibrations, from that producing a high note to those in which not more than five or six occurred in a second, could be obtained. The crispations were formed upon a glass plate attached to the middle of the lath, by two or three little pellets of soft cement.[5]

70. Obtained in this way the appearances were very beautiful, and the facilities very great. A glass plate, from four to eight inches square, could be covered uniformly with crispations of the utmost regularity; for, by attaching the plate with a little method, and at points equidistant from the centre of the bar, it was easy to make every part travel with the same velocity, and in that respect differ from and surpass the bar which sustained it. The conoidal heaps constituting the crispation could be so enlarged by slowness of vibration, that three or four occupied a linear inch. The glass plate could be removed, and another of different form or substance, and with other fluids, as mercury, &c., substituted in an instant.

71. In using laths, it is necessary to confine the parts bearing upon the bridges, either by slight pressure of the fingers, or by loops of string, or by weights. The exciting glass rod need not necessarily rest upon the middle of the bar or plate, but

may be applied with equal effect at some distance from it. Long laths may be made to subdivide in their mode of vibration, according as the rod is applied to different places, and the pressure given by the exciting moist fingers is varied; with each change of this kind an immediate change of the crispation is observed.

72. This form of apparatus was enlarged until a board eighteen feet long was used, the layer of water being now three fourths of an inch in depth and twenty-eight inches by twenty inches in extent. The sides of the cistern were very much inclined, so that the water should gradually diminish in depth, and thus reflected waves be prevented. The vibrations were so slow as to be produced by the direct application of the hand, and the heaps were each from an inch to two inches in extent. Though of this magnitude, they were identical in their nature with those forming crispations on so small a scale as to appear merely like a dullness on the surface of the water.

73. In these experiments the proportion of water requires a general adjustment, the crispations being produced more readily and beautifully when there is a certain quantity than when there is less. For small crispations, the water should flow upon the surface freely. Large crispations require more water than small ones. Too much water sometimes interferes with the beauty of the appearance, but the crispation is not incompatible with much fluid, for the depth may amount to eight, ten, or twelve inches (111), and is probably unlimited.

74. These crispations are equally produced upon the under with the upper surface of vibrating plates. When the lower surface is moistened, and the bow applied (65), the drops which hang down by the force of gravity are rippled; but being immediately gathered up as described in the former paper (44), a certain definite layer is produced, which is beautifully rippled or crispated at the centres of vibration.

75. Most fluids, if not all, may be used to produce these crispations, but some with particular advantages; alcohol, oil of turpentine, white of egg, ink, and milk produce them. White of egg, notwithstanding its viscosity, shows them readily and beautifully. Ink has great advantages, because, from its colour and opacity, the surface form is seen undisturbed by any reflection from the glass beneath; its appearance in sunshine is exceedingly beautiful. When diluted ink is used for large crispations, upon tin plate or over white paper, or mercury, the different degrees of colour or translucency corresponding to different depths of the fluid, give important information relative to the true nature of the phenomena (78. 85. 97). Milk is, for its opacity, of similar advantage, especially when a light is placed beneath, and being more viscid than water is better for large arrangements (72. 98), because it produces less splashing.

76. Oil does not show small crispations readily (120), and was supposed to be incapable of forming them, but when warmed (by which its liquidity is increased) it produces them freely. Cold oil will also produce large crispations, and for very large ones would probably be better than water, because of its cohesion. The difference between oil and white of egg is remarkable; for the latter, from common observation, would appear to be a thicker fluid than oil: but the qualities of

cohesion differ in the two, the apparent thickness of white of egg depending upon an elastic power (probably due to an approach to structure), which tends to restore its particles to their first position, and coexisting with great freedom to move through small spaces, whilst that of oil is due to a real difficulty in removing the particles one by another. It is possible that the power of assuming, more or less readily, the crispated state may be a useful and even important indication of the internal constitution of different fluids.

77. With mercury the crispations are formed with great facility, and of extreme beauty, when a piece of amalgamated tin or copper plate being fixed on a lath (69), is flooded with the fluid metal, and then vibrated. A film quickly covers the metal, and then the appearances are not so regular as at first; but on removing the film by a piece of paper, their regularity and beauty are restored. It is more convenient to cover the mercury with a little very dilute acetic or nitric acid; for then the crispations may be produced and maintained for any length of time with a surface of perfect brilliancy.

78. When a layer of ink was put over the mercury, the acid of the ink removed all film, and the summits of the metallic heaps, by diminishing the thickness of the ink over them, became more or less visible, producing the appearance of pearls of equal size beautifully arranged in a black medium. When mercury covered with a film of dilute acid was vibrated in the sunshine, and the light reflected from its surface received on a screen, it formed a very beautiful and regular image; but the screen required to be placed very near to the metal, because of the short focal lengths of the depressions on the mercurial surface.

79. It is sometimes difficult to arrive by inspection at a satisfactory conclusion of the forms and arrangements thus presented, because of multiplied reflection and the particular condition of the whole, which will be described hereafter (95). When observed, well formed with vibrations so slow as to produce three or four elevations in a linear inch (70), they are seen to be conoidal heaps rounded above, and apparently passing into each other below by a curvature in the opposite direction. When arranged regularly, each is surrounded by eight others, so that, a single light being used, nine images may be sent from each elevation to the eye. These are still further complicated, when transparent fluids are used, by reflections from the glass beneath. The use of ink (75) removes a good deal of the difficulty experienced, and the production of slow, regular, sustained vibrations, more (67. 69).

80. These elevations I will endeavour to distinguish henceforth by the term *heaps*.

81. The crispation on the long plate of glass described (67) always ultimately assumed a rectangular arrangement, i. e. the heaps were equidistant, and in rows parallel or at right angles to each other. The rows usually form angles of 45° to the sides of the plate at the commencement; but if the vibration be continued, the whole system usually wheels round through 45° until the rows coincide with the edges of the plate.

82. The lateral dimension of the heaps remained constant notwithstanding considerable variations in the force of vibration. But it was soon found that variation

in the depth of water affected their number; that with less water the heaps were smaller, and with more water larger, though the sound and therefore the number of vibrations in a given period remained the same. The number of heaps could be reduced to eight or increased to eleven and a half in the three inches by a change in no other condition than the depth of fluid.

83. With the above plate (67. 81) the appearances were usually in the following order, the pool of water being quadrangular or nearly so, and the exciting rod resting in the middle of it. Ring-like linear heaps concentric to the exciting rod first form to the number of six or seven; these may be retained by a moderated state of vibration, and produce intervals which measured across the diameter of the rings are to the number of ten in three inches, with a certain constant depth of water. By increasing the force of vibration the altitude of these elevations increases, but not their lateral dimension, and then linear heaps form across these circles and the plate, and parallel to the bridges, having an evident relation to the manner in which the whole plate vibrates. These, which like all other of these phenomena are strongest at the part most strongly vibrating, soon break up the circles, and are themselves broken up, producing independent heaps, which at first are irregular and changeable, but soon become uniform and produce the quadrangular order; first at angles of 45° to the edges of the plate, but gradually moving round until parallel to them. So the arrangement continues, unless the force be so violent as to break it up altogether: if the vibratory force be gradually diminished, then the heaps as gradually fall, but without returning through the order in which they were produced. The following lines may serve to indicate the course of the phenomena.

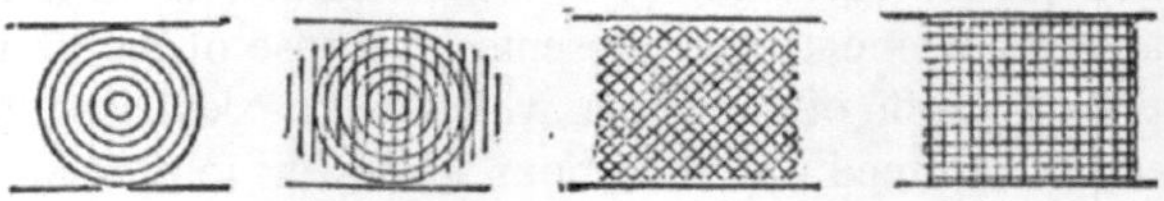

Figure 13.14

When perfectly formed, the heaps are also to the number of ten in three inches with the same depth of water as that which produced the rings. The intervals between the rings and the heaps are the same, other influential circumstances remaining unaltered.

84. Then another form of heaps occasionally occurred, but always passing ultimately into those described. These heaps were grouped in an arrangement still very nearly rectangular, and at angles of 45° to the sides of the plate, but were contracted in one direction, and elongated in the other; these directions being parallel to the sides and ends of the plate. If the marks in Figure 13.15 be supposed to represent the tops of the heaps, an idea of the whole will be obtained. Three inches along these heaps included eight, but across them it included fifteen nearly. These numbers are therefore the relation of length to breadth. But along the lines of the quadrilateral arrangement three inches included eleven heaps, which, notwithstanding the difference in form, is the same number that was produced by the

Figure 13.15

same plate, with the same depths of water, when the heaps were round; therefore an equal number of heaps existed in the same area in both cases; and the departure from perfect rectangular arrangement, and also the ratio of 1:2, is probably due to some slight influence of the sides of the plate.

85. When mercury covered with a film of very dilute nitric acid is vibrated (77), the rectangular arrangement is constantly obtained. When vibrated under dilute ink (78), it is still more beautifully seen and distinguished. The tin plate sustaining the mercury was square, and when the whole surface was covered with crispations, the lines of the rectangular arrangement were always at angles of 45° to its edges.

86. When sand is sprinkled uniformly over a plate on which large water crispations are produced, i. e. four, five or six in the inch, it gives some very important indications. It immediately becomes arranged under the water, and with a little method may be made to yield very regular forms. It is always removed from under the heaps, passing to the parts between them, and frequently producing therefore the accompanying form, Figure 13.16, of great regularity. As the sand figure

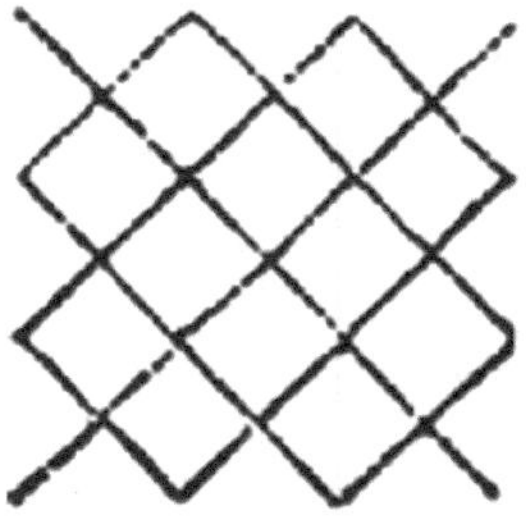

Figure 13.16

remains when the vibration has ceased, it allows of the determination of position, the measurement of intervals, &c. very conveniently.

87. Very often the lines of sand are not continuous, but separated with extreme regularity into portions as represented Figure 13.17. The portions of these lines were sometimes, with little sand on the plate, very small, Figure 13.18; and when more sand was present they were thickened occasionally, Figure 13.19; then

Figure 13.17

Figure 13.18

Figure 13.19

assuming the appearance of heaps arranged in straight lines at angles of 45° to the lines regulating the position of the water-heaps which formed them, and just double in number to the latter. At other times the sand instead of being deficient at the intersecting angle would accumulate there only, Figure 13.20; and at other times would accumulate there principally, but still show the original form by a few connecting particles, Figure 13.21.

88. When the heaps were of the form described (84), the sand was still washed from under them; it did not however assume lines parallel to the rectangular arrangement of the heaps, but was arranged as in Figure 13.22.

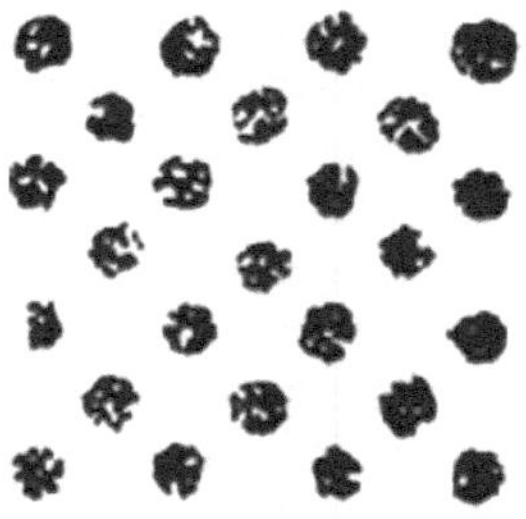

Figure 13.20

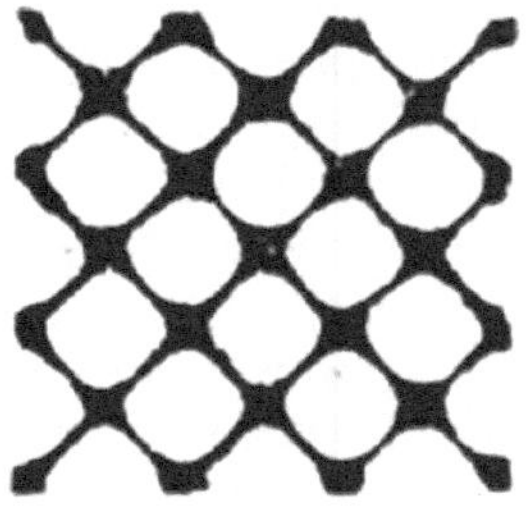

Figure 13.21

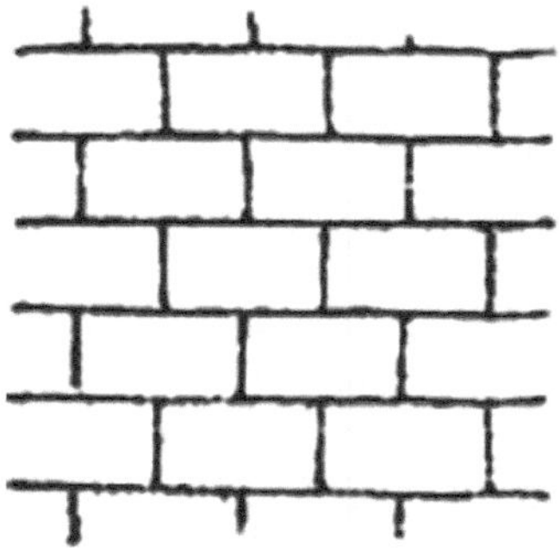

Figure 13.22

89. When only the circular linear heaps (83) were produced, the sand assumed similar circular forms, concentric and alternating with the water elevations.

90. On strewing a little lycopodium over the water for the purpose of gaining information relative to what occurred at the surface during the crispation, it moved about over the fluid in every possible direction, whilst the crispations existed of the utmost steadiness beneath. The same thing occurred with pieces of cork on very large crispations (98). But when much lycopodium was put on, so that the particles retained each other in a steady position, then it formed lines parallel to

the arrangement of the heaps, the powder being displaced from the parts over the heaps, and taking up an arrangement perpendicularly over the sand beneath. As the lycopodium forms float on the water they are easily disturbed, and in no respect approach as to beauty and utility to the forms produced by the sand; but lycopodium may be used with smaller crispations than sand.

91. The crispations are much influenced by various circumstances. They tend to commence at the place of greatest vibration; but if the quantity of fluid is too little there, and more abundant elsewhere, they will often commence at the latter place first. Their final arrangement is also much affected by the form of the plate, or of the pool of water on which they occur. When the plates or pools are rectangular, and all parts vibrate with equal velocity, the lines of heaps are at angles of 45° to the edges. But when semicircular and other plates were used, the arrangement, though quadrangular, was unsteady, often breaking up and starting by pieces into different and changing positions.

92. When mercury was used (77), the film formed on it after a few moments had great power, according to the manner in which it was puckered, of modifying the general arrangement of new crispations.

93. When a circular plate, supported by cork feet attached where a single nodal line would occur, was covered with water and vibrated by a rod resting upon the middle, the crispations extended from the middle towards the nodal line; these were sometimes arranged rectangularly, but had no steadiness of position, and changed continually. At other times the heaps appeared as if hexagonal, and were arranged hexagonally, but these also shifted continually. This and many other experiments (83) showed that the direction and nature of the vibration of the plate (i. e. of the lines of equal or varying vibrating force), had a powerful influence over the regularity and final arrangement of the crispations.

94. The beautiful appearance exhibited when the crispations are produced in sunshine, or examined by a strong concentrated artificial light, has been already referred to (78. 79). When the reflected image from any one heap is examined, (for which purpose ink (75) or mercury (77) is very convenient,) it will be found not to be stationary, as would happen if the heap was permanent and at rest, nor yet to form a vertical line, as would occur if the heap were permanent but travelled to and fro with the vibrating plate; but it moves so as to re-enter upon its course, forming an endless figure, like those produced by Dr. YOUNG's piano-forte wires, or WHEATSTONE's kaleidophone, varying with the position of the light and the observer, but constant for any particular position and velocity of vibration. Upon placing the light and the eye in positions nearly perpendicular to the general surface of the fluid, so as to avoid the direct influence of the motion of vibration, still the luminous, linear, endless figure was produced, extending more or less in different directions, according to the relation of the light and eye to the crispated surface, and occasionally corresponding in its extent one way to the width of the heap, i. e. to the distance between the summit of one heap and its neighbours, but never exceeding it. The figure produced by one heap was accurately repeated by all the heaps when the vibrating force of the plate was equal (70) and the arrangement regular.

95. The view which I had been led to anticipate of the nature of the heaps, from the effects described in the former paper, were, that each heap was a permanent elevation, like the cones of lycopodium powder (53. 58), the fluid rising at the centre, but descending down the inclined sides, the whole system being influenced, regulated, and connected by the cohesive force of the fluid. But these characters of the reflected image, with others of the effects already described, led to the conclusion, that notwithstanding the apparent permanency of the crispated surface, especially when produced on a small scale, as by the usual method, the heaps were not constant, but were raised and destroyed with each vibration of the plate; and also that the heaps did not all exist at once, but (referring to locality) formed two sets of equal number and arrangement, Figure 13.23, never existing

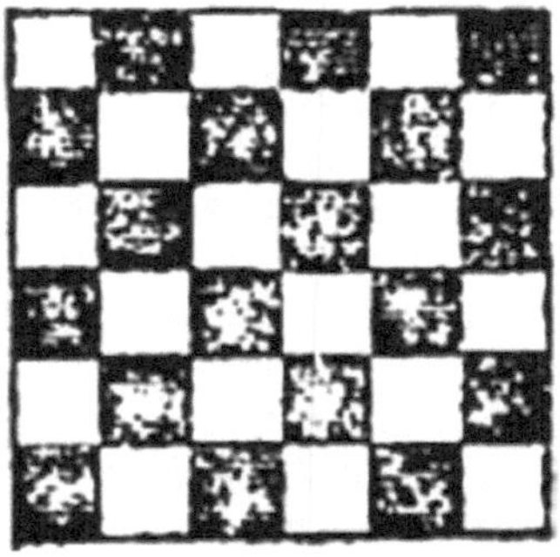

Figure 13.23

together, but alternating with, and being resolved into each other, and by their rapidity of recurrence giving the appearance of simultaneous and even permanent existence. Provided this view were confirmed, it seemed as if it would be easy to explain the production of the heaps, their regular arrangement, &c., and to deduce their recurrence, dimensions, and many other points relative to their condition.

96. On producing a water crispation, having four or five heaps in a linear inch, placing a candle beneath, and a screen of French tracing paper above it, the phenomena were very beautiful, and such as supported the view taken. By placing the screen at different distances, it could be adapted to the focal length due to the curvature at different parts of the surface of fluid, so that by observing the luminous figure produced and its transitions as the screen was moved nearer or further, the general form of the surface could be deduced. Each heap with a certain distance of screen gave a star of light ⊕, Figure 13.24, which twinkled, i. e. appeared and disappeared alternately, as the heap rose and fell. At the corners × equidistant from these, fainter starred lights appeared; and by putting the screen nearer to or further from the surface, lines of light, in two or even four directions, appeared intersecting the luminous centres and apparently permanent, whilst circumstances remained unchanged. These effects could be magnified to almost any scale (72).

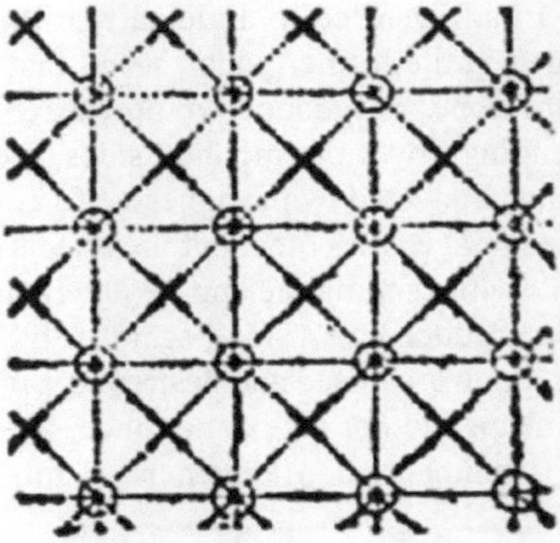

Figure 13.24

97. When heaps of similar magnitude were produced, with diluted ink on glass (75), and white paper or an illuminated screen looked at through them, a chequered appearance was observed. In one position, lines of a certain intensity separated the heaps from each other, but the square places representing the heaps looked generally lighter. In another position, when but little reflected light came from the surface of the heaps, their places could be perceived as dark, from the greater depth of ink there. By care, another position could be found in which the whole surface looked like an alternate arrangement of light and dark chequers, Figure 13.25, not steady, but with a quivering motion, which further attention

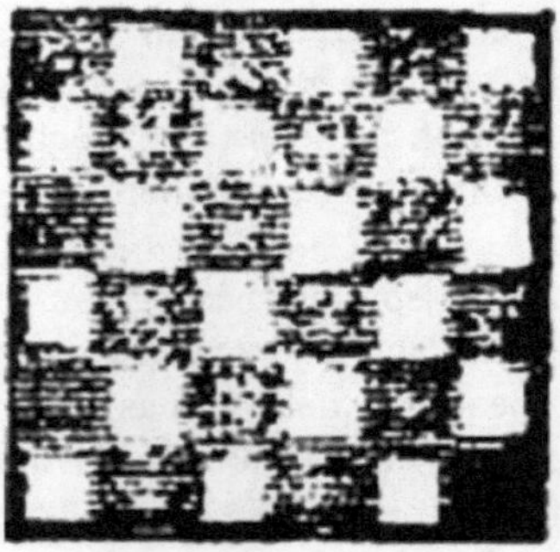

Figure 13.25

could trace as due to a rapid alternation in which the light spaces became dark and the dark light, simultaneously. When, instead of glass, a bright tin plate was used under the diluted ink, the chequered spaces and their alternations could be seen still more beautifully.

98. It was in consequence of these effects that very large arrangements were made (72), giving heaps that were two inches and a half wide each[6]; and now it

was evident, by ordinary inspection, that the heaps were not stationary, but rose and fell; and also that there were two sets regularly and alternately arranged, the one set rising as the other descended.

99. Sand gave no indications of arrangement with these large heaps (86); but when some coarse saw-dust was soaked, so as to sink in water, and then distributed in the fluid, its motions were beautifully illustrative of the whole philosophy of the phenomena. It was immediately washed away from under the rising and falling heaps, and collected in the places equidistant between these spots, as the sand did in the former experiments (86), and by its vibratory motion to and fro, it showed distinctly how the water oscillated from one heap towards another, as the heaps sunk and rose.

100. When milk (75) was used instead of water for these large arrangements in a dark room, and a candle was placed beneath, the appearances also were very beautiful, resembling in character those described (97).

101. Each heap (identified by its locality) recurs or is re-formed in two complete vibrations of the sustaining surface[7]; but as there are two sets of heaps, a set occurs for each vibration. The maximum and minimum of height for the heaps appears to be alternately, almost immediately after the supporting plate has begun to descend in one complete vibration.

102. Many of these results are beautifully confirmed by the appearances produced, when regular crispations have been sustained for a short time with mercury, on which a certain degree of film has been allowed to form (77). On examining the film afterwards in one light, lines could be seen on it, coinciding with the intervals of the heaps in one direction; in another light, lines coinciding with the other direction came into sight, whilst the first disappeared; and in a third light, both sets of lines could be seen cutting out the square places where the heaps had existed: in these spaces the film was minutely wrinkled and bagged, as if it had there been distended; at the lines it was only a little wrinkled, giving the appearance of texture; and at the crossing of the lines themselves, it was quite free from mark, and fully distended. All these are natural consequences, if the film be considered as a flexible but inelastic envelope formed over the whole surface whilst the heaps were rising and falling.

103. The mode of action by which these heaps are formed is now very evident, and is analogous in some points to that by which the currents and the involving heaps already described are produced. The plate in rising tends to lift the overlying fluid, and in falling to recede from it; and the force which it is competent to communicate to the fluid can, in consequence of the physical qualities of the latter, be transferred from particle to particle in any direction. The heaps are at their maximum elevation just after the plate begins to recede from them; before it has completed its motion downwards, the pressure of the atmosphere and that part of the force of the plate which through cohesion is communicated to them, has acted, and by the time the plate has begun to return, it meets them endowed with momentum in the opposite direction, in consequence of which they do not rise as a heap, but expand laterally, all the forces in action combining to raise a similar set of heaps, at exactly intermediate distances, which attain their maximum height just after the plate again begins to recede; these therefore undergo a similar

process of demolition, being resolved into exact duplicates of the first heaps. Thus the two sets oscillate with each vibration of the plate, and the action is sustained so long as the plate moves with a certain degree of force; much of that force being occupied in sustaining this oscillation of the fluid against the resistance offered by the cohesion of the fluid, the air, the friction on the plate, and other causes.

104. A natural reason now appears for the quadrangular and right-angled arrangement which is assumed, when the crispation is most perfect. The hexagon, the square, and the equilateral triangle are the only regular figures that can fill an area perfectly. The square and triangle are the only figures that can allow of one half alternating symmetrically with the other, in conformity with what takes place between the two reciprocating sets of heaps, Figure 13.26; and of these two the boundary lines between squares are of shorter extent than those between equilateral triangles of equal area. It is evident therefore that one of these two will be finally assumed, and that that will be the square arrangement; because then the fluid will offer the least resistance in its undulations to the motions of the plate, or will pass most readily to those positions into which the forces it receives from the plate conspire to impel it.

105. All the phenomena observed and described may, as it appears to me, be now comprehended. The fluid may be considered as a pendulum vibrating to and

Figure 13.26

fro under a given impulse; the various circumstances of specific gravity, cohesion, friction, intensity of vibrating force, &c. determining the extent of oscillation, or, what is the same thing, the number of heaps in a given interval. When the number of vibrations in a given time is increased, these heaps are more numerous, because the oscillation, to be more rapid, must occur in a shorter space. The necessity of a certain depth of fluid (73) is evident, and also the reason why, by varying the depth (82), the lateral extent of the heaps is changed. The arrangement of the sand and lycopodium, by the crispations, and the occurrence of the latter at centres of vibration, and only upon surfaces vibrating normally, are all evident consequences. The permanency of the lateral extension of the heaps, when the velocity of the vibrating plate varies, is a very marked effect, and it is probable that the investigation of these phenomena may hereafter importantly facilitate inquiries into the undulations of fluids, their physical qualities, and the transmission of forces through them.

106. As to the origin or determination of crispations, no difficulty can arise; the smallest possible difference in almost any circumstance, at any one part, would, whilst the plate is vibrating, cause an elevation or depression in the fluid there; the smallest atom of dust falling on the surface, or the smallest elevation in the plate, or the smallest particle in the fluid of different specific gravity to the liquid itself, might produce this first effect; this would, by each vibration of the plate, be increased in amount, and also by each vibration extended the breadth of a heap, in at least four directions: so that in less than a second a large surface would be affected, even under the improbable supposition that only one point should at first be disturbed.

107. I have thought it unnecessary to dwell upon the explanation of the circular linear heaps (83. 93. 110) produced on long or circular plates by feeble vibration. They are explicable upon the same principles, account being at the same time taken of the arrangement and proportion of vibrating force in the various parts of the plates.

108. The heaps which constitute crispation (as the word has been used in this paper) are in form, quality, and motion of their parts, the same with what are called stationary undulations; and if the mercury in a small circular basin be tapped at the middle, stationary undulations, resembling the ring-like heaps (83. 110), will be obtained; or if a rectangular frame be made to beat at equal intervals of time on mercury or water, heaps like those of the crispations, arranged quadrangularly at angles of 45° to the frame, will be produced. These effects are in fact the same with those described, but are produced by a cause differing altogether. The first are the result of two progressing and opposed undulations, the second of four: but the heaps of crispations are produced by the power impressed on the fluid by the vibrating plate; are due to vibrations of that fluid occurring in twice the time of the vibrations of the plate; and have no dependence on progressive undulations, originating laterally, as many of the phenomena described prove. Thus, when the edges were bevelled (72. 110), or covered with cloth, or wet saw-dust, so that waves reaching the side should be destroyed, or when the limits of the water or plates were round (91) or irregular, still the heaps were produced, and their arrangement square. When the round plate (93) was used, regular crispations were still produced, though, as the water extended over the nodal line, and was there

perfectly undisturbed, no progressing and opposed undulations could originate to produce them. Vellum stretched over a ring, and rendered concave by the pressure of the exciting rod, produced the same effect.

109. When a plate of tin, rendered very slightly concave, was attached to a lath (69), so as to have equality of vibratory motion in all its parts, and a little dilute alkali (which would wet the surface) put into it, the crispations formed in the middle, but ceased towards the sides, where, though well wetted, there was not depth enough of water, and from whence also no waves could be reflected to produce stationary undulations in the ordinary manner.

110. When a similar arrangement was made with mercury on a concave tin plate, the effects were still more beautiful and convincing. The centre portion was covered with one regular group of quadrangular crispations; at some distance from the centre, and where the mercury was less in depth, these passed into concentric, ring-like heaps, of which there were a great many; and outside of these there was a part wet with mercury, but with too little fluid to give either lines or heaps. Here there could be no reflected waves; or, if that were thought possible, those waves could not have formed both the circular rings and the square crispation. When this plate was vibrated, the mercury spread in all directions up the side, a natural consequence of the production of powerful oscillations at the middle, which would extend their force laterally, but quite against their being due to the opposition and crossing of waves originating at the sides.

111. A limited depth of fluid is by no means necessary to produce crispations on the surface (73). A circular glass basin about five inches in diameter and four inches deep was attached to a lath (69), filled with water and vibrated, the exciting rod being applied at the side (71). The surface of the water was immediately covered with the most regular crispations, i. e. heaps arranged quadrangularly. On taking out part of the water and filling it up with oil, the oil assumed the same superficies. On putting an inch in depth of mercury under the water, the mercury became crispated. The experiment was finally made with water fourteen inches in depth. Particles at a very moderate depth in the water seemed to have no motion except the general motion of the fluid, and the whole of the lower part of the water may be considered as performing the part of a solid mass upon which the superficial undulating portion reposed. In fact it matters not to the fluid, what is beneath, provided it has sufficient cohesion, is uniform in relation to the surface fluid, and can transmit the vibrations to it in an undisturbed manner.[8]

112. The beautiful action thus produced at the limits of two immiscible fluids, differing in density or some other circumstances, by which the denser was enabled most readily to accommodate itself to rapid, regular and alternating displacements of its support when that support was horizontal, suggested an inquiry into the probable arrangement of the fluid when the displacements were lateral or even superficial.

113. On arranging the long plate (67. 81) vertically, so that the lower extremity dipped about one third of an inch into water, Figure 13.27, and causing it to vibrate by applying the rod at ×, or by tapping the plate with the finger, undulations of a peculiar character were observed: those passing from the plate towards the sides of the basin were scarcely visible though the plate vibrated strongly, but in place of such appeared others, in the production of which the mechanical force

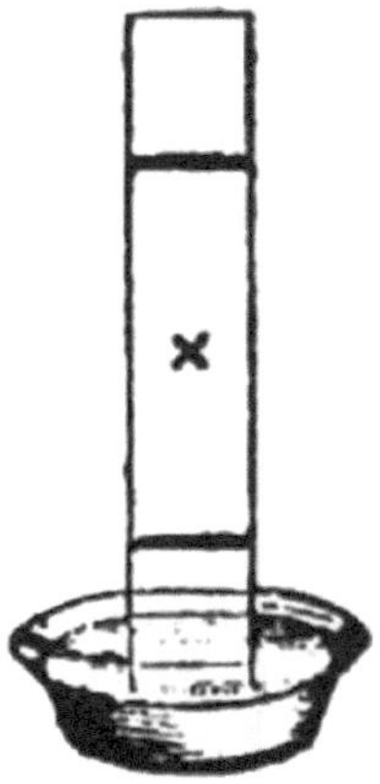

Figure 13.27

of the vibrating plate exerted upon the fluid was principally employed. These were apparently permanent elevations, at regular intervals, strongest at the plate, projecting directly out from it over the surface of the water, like the teeth of a coarse comb gradually diminishing in height, and extending half or three-quarters of an inch in length. These varied in commencing at the glass, or having intervening ridges, or in height, or in length, or in number, or in breaking up into violently agitated pimples and drops, &c. according as the plate dipped more or less into the water, or vibrated more or less violently, or subdivided whilst vibrating into parts, or changed in other circumstances. But when the plate (sixteen or seventeen inches long) dipped about one sixth of an inch, then four of these linear heaps occupied as nearly as possible the same space was four heaps formed with the same plate in the former way (83) and accompanied with the same sound.

114. By fixing a wooden lath (69) perpendicularly downwards in a vice, plates of any size or form could be attached to its lower end and immersed more or less in water; and by varying the immersion of the plate, or the length of the lath, or the place against which the exciting rod (71) was applied, the vibrations could be varied in rapidity to any extent.

115. On using a piece of board at the extremity of the lath, eight inches long and three inches deep, with pieces of tin plate four inches by five, fixed on at the ends in a perpendicular position to prevent lateral disturbance at those parts, very regular and beautiful ridges were obtained of any desired width, Figure 13.28.

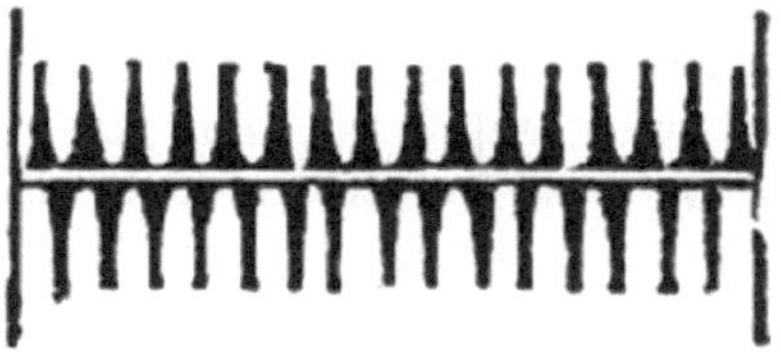

Figure 13.28

These ridges, as before, formed only on the wood, and were parallel to the direction of its vibration. They occurred on each side of the vibrating plane with equal regularity, force and magnitude, but seemed to have no connection, for sometimes they corresponded in position, and at other times not; the one set shifting a little, without the others being displaced.

116. It could now be observed that the ridges on either side the vibrating plane consisted of two alternating sets; the one set rising as the other fell. For each fro and to motion of the plane, or one complete vibration, one of the sets appeared, so that in two complete vibrations the cycle of changes was complete. Pieces of cork and lycopodium powder showed that there was no important current setting in the direction of the ridges; towards the heads of the ridges pieces of cork oscillated from one ridge towards its neighbour, and back again. The lycopodium sometimes seemed to move on the ridges from the wood, and between them to it; but the motion was irregular, and there was no general current outwards or inwards. There was not so much disturbance as amongst the heaps (90).

117. A very simple arrangement exhibits these ripples beautifully. If an oval or circular pan, fifteen or eighteen inches in diameter, be filled with water, and a piece of lath (69) twelve or fifteen inches long be held in it, edge upwards, so as to bear against the sides of the pan as supporting points, and cut the surface of the water, then on being vibrated horizontally by the glass rod and wet finger, the phenomenon immediately appears with ripples an inch or more in length. When the upper edge of the lath was an inch below the surface, the ripples could be produced. When the vessel had a glass bottom, the luminous figures produced by a light beneath and a screen above, were very beautiful (96). Glass, metal and other plates could thus be easily experimented with.

118. These ripple-like stationary undulations are perfectly analogous as to cause, arrangement and action with the heaps and crispations already explained, i. e. they are the results of that vibrating motion in directions perpendicular to the force applied (105), by which the water can most readily accommodate itself to rapid, regular, and alternating changes in bulk in the immediate neighbourhood of the oscillating parts.

119. From this view of the effect it was evident that similar phenomena would be produced if a substance were made to vibrate in contact with and normally to the surface of a fluid, or indeed in any other direction. A lath was therefore fixed horizontally in a vice by one end, so that the other could vibrate vertically; a cork was cemented to the under surface of the free end, and a basin of water placed beneath with its surface just touching the cork; on vibrating the lath by means of the glass rod and fingers (67), a beautiful and regular star of ridges two, three, or even four inches in length, was formed round the cork, Figure 13.29. These ridges were more or less numerous according to the number of vibrations, &c. As the water was raised, and more of the cylinder immersed, the ridges diminished in strength, and at last disappeared: when the cylinder of cork just touched the surface, they were most powerfully developed. This is a necessary consequence of the dependence of the ridges upon the portion of water which is vertically

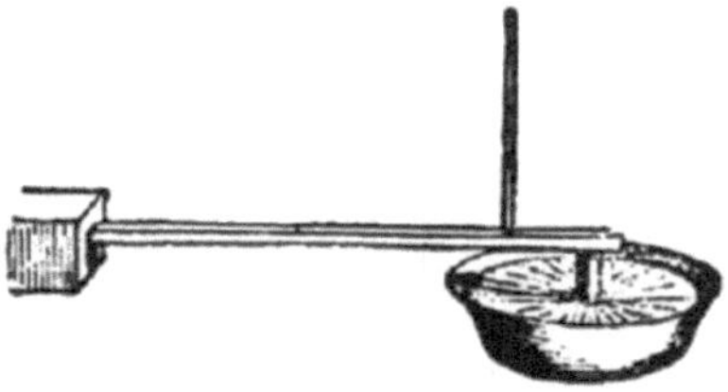

Figure 13.29

displaced and restored at each vibration. When that, being partial in relation to the whole surface, is at or near the surface, the ridges are freely formed in the immediate vicinity; when at a greater depth (being always at the bottom of the cork), the displacement is diffused over a larger mass and surface, each particle moves through less space and with less velocity, and consequently the vibrations must be stronger or the ridges be weaker or disappear altogether. The refraction of a light through this star produces a very beautiful figure on a screen.

120. A heavy tuning-fork vibrating, but not too strongly, if placed with the end of one limb either vertical, inclined, or in any other position, just touching the surface of water, ink, milk, &c. (75), shows the effect very well for a moment. It also shows the ridges on mercury, but the motion and resistance of so dense a body quickly bring the fork to rest. It formed ridges in hot oil, but not in cold oil (76). With cold oil a very inclined fork produced a curious pump-like action, throwing up four streams, easily explained when witnessed, but not so closely connected with the present phenomena as to require more notice here.

121. There is a well known effect of crispation produced when a large glass full of water is made to sound by passing the wet finger round the edges. The glass divides into four vibrating parts opposite to which the crispations are strongest, and there are four nodal points considered in relation to a horizontal section, at equal distances from each other, the finger always touching at one of them. If the vessel is a large glass jar, and soft sounds are produced, the surface of the water exhibits the ridges at the centres of vibration; as the sound is rendered louder, these extend all round the glass, and at last break up at the centres of vibration into irregular crispations, but both the ridges and crispations are effects of the kind already described, and require no further explanation.

122. There are some other effects, one of which I wish here briefly to notice, as connected more or less with the vibratory phenomena that have been described. If, during a strong steady wind, a smooth flat sandy shore, with enough water on it, either from the receding tide or from the shingles above, to cover it thoroughly, but not to form waves, be observed in a place where the wind is not broken by pits or stones, stationary undulations will be seen over the whole of the wet surface, forming ridges like those already described, and each several inches long. These are not waves of the ordinary kind; they are accurately parallel to the course of the wind; they are of uniform width whatever the extent

of surface, varying in width only as the force of the wind and the depth of the stratum of water varies. They may be seen at the windward side of the pools on the sand, but break up so soon as waves appear. If the waves be quelled by putting some oil on the water to windward, these ripples then appear on those parts. They are often seen, but so confused that their nature could not be gathered from such observations, on the pavements, roads, and roofs when sudden gusts of wind occur with rain. The character of these ripples, and their identity with stationary undulations, may be ascertained by exerting the eye and the mind to resolve them into two series of ordinary advancing waves moving directly across the course of the wind in opposite directions. But as such series could not be caused by the wind exerted in a manner similar to that by which ordinary waves are produced, (the direction being entirely opposed to such an idea,) I think the effect is due to the water acquiring an oscillatory condition similar to those described, probably influenced in some way by the elastic nature of the air itself (124) and analogous to the vibration of the strings of the Æolian harp, or even to the vibration of the columns of air in the organ-pipe and other instruments with embouchures.

These ridges were strong enough to arrange the sand beneath where ordinary waves had not been powerful enough to give form to the surface.

123. All the phenomena as yet described are such as take place at the *surfaces* of those fluids in common language considered as inelastic, and in which the elasticity they possess performs no necessary part; nor is it possible that they could be produced within their mass. But on extending the reasoning, it does not seem at all improbable that analogous effects should take place in gases and vapour, their elasticity supplying that condition necessary for vibration which in liquids is found in an abrupt termination of the mass by an unconfined surface.

124. If this be so, then a plate vibrating in the atmosphere may have the air immediately in contact with it separated into numerous portions, forming two alternating sets like the heaps described (95); the one denser, and the other rarer than the ordinary atmosphere; these sets alternating with each other by their alternate expansion and condensation with each vibration of the plate.

125. With the hope of discovering some effect of this kind, a flat circular tin plate had a raised edge of tin three quarters of an inch high fixed on all round, and the plate was then attached to a lath (69), a little lycopodium put on to it, and vibrated powerfully, so that the powder should form a mere cloud in the air, which, in consequence of the raised edge and the equal velocity (70) of all parts of the plate, had no tendency to collect. Immediately it was seen that in place of a uniform cloud it had a misty honeycomb appearance, the whole being in a quivering condition; and on exerting the attention to perceive waves as it were travelling across the cloud in opposite directions, they could be most distinctly traced. This is exactly the appearance that would be produced by a dusty atmosphere lying upon the surface of a plate and divided into a number of alternate portions rapidly expanding and contracting simultaneously.

126. But the spaces were very many times too small to represent the interval through which the air by its elasticity would vibrate laterally once for two vibrations of the plate, in analogy with the phenomena of liquids; and this forms a strong objection to its being an effect of that kind. But it does not seem impossible that the air may have vibrated in subdivisions like a string or a long column of air; and the air itself also being laden with particles of lycopodium would have its motions rendered more sluggish thereby. I have not had time to extend these experiments, but it is probable that a few, well chosen, would decide at once whether these appearances of the particles in the air are due to real lateral vibrations of the atmosphere, or merely to the direct action of the vibrating plate upon the particles.

127. If the atmosphere vibrates laterally in the manner supposed, the effect is probably not limited to the immediate vicinity of the plate, but extends to some distance. The vertical plates intersecting the surface of water and vibrating in a horizontal plane (117) produced ripples proceeding directly out from them five or six inches long; whilst the waves parallel to the vibrating plate were hardly sensible; and something analogous to this may take place in the atmosphere. If so, it would seem likely that these vibrations occurring conjointly with those producing sound, would have an important influence upon its production and qualities, upon its apparent direction, and many other of its phenomena.

128. Then by analogy these views extend to the undulatory theory of light, and especially to that theory as modified by M. FRESNEL. That philosopher, in his profound investigations of the phenomena of light, especially when polarized, has conceived it necessary to admit that the vibrations of the ether take place transversely to the ray of light, or to the direction of the wave causing its phenomena. "In fact we may conceive direct light to be an assemblage, or rather a rapid succession, of an infinity of systems of waves polarized (i. e. vibrating transversely) in all azimuths, and so that there is as much polarized light in any one plane as in a plane perpendicular to it." HERSCHEL says that FRESNEL supposes the eye to be affected *only* by such vibrating motions of the ethereal molecules as are performed in planes perpendicular to the direction of the rays. Now the effects in question seem to indicate how the direct vibration of the luminous body may communicate transversal vibration in every azimuth to the molecules of the ether, and so account for that condition of it which is required to explain the phenomena.

129. When the star of ridges formed by a vibrating cylinder (119) upon the surface of water is witnessed instead of the series of circular waves that might be expected, it seems like the instant production of the phenomena of radiation by means of vibratory action. Whether the contiguous rarified and condensed portions which I have supposed in air, gases, vapour and the ether, are arranged radially like the ridges in the experiment just quoted, or whether rare and dense alternate in the direction of the radii as well as laterally, is a question which may perhaps deserve investigation by experiment or calculation.

Royal Institution,
July 30*th*, 1831.

NOTES

1 LIEBER's Hist. of Natural Phenomena for 1813.
2 Annals of Philosophy, N. S. vi. p. 82.
3 Wellenlehre, p. 414.
4 WEBER's Wellenlehre, p. 414.
5 Equal parts of yellow wax and turpentine.
6 This estimate is given in accordance with the mode of estimating the former and smaller heaps, as if the heaps were formed simultaneously; but it is evident that if only half the number exist at once, each heap will have twice the width or four times the area of those which can be formed if all exist together.
7 A vibration is here considered as the motion of the plate, from the time that it leaves its extreme position until it returns to it, and not the time of its return to the intermediate position.
8 I have seen the water in a pail placed in a barrow, and that on the head of an upright cask in a brewer's van passing over stones, exhibit these elevations.

Editorial Headnote

14. Charles Wheatstone, "On the Figures Obtained by Strewing Sand on Vibrating Surfaces, Commonly Called Acoustic Figures", *Philosophical Transactions of the Royal Society* 123 (London, England, 1833), pp. 593–633.

Having performed Chladni's acoustic plates as both private inquiry and popular exhibition for well over a decade, Wheatstone finally wrote up an extensive account of these experiments in 1833. Like Faraday's work on vibrating fluids, or "crispations", Wheatstone published his study of acoustic figures in the country's most eminent scientific journal, the *Philosophical Transactions of the Royal Society*. This choice of publication, and its acceptance, emphasized the importance contemporaries attached to Chladni's figures and their implications on understanding sound's behaviour. Wheatstone's paper therefore marked something of a culmination of his efforts to promote Chladni's acoustic figures, which the German natural philosopher had first published back in 1787 in his *Entdeckungen über die Theorie des Klanges* and later developed in his *Die Akustik* (1802), both of which had been little known among British audiences until the 1820s. Wheatstone argued that as the pitch of sound produced by a plate increased, the patterns formed by the sound became ever more complex: a higher frequency seemed to produce a greater number of nodal points throughout a solid body. The purpose of Wheatstone's publication was not just to extend Chaldni's experiments but to account for why these patterns formed. As he explained, a year later, while Chladni had identified and explained the figures produced through sonorous impulse, Chladni had said nothing on "the causes of these phenomena". Wheatstone's task then, was to unlock these "hieroglyphics", and he believed he had indeed "at last succeeded in deciphering them . . . and the results of which are in such strict accordance with experiment that there is not the slightest bend in a line which it does not fully account for".[1] Wheatstone's paper, especially its illustrations, secured a considerably positive response. Scottish natural philosopher David Brewster (1781–1868), for one, was impressed. When reviewing the first edition of Mary Somerville's (1780–1872) *On the Connexion of the Physical Sciences* in 1834, Brewster regretted that she had omitted to include illustrations of Wheatstone's acoustic plates. In her second edition, published in 1835, Somerville took Brewster's advice, replicating the pictures from Wheatstone's *Philosophical Transactions* contribution.[2] It is telling of sound's ability to transcend popular and elite scientific culture that accounts of acoustic figures were just as at home in eminent journals like the *Philosophical Transactions*, as they were in magazines and publications aimed at more diverse audiences. Nevertheless, the mid-1830s marked a shift in Wheatstone's work away from questions of acoustics. Although he continued to manage his family's commercial operations, designing and building a series of financially lucrative instruments, notably his Concertina and Symphonium, Wheatstone's main line of experimental inquiry would turn ever more to electromagnetic phenomena and the development of telegraphic apparatus.[3]

NOTES

1 King's College London Library: Wheatstone 5, "Lecture IV" in "Various: 8 lectures on sound" (ca. 1835).

2 Mary Somerville, *On the Connexion of the Physical Sciences*, 2nd ed. (London: John Murray, 1835), pp. 168–70.

3 Allan W. Atlas, *The Wheatstone English Concertina in Victorian England* (Oxford: Clarendon Press, 1996); Allan W. Atlas, "Ladies in the Wheatstone Ledgers: the gendered concertina in Victorian England, 1835–1870", *Royal Musical Association Research Chronicle* 39 (2006), pp. 9–10.

14

ON THE FIGURES OBTAINED BY STREWING SAND ON VIBRATING SURFACES, COMMONLY CALLED ACOUSTIC FIGURES

Charles Wheatstone

Source: Charles Wheatstone, "On the Figures Obtained by Strewing Sand on Vibrating Surfaces, Commonly Called Acoustic Figures", *Philosophical Transactions of the Royal Society* 123 (London, England, 1833), pp. 593–633.

Received February 12, – Read March 14, 1833.

§ 1.

HALF a century has nearly elapsed since the attention of philosophers was first called to the curious phenomena exhibited when sand is strewed on vibrating surfaces. Long before this time, GALILEO had noticed that small pieces of bristle laid on the sounding-board of a musical instrument, were violently agitated on some parts of the surface, whilst on other parts they did not appear to move; and our own countryman Dr. HOOKE, whose sagacity in anticipating many of the discoveries of later times has been so frequently remarked, had proposed to observe the vibrations of a bell by strewing flour upon it. But to CHLADNI is due the sole merit of having discovered the symmetrical figures exhibited on plates of regular forms when caused to sound. His first investigations on this subject, Entdeckungen über die Theorie des Klanges, were published in 1787; this work was followed by his Akustik in 1802, and his Neue Beyträge zur Akustik, 1817. A French translation, by himself, of his second work was published at Paris in 1809.

All the figures obtained by CHLADNI on square surfaces are delineated in pages 611, 613, 615; they are copied from the Neue Beyträge, which work contains his most mature experiments; but not having been translated either into French or English, it is but little known in this country. The following are the general results deduced by CHLADNI from his observations respecting these figures: his works may be referred to for the details omitted, and for those concerning the vibrations of plates in general.

In all the modes of vibration of a square or rectangular plate, the figures, even if they consist of diagonal or tortuous lines, may all be referred to a certain number of nodal lines in the two directions parallel to the sides.

To establish a convenient notation for these figures, he represents the lines in the two directions by numbers separated by a vertical line. Thus, for example, 3|0 signifies the mode of vibration, in which there are three lines in one direction and none in the other; 5|2 denotes that in which there are five lines parallel to one side, and two to the other, &c.

The nodal lines, which may always be considered as having been originally straight, may curve themselves more or less; and in general the flexions of these lines, whether they adjoin each other or are separated by a straight line, mutually approach to or recede from each other. In some modes of vibration the nodal lines are never straight.

In some instances, the same mode of vibration may manifest itself in two essentially different ways, according as the flexions of the lines, or the greater number of them, are inward or outward; in the first case, the sound is usually graver than in the second. This difference is remarked in those figures where there is an entire number of flexions, as in 2|0, 3|1, 4|0, 5|3, 6|2, &c.; but never in those figures where there are 1½, 2½, &c., as in 3|0, 4|1, 5|0, 5|2, &c. To distinguish the first from the second figure, Chladni places a horizontal line above the numbers in the first case, and below them in the second case, thus, $\overline{4|2}$, $\underline{4|2}$.

When two or more figures having the same notation, occur in the Table, the others are to be considered as distortions of the first, occasioned by altering the fixed points, and the place at which the bow is applied.

If four plates of the same size, and upon which the same figure has been produced, be placed together so as to form a larger square, this compound figure may also be more or less accurately produced on a single larger plate. Several instances of this may be seen by reference to the Table of figures.

The following Table contains the relative sounds (expressed both by their musical names and the number of their vibrations,) of all the modes of vibration of a square plate, experimentally ascertained by Chladni. The horizontal series of numbers denotes the lines parallel to one of the sides, and the vertical series those parallel to the other.

The figures placed as exponents indicate the octaves in which the sounds occur; and the characters + and – denote that the sounds to the characters of which they are affixed are respectively sharper and flatter than the true intervals.

Having thus briefly stated the general results deduced by Chladni from his experimental researches, I shall proceed to class and analyse the phenomena; and I shall

TABLE.

1		G, 6					
2	D–, 9– – E..F, 10+	B, 15	A^1..B^1 27+, 28–				
3	G♯, 25	B′, 30 – C^2+, 32? 33?	F♯^{2}, 45	C^3+ 64, 65			
4	G♯^{2}–, 49– – G♯^{2}, 50	B^2, –, 55, 56–	C♯^{3}, 70 – D^3, 72	F♯^{3} 90, 91	B^3–, 110+, 112		
5	E^3+, 81	F^3, 84 – F♯^{3}, 90 91	G♯^{3}, 98, 99, 100–	B^3, 119 120 – C^4, 125 126 128–	D♯^{4}, 150, 153	F♯^{4}, 180	
6	B^3, 120 121 – C^4–, 125 126	C^4, 128	C♯^{4}, 135 140 – D^4, 144	E^4, 160 162	G^4, 189 192 – G♯^{4}, 196 198 200–	B♯^{4}, 224+ 231–	C^5, 256 264

(continued)

(*Continued*)

7	F^4, 169	F^4 $F\sharp^4$ 1 7 5 ? 180?	$F\sharp^4$+ 180+? 189–?	$G\sharp^1$+, 209 210 – A^4, 216 220	B^4, 240 240	$C\sharp^5$, 275 280 – D^5, 286 288	E^5, 320 324 325	$F\sharp^5$, 360 364		
8	$B\sharp^4$, 224 225 – $B\sharp^4$+ 225? 231?		B^4, 240 242 – C^5– 245 250	C^5+, 256+ 264–	D^5, 286 288 – $D\sharp^5$, 294 299	F^5, 336 338	G^5, 377 + 384 – $G\sharp^5$, 390 392	A^5, 432 435	B^5, 480	
9	D^5, 289		E^5– 315	F^5 330 336 – F^5+, 343 345	$F\sharp^5$, 360	$G\sharp^5$– 390 392	$B\sharp^5$, 450	C^5–, 495 – C^5 510 512	$C\sharp^6$+ 561	$D\sharp^6$, 612
	0	1	2	3	4	5	6	7	8	9

endeavour to show, that all the figures of vibrating surfaces are the resultants of very simple modes of vibration, oscillating isochronously, and superposed upon each other; the resultant figure varying with the component modes of vibration, the number of the superpositions, and the angles at which they are superposed. In this first part of the investigation I shall confine myself to the figures of square and other rectangular plates.

§ 2.

The most simple modes of vibration of a rectangular surface are those which exhibit quiescent lines parallel to one of its edges. EULER has theoretically established, that a rod or band, having both its ends unfixed, can vibrate with 2, 3, 4, 5, 6, &c. quiescent lines parallel to the ends, and that the corresponding numbers of vibration are very nearly as the squares of the arithmetical progression 3, 5, 9, 11, &c. These conclusions are fully confirmed by experiment. He has proved, moreover, that when the same mode of vibration of different plates is compared, the number of vibrations is inversely as the square of the length of the plate, but that increase of breadth occasions no difference in the sound; and that the distance from a free end to a quiescent line is rather less than half the distance between two quiescent lines.

Fig. 1. *a.* page 617, shows the number and situations of the quiescent lines in the first four modes of vibration of this series. Fig. 1. *b.* and *c.* are profiles of the preceding, and represent the curvature of each parallel fibre perpendicular to the quiescent lines at the two opposite limits of their vibration. The quantity of motion at each point is indicated by the corresponding ordinate of the curve, and its direction by its situation above or below the horizontal line. It will be convenient to distinguish these states of motion, in which every corresponding point is moving in direct opposition; and I shall therefore call the first, *b.* positive states of vibration, and the second, *c.* negative states of vibration. When there is an even number of quiescent points, the positive state of vibration may be considered as that in which the motion at the *central* part is above the plane of equilibrium, and the negative, that in which it is below it. If we suppose two similar surfaces with the same number of quiescent lines to be superposed, and both to vibrate in concurrence, i. e. both either positively or negatively, they will mutually assist each other's effects; but if they vibrate in opposing directions, they will destroy each other's motions, and the entire surface will be at rest.

§ 3.

When the rectangular surface is equilateral, it is obvious that it may vibrate in two different rectangular directions, so as to give the same sound, and present the same arrangement of quiescent lines. Now this plate may be excited at various points where the motion of each mode of vibration is at its maximum, in the same direction, and of equal intensity: such being the case, there is no reason why one mode of vibration should be produced in preference to the other; and on calculating the effect of such coexistence, it will be found that the resultants of these combined modes of vibration, similar in everything but in their direction with regard to the sides of the plate, give rise to new quiescent lines which accurately correspond with figures described by CHLADNI; while the number of vibrations does not materially differ from that of the component modes of vibration.

The principal results of the superposition of two similar modes of vibration are these: 1st, The points where the quiescent lines of each figure intersect each other, remain quiescent points in the resultant figure; 2ndly, The quiescent lines of one figure are obliterated when superposed by the vibrating parts of the other; 3rdly, New quiescent points, which may be called points of compensation, are formed wherever the vibrations in opposite directions neutralize each other; and, lastly, At all other points the motion is as the sum of the concurring, or the difference of the opposing vibrations.

A primary figure, having an even number of quiescent lines, may be superposed two ways, and may consequently give rise to two distinct resultant figures: one, when the central vibrating parts concur; and the other, when they are in opposition; but if the number of the quiescent lines in the primary figure be uneven, there can be only one resultant figure.

The quiescent lines which thus result may be very easily ascertained. I will take as an example the first mode of vibration, having two parallel quiescent lines; this being superposed in two rectangular directions, and so that the states of vibration are opposing (page 617, fig. 2.), it is obvious that no lines of compensation can exist in the four rectangular segments *a a a a*, as every point included within them is actuated by concurrent motions; but in all the other rectangles they must necessarily be formed, as every point within them is affected by the two opposing motions, and if the two modes be of equal intensity, the compensations must occur at every point equally distant from the two rectangular quiescent lines, each appertaining to a different mode of vibration. The resultant figure will thus be found to consist of two diagonal lines perpendicular to each other, and passing through the centre of the plate.

But if the two superpositions vibrate in concurrence, the rectangles *b b b b* will be free from compensating points; but these will occur in the other rectangles, and form the figure represented (fig. 3.), which also consists of diagonal lines.

In the same manner the resultant of any two similar modes of vibration with nodal lines, parallel to the sides, may be proved to consist of lines parallel to the diagonals.

§ 4.

It is not a necessary condition for the vibrations of a square plate that the primary nodal lines shall be parallel to a side; they may also be parallel to a diagonal, or to any line intermediate between a transverse and a diagonal line. In these cases the superpositions take place according to the following rule: That the axes of the superposed modes of vibration must make equal angles with a transversal line passing through the centre; for otherwise the modes of vibration would not be similar. By the axis of a primary mode of vibration, I mean a straight line passing through the centre of the plate and parallel to the quiescent lines. Considerations of the kind already employed will show that in all these instances the resultant figures consist of lines parallel to the edges of the plate, and that they are always the same in number as the nodal lines of a component mode of vibration, but differently distributed in the two directions, according as the angle of superposition varies.

The various primary modes of vibration, transverse, intermediate, and diagonal, and the angles which the quiescent lines of two similar figures make with each other when they are superposed, are represented in the first column of the general Table, page 619–633; in the second column of this Table are placed the figures resulting from their opposing superpositions, and in the third column those which arise from their concurring superpositions.

We obtain by experiment a limited number only of figures which can be considered the resultants of primary modes of vibration consisting of any given number of oblique lines; but it would seem, that as the various degrees of obliquity are infinite, so there should be an infinite number of resultant figures passing into each other by insensible gradations: by calculation this should be so, but there are causes of limitation which I shall proceed to explain.

It appears that no resultant figure is maintainable unless the greatest excursions of the external vibrating parts occur at the edges of the plate. In the concurring superpositions of eight oblique lines, this condition can only be fulfilled when the angles they make with each other are either 90°, or 143° 8′; in the first case the resultant figure consists of four lines in each transverse direction, in the second of six lines in one direction and two in the other. In the opposing superpositions of the same number of lines, the condition is fulfilled when the angles at which the lines are inclined are 118° 4′ and 163° 44′; the resultant figure of the former consists of five lines in one direction and three in the other, and that of the latter of seven in one direction and one in the other.

§ 5.

I have, in the preceding sections, described the various modes of binary superpositions which may take place on a square surface. But there are numerous cases in which four superpositions may coexist, and these I shall now proceed to take into consideration.

When the axis of a primary figure corresponds either with a diagonal or with a transverse line, passing through the centre of the plate, it is obvious that there can be only one other line of equal length, which can be considered as the axis of a similar and isochronous mode of vibration; in these cases it is evident, therefore, that there can only be two superpositions. But in every intermediate direction of an axis, there are three other lines of equal length which constitute axes of similar modes of vibration; and four superpositions can therefore take place whenever the axis of a component mode of vibration is neither a diagonal nor a transverse line.

It would be a tedious and laborious process to ascertain a resultant figure by combining its four component modes of vibration; but the same purpose will be effectually answered by combining them first in pairs, as explained in the preceding section, and then combining two of these first resultants rectangularly together.

The following process affords great facility for ascertaining the second resultant figure, which arises from the superposition of two first resultants. *a.* and *b.* (fig. 4. page 617,) are the two component first resultants, the similar lines of one being placed rectangularly to those of the other; the vibrating parts are indicated by the letters P and N, according as the vibrations are positive or negative. At C the two figures are superposed, A being represented by the continuous and B by the dotted lines. The surface is now subdivided into a number of unequal rectangles, and by comparing the two component figures together, it is easy to see which of these rectangles are influenced by conspiring, and which by opposing motions; if the motions are found to conspire, the letters P or N must be placed in these rectangles according as the coexisting motions are positive or negative; if the motions are in opposition, a mark may be made to indicate that a quiescent line passes through this rectangle. Wherever a continuous line intersects a dotted line, a mark is to be made, to indicate that a quiescent point is formed; and as in every other part, the quiescent lines of one figure pass over vibrating parts of the other, the boundary lines of all the rectangles must be marked with the letters indicating the motions of the vibrating parts they superpose. The figure C being thus marked, the resultant figure is easily described by joining the fixed points by lines drawn through the rectangles shown to be actuated by opposing motions; carefully avoiding to encroach upon the rectangles of conspiring motion marked P or N.

That the diagonal line is perfectly straight may be proved in the following manner. It must first be premised that the rectangles included within the quiescent lines of each of the first resultant figures, have precisely the same quantity of motion in the same relative points with respect to the surrounding sides, and that the vibrating parts at the edges and corners of the plate must be considered respectively as exact halves and quarters of a complete vibrating part. This being understood, if two similar first resultants be laid alongside each other in the directions in which they are to be superposed, and if a diagonal line be similarly drawn through each of them, it will be obvious that every corresponding point of each line must possess the same intensity of motion. If the successive segments (equal in each figure,) through which the lines simultaneously pass, be in opposite states of vibration, they will neutralize each other's effects, and a diagonal quiescent line will be formed; and if they be concurring, all the parts between the coincident or fixed points will be in motion. In the diagram (fig. 4. *c.*) one diagonal is in the first state and the other in the second.

If the number of lines in the component first resultant figures be uneven, they admit of only one mode of superposition. But first resultants having an even number of quiescent lines, admit of being superposed in two ways, according as they are vibrating in concurrence or in opposition.

It frequently occurs that entire quiescent lines superpose each other; thus, when in the component figure there is an uneven number of lines in each direction, the two rectangular central lines of each superposed figure must coalesce, and consequently they continue in both the resultant figures; examples of this are seen in the resultants of 3|1, 5|1, 5|3, 7|1, 7|3, 7|5, &c. Again, when the number of quiescent lines in one direction of the component figure is three times greater than that in the other, the number of coinciding fixed lines in each direction of the resultant figure is equal to the smallest number in the first resultant, as in 6|2, 9|3, 12|4, 15|5, &c. See the general Table, page 619–633.

The following general results are obtained from constructing the second resultant figures according to the rules above given.

In superposing first resultants consisting of an even number of lines: 1st, When the number of lines in each direction of the first resultants is even, and the modes of vibration are concurring, no line passes through the centre of the second resultant figure. When the modes of vibration are opposing, two rectangular diagonal lines of compensation occur. 2ndly, When the number of lines in each direction is uneven, and the modes of vibration are concurring, there are always two perpendicular transversal lines passing through the centre; and when the modes are opposing, there are, in addition to these fixed lines, the two diagonal lines of compensation.

When the first resultants consist of an uneven number of lines, in which case there is no distinction of concurring and opposing vibrations, one diagonal line only invariably occurs.

In no case is it necessary to calculate an entire figure. When the number of nodal lines in the primary mode of vibration is even, only one quarter of the figure is required to be calculated, as it is obvious that every second resultant of this kind consists of four symmetrical, and as it were reflected, portions. But when the primary number of quiescent lines is uneven, it is necessary to calculate one half; the other half is symmetrical and inverted.

Some of the first resultants are never obtained by experiment. When the number of quiescent lines in the primary mode of vibration is uneven, either the first or the second resultant may be obtained at pleasure; thus in 3|2 if the impulses be made at a corner, where the motion of both superpositions is at its maximum, the second resultant must arise; but if they be made at the middle of a side, the first resultant only will appear, because the point of excitation is a quiescent point of the other. But in all cases where there is an *even* number of lines, it is impossible to obtain the first resultant, because each maximum of vibration equally belongs to both superpositions.

§ 6.

I have given in pages 619–633, a Table which shows every perfect resultant figure of a square surface when the number of quiescent lines in the primary modes of vibration does not exceed twelve. I have carefully calculated each figure by the rules laid down in the preceding sections, and have shown in the Table the successive processes of superposition. The first vertical row exhibits the two primary modes of vibration superposed at the required angles; one figure being represented by the continuous lines, and the other by the dotted lines. The second row contains the first resultants which arise from the opposing superpositions of the preceding; and the third row, those which result from their concurring superpositions. The fourth and fifth rows exhibit the perfect second resultants which are formed, the former by the opposing and the latter by the concurring superpositions of the first resultant which the plate has been already found competent to produce.

On comparing the calculated figures with those obtained experimentally by CHLADNI, the greater number are found exactly to agree; there are, however, some differences which it will be necessary to explain. In the first place, there is an obvious cause of error in delineating figures from experiment, from this circumstance, – that the sand accumulates in the spaces where two convex curves are near and opposite to each other, the motion being there very small, so that it is difficult to ascertain whether the curves join, or not. Secondly, Inequalities in the plate will sometimes occasion lines which ought to intersect each other, so as to appear separated curves. On comparing together the figures CHLADNI has marked as $\underline{6|4}$, $\overline{6|4}$, 7|2, 7|4, 8|3 *c.* &c. with those of the calculated Table, they will be found to differ only in these respects.

Another cause of difference is this: When the lines of one component figure very nearly coincide with those of the other, but without actually doing so, the resultant figure may be such as would arise from their actual superposition, instead of that which accurate calculation would give. In CHLADNI's Table there are two instances of this alteration, 7|2 *a.* and 8|3 *a.*

A few of the figures delineated by CHLADNI are irregular resultants formed by the superpositions of dissimilar modes of vibration. These irregular resultants can be formed only when the dissimilar component modes of vibration give the same sound, and have a maximum point of vibration in common, at which they can be simultaneously excited. The figure marked by CHLADNI 6|1, I find to be an irregular resultant formed by the combination of 6|1 with 3|5, which both give the same sound C^4. The irregular figure marked in his Table 5|1 *a.* is a compound of 5|1 and 2|5.

The calculated figures 6|1, 7|1, 8|1, 9|1, 10|1, 10|2, and 11|1, are not to be found in CHLADNI's Table. The near approach of the inclined lines of their primary component figures to parallelism is the cause of the great difficulty in obtaining these figures by experiment.

The figures marked by CHLADNI 10|3, 9|4, 8|5, 7|6, 10|4, 9|5, 8|6, and 9|7, exceed the limits within which I have calculated the Table, and are therefore not to be found in it.

§ 7. IMPERFECT RESULTANT FIGURES

I have hitherto considered those resultant figures only which arise from the superpositions of similar modes of vibration, each exactly equal in intensity; these I shall in future call *perfect* resultant figures. But when the vibrations of the superposed modes are unequal in intensity, then a figure intermediate between the perfect resultant and one of its components is formed; these intermediate figures I shall call *imperfect* resultants. They are experimentally obtained by varying in a slight degree the places at which the plate is held or touched, from those necessary to determine the corresponding perfect resultant figure; the place at which the bow is applied remaining in both cases the same.

Fig. 6. *a. b. c. d. e.* in CHLADNI's Table, page 611, represents the successive transformations of figure which take place when each component mode of vibration presents three transversal lines: *a.* and *e.* are the two components; *c.* the perfect resultant; *b.* an imperfect resultant, in which the excursions of *a.* are the greatest: and *d.* an imperfect resultant, in which *e.* has the greatest energy.

Fig. 11. *a. b.* page 611, exhibits the transitions of the opposing superposition of two primary figures, each presenting four transversal lines; and fig. 12. *a.* and *b.* the changes of the concurring superposition of the same.

These are the principal types of the transformation of primary figures into first resultants. In each of these series of transitions there are certain points which are invariable during every change: these are, the quiescent points formed by the nodal lines of one figure intersecting those of the other, and the centres of vibration where the maxima of positive or negative vibration agree in each component mode of vibration. The points of compensation are changeable.

Figs. 29. *a.* and 30. *a.* page 613, represent imperfect second resultants, formed by two superpositions of the first resultant figure 6|2. Fig. 29. *a.* arises from concurring, and fig. 30. *a.* from opposing superpositions. The straight lines in these imperfect figures arise from the coincidence of entire quiescent lines in each component figure, and, consequently, they remain unaltered whatever may be the relative intensities of the superposed modes of vibration. But the curved lines, which are formed of compensating points, change with the varying intensities.

§ 8. FIGURES OF IRREGULAR PLATES

If the sides of the square be nearly, but not exactly, equal, the superpositions of two similar modes of vibration with transversal lines still take place; but instead of exhibiting perfect resultants, figures resembling transitional figures appear. Thus in the binary superposition (page 611, fig. 1.) of the figure with two transversal lines, if the sides be unequal, the crossed lines separate at their point of intersection and are converted into two curves, the summits of which recede from each other as the difference in the lengths of the sides becomes greater.

Also, if the diagonals of the square be unequal, the resultant figure (page 611, fig. 2.) arising from two superposed modes of vibration with diagonal axes, will be modified in a similar manner.

Corresponding modifications are occasioned, through accidental differences of elasticity, &c. in the directions of the axes of the superposed modes of vibration, even when the dimensions of the plate are apparently equal.

If a plate of glass be covered on one of its sides with leaf-gold, or if a plate of ground glass be substituted for an ordinary glass plate with smooth surfaces, the figures may be obtained distinctly delineated by lines consisting of a single row of grains of sand. Experiments made in this manner, upon square plates carefully prepared, induced Professor STREHLKE to conclude, after many minute measurements of these lines, that all acoustic figures are formed of hyperbolic curves, and that the quiescent lines never intersect each other. But however correct these experiments may have been, the conclusions drawn from them are unwarranted; were it possible to obtain plates of a perfectly homogeneous substance, and of accurately equal dimensions, there can be no doubt that the lines, however finely defined, would actually intersect each other.

§ 9.

I have already, § 1., given CHLADNI's Table of the comparative sounds, and numbers of vibrations of the figures of square surfaces obtained from experiment. In the following Table these results are arranged so as to correspond with the views taken in this paper. The numbers in the first vertical row indicate the number of parallel quiescent lines in the primary figure. In the horizontal rows the angles at which the lines of the primary modes of vibration intersect each other are shown; and the figures between brackets give the notation of the first resultant produced by their superposition; below these its number of vibrations, and the character representing its musical sound, are given; when there are two of these lower lines separated by a horizontal dash, the one above indicates the sound of the opposing superposition, and that below it the concurring superposition.

Thus it appears, that every figure of a square surface which experiment can give, may be reduced to a primary figure with parallel lines, giving the same sound: if, therefore, the analytical investigation be confined to these, many of the difficulties will disappear. EULER has investigated the subject when lines parallel to a side only are concerned; it remains to extend the inquiry to modes of vibration the nodal lines of which are perpendicular to any line passing through the centre of the surface. An analytical expression for all the sounds of a square plate may probably be obtained, which shall be a function of the number of quiescent lines, and the length of the axis of the mode of vibration.

TABLE.

<table>
<tr><td>2</td><td>90° (1|1)</td><td>180° (2|0)</td><td></td><td></td><td></td><td></td><td></td></tr>
<tr><td></td><td>G, 6</td><td>D–, 9</td><td></td><td></td><td></td><td></td><td></td></tr>
<tr><td></td><td></td><td>– –</td><td></td><td></td><td></td><td></td><td></td></tr>
<tr><td></td><td></td><td>E..F, 10+</td><td></td><td></td><td></td><td></td><td></td></tr>
<tr><td>3</td><td>126° 52′ (2|1)</td><td>180° (3|0)</td><td></td><td></td><td></td><td></td><td></td></tr>
<tr><td></td><td>B, 15</td><td>G♯^{1}, 25</td><td></td><td></td><td></td><td></td><td></td></tr>
<tr><td>4</td><td>90° (2|2)</td><td>143° 8′, (3|1)</td><td>180° (4|0)</td><td></td><td></td><td></td><td></td></tr>
<tr><td></td><td>A^{1}..B^{1}, 27+, 28–</td><td>B^{1}, 30</td><td>G♯^{2}–, 49–</td><td></td><td></td><td></td><td></td></tr>
<tr><td></td><td></td><td>– –</td><td>– –</td><td></td><td></td><td></td><td></td></tr>
<tr><td></td><td></td><td>C^{2}+, 32.?.33.?</td><td>G♯^{2}, 50</td><td></td><td></td><td></td><td></td></tr>
<tr><td>5</td><td>112° 38′, (3|2)</td><td>151° 56′, (4|1)</td><td>180°, (5|0)</td><td></td><td></td><td></td><td></td></tr>
<tr><td></td><td>F♯^{2}, 45</td><td>B^{2}–, 55..56–</td><td>E^{3}+, 81</td><td></td><td></td><td></td><td></td></tr>
<tr><td>6</td><td>90°, (3|3)</td><td>126° 52′, (4|2)</td><td>157° 22′ (5|1)</td><td>180° (6|0)</td><td></td><td></td><td></td></tr>
<tr><td></td><td>C^{3}+, 64.65</td><td>C♯^{3}, 70</td><td>F^{3}, 84</td><td>B^{3}, 120..121</td><td></td><td></td><td></td></tr>
<tr><td></td><td></td><td>–</td><td>– –</td><td>– –</td><td></td><td></td><td></td></tr>
<tr><td></td><td></td><td>D^{3}, 72</td><td>F♯^{3}, 90.91</td><td>C^{4}–, 125..126</td><td></td><td></td><td></td></tr>
<tr><td>7</td><td>106° 16′ (4|3)</td><td>136° 22′ (5|2)</td><td>161° 4′ (6|1)</td><td>180° (7|0)</td><td></td><td></td><td></td></tr>
<tr><td></td><td>F♯^{3}, 90..91</td><td>G♯^{3}, 98.99.100–,</td><td>C^{4}, 128</td><td>F^{4}, 169</td><td></td><td></td><td></td></tr>
<tr><td>8</td><td>90° (4|4)</td><td>118° 4′, (5|3)</td><td>143° 8′ (6|2)</td><td>163° 44′ (7|1)</td><td>180° (8|0)</td><td></td><td></td></tr>
<tr><td></td><td>B♯^{3}–, 110+.112</td><td>B^{3}.119.120</td><td>C♯^{4} 135.140</td><td>F^{4}..F♯^{4} 175.? 180.?</td><td>B♯^{4} 224.225</td><td></td><td></td></tr>
<tr><td></td><td></td><td>– –</td><td>– –</td><td></td><td>– –</td><td></td><td></td></tr>
<tr><td></td><td></td><td>C^{4}, 125.126.128–</td><td>D^{4}.144</td><td></td><td>B^{4}+225+231.?</td><td></td><td></td></tr>
<tr><td>9</td><td>102° 40′ (5|4)</td><td>126° 52′ (6|3)</td><td>148° 6′ (7|2)</td><td>165° 44′ (8|1)</td><td>180° (9|0)</td><td></td><td></td></tr>
<tr><td></td><td>D♯^{4} 150.153</td><td>E^{4} 160.162</td><td>F♯^{4}+.180+.?189–</td><td></td><td>D^{5} 289</td><td></td><td></td></tr>
<tr><td>10</td><td>90° (5|5)</td><td>112° 38′ (6|4)</td><td>133° 26′ (7|3)</td><td>151° 56′ (8|2)</td><td>167° 18′ (9|1)</td><td>180° (10|0)</td><td></td></tr>
<tr><td></td><td>F♯^{4}, 180</td><td>G^{4} 189..192</td><td>G♯^{4}+ 209.210</td><td>B^{4} 240.242</td><td>unascertained.</td><td></td><td></td></tr>
<tr><td></td><td></td><td>– –</td><td>– –</td><td>– –</td><td></td><td></td><td></td></tr>
<tr><td></td><td></td><td>G♯^{4}196.198.200–</td><td>A^{4}, 216.220</td><td>C^{5}– 245.250</td><td></td><td></td><td></td></tr>
<tr><td>11</td><td>100° 24′ (6|5)</td><td>120° 30′ (7|4)</td><td>138° 52′ (8|3)</td><td>154° 56′ (9|2)</td><td>168° 34′ (10|1)</td><td>180° (11|0)</td><td></td></tr>
<tr><td></td><td>B♯^{4}, 224+, 231–</td><td>B^{4} 240.242</td><td>C^{5}+, 256+.264–</td><td>E^{5}– 315</td><td></td><td></td><td></td></tr>
<tr><td>12</td><td>90° (6|6)</td><td>108° 56′ (7|5)</td><td>126° 52′ (8|4)</td><td>143° 8′ (9|3)</td><td>157° 22′ (10|2)</td><td>169° 36′ (11|1)</td><td>180° (12|0)</td></tr>
<tr><td></td><td>C^{5} 256+.264–</td><td>C♯^{5}, 275.280</td><td>D^{5}, 286.288</td><td>F^{5}, 330.336</td><td></td><td></td><td></td></tr>
<tr><td></td><td></td><td>– –</td><td>– –</td><td>– –</td><td></td><td></td><td></td></tr>
<tr><td></td><td></td><td>D^{5}, 286.288</td><td>D♯^{5} 294.299</td><td>F^{5}+ 343, 345</td><td></td><td></td><td></td></tr>
</table>

§ 10.

Immediately after the publication of CHLANDI's experiments on square plates, JAMES BERNOUILLI attempted to demonstrate them analytically; but his investigation was entirely unsuccessful; his conclusions were founded on erroneous data, and the results he obtained were at variance with experiment. His assumptions were these: that the primary figures consisted of 2, 3, 4, 5, 6, 7, &c. lines parallel to the side only, these being the modes of vibration of a lamina as investigated by EULER; that any two similar or dissimilar modes of vibration might superpose each other rectangularly, the nodal lines of the two components appearing together in the resultant mode of vibration; and that the sound of the resultant differed from that of either of the two components, being much higher. The figures given by experiment he considered accidental distortions of these compound figures. This theory gave no account of those figures in which a single line in one direction coexists with any number in the other, as there is no primary figure consisting of one nodal line only; and BERNOUILLI acknowledged his theory to be imperfect in this respect.

The failure of BERNOUILLI led CHLADNI inconsiderately to state, that "the supposition of regarding such a rigid membraniform body as a network formed by curved lines in one direction applied upon curved lines in another direction, is not conformable to nature, and will never give, either results agreeing with experiment, or an appearance of explanation of some of the most simple vibrations." That this assertion is erroneous, the considerations in the present paper have, I conceive, fully proved. The error of BERNOUILLI did not consist in assuming that the observed acoustic figures were formed by superposing simple modes of vibration on each other, for this has been shown to be true; but in his assumptions of the manner in which these superpositions were made, and the effects which resulted from such hypothetical superpositions.

The various mathematicians who have more recently undertaken to investigate the laws of vibrating surfaces, as POISSON, CAUCHY, Mademoiselle GERMAIN, &c., do not appear to have taken into consideration anything resembling the theory of superposition.

Dr. YOUNG seems to have had a correct notion of the origin of the acoustic figures; for in his Lectures, when slightly noticing CHLADNI's experiments, he remarks, "The vibrations of plates differ from those of rods in the same manner as the vibrations of membranes differ from those of chords, the vibrations which cause the plate to bend in different directions being combined with each other, and sometimes occasioning singular modifications."

The brothers WEBER, in their excellent work the Wellenlehre, published in 1825, have advanced a step nearer the truth than any of their predecessors. They have shown, that in a square vessel containing water or mercury, two series of stationary waves, one parallel to each side, may be made to intersect each other; and that the compound wave formed by their interferences assumes the form shown in § 3. to be the resultant of two superpositions of parallel transverse lines. Their observations are confined to those modes of undulation, analogous to the first resultants of primary modes of vibration with lines parallel to a side. Though I had advanced considerably in the present inquiry before I saw this work, yet I should be wanting in justice to these philosophers did I not here state that theirs is the merit of having shown, in the most simple case, the way in which the superpositions of modes of vibration or undulation actually do take place.

§ 11. PLATES OF WOOD

From the rules already laid down, it is obvious that the series of figures presented by a square plate of any homogeneous material ought not to be obtained on a square plate of wood, in which substance the elasticity is not the same in all directions. If a square plate of wood be prepared with its fibres parallel to one of the sides of the square, the axes of greatest and least elasticity will be disposed rectangularly, and parallel to the adjacent sides; in this case the same primary mode of vibration in the two directions will not give the same sound, although the dimensions of the vibrating parts are the same in both; consequently they cannot coexist, and the resultant figures with diagonal lines will be wanting on such a plate. But if the axes of the two component modes of vibration be equally inclined to either of the axes of elasticity, these modes of vibration will be necessarily similar and isochronous, and therefore capable of superposition; on a square plate of wood, consequently, all those first resultants which consist of any number of lines parallel to the sides intersecting each other rectangularly, may be obtained; and the same figure will be accompanied by different sounds, according as the axes of the modes of vibration are inclined to the axes of least or greatest elasticity. It is easy to foresee that none of the second resultants, which consist of four isochronous superpositions, can be obtained on such a plate.

But if the wooden plate, instead of being square, be a rectangle, the sides of which are inversely as the squares of their resistance to flexion, the two modes of vibration parallel to the sides, though differing in length, will be isochronous; and their coexistence will give rise to a resultant figure with lines parallel to the diagonal. Thus on a rectangular plate of straight-fibred deal wood, in which the proportion of the sides were as 28 to 59, I obtained the two crossing diagonal lines, corresponding to the second figure of a homogeneous square plate of glass or metal.

Savart has made a series of numerous and accurate experiments on the changes which take place in the sound, and also in the form and position of the figure of the first mode of vibration on circular plates of wood of similar dimensions, cut in different directions with respect to the three principal axes of elasticity. All the results he has obtained are in perfect accordance with the rules stated in this paper, and might have been predicted by them. He has extended his investigations to circular slices of crystals, cut in various directions with respect to their axes, and has obtained in this way much valuable information. These researches of Savart point out a new direction to our inquiries respecting the structure of bodies; and the utility of his experiments will be greatly enhanced by the knowledge we now possess of the causes on which these phenomena depend. I shall shortly return to this subject.

Chladni's Table

Of the figures of square surfaces, obtained by experiment.

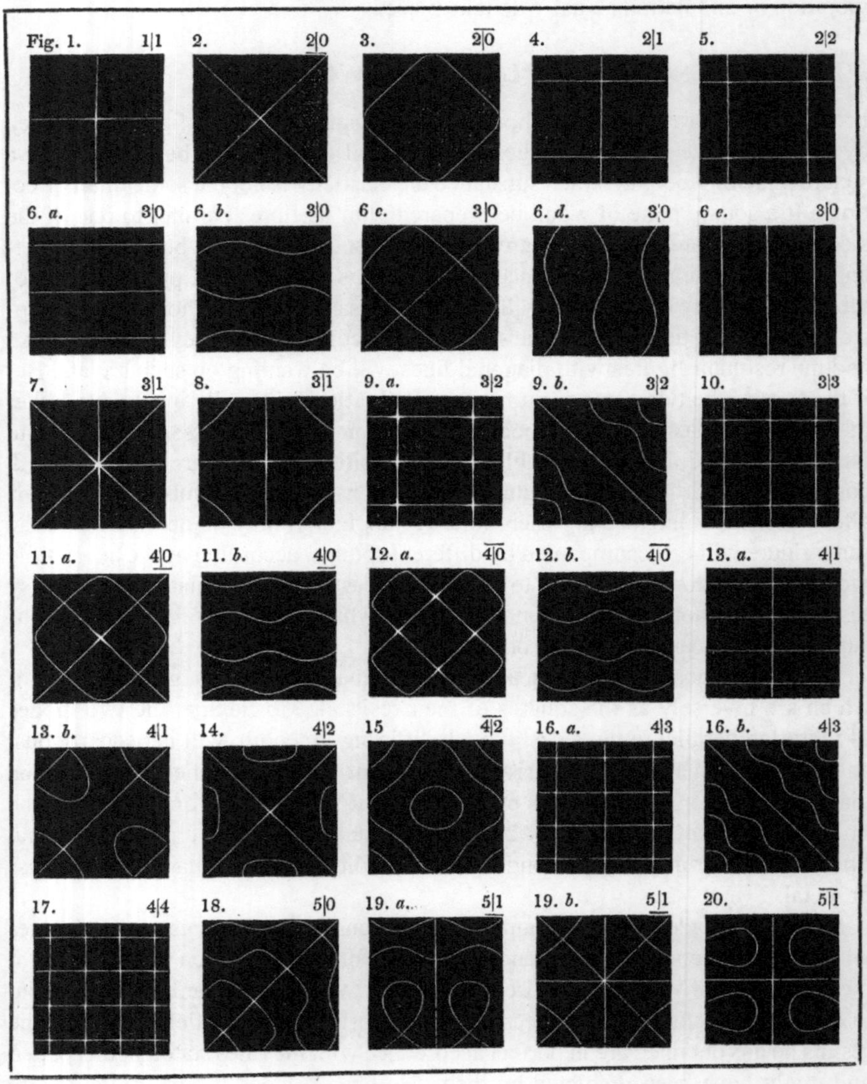

(Continuation of CHLADNI'S TABLE)

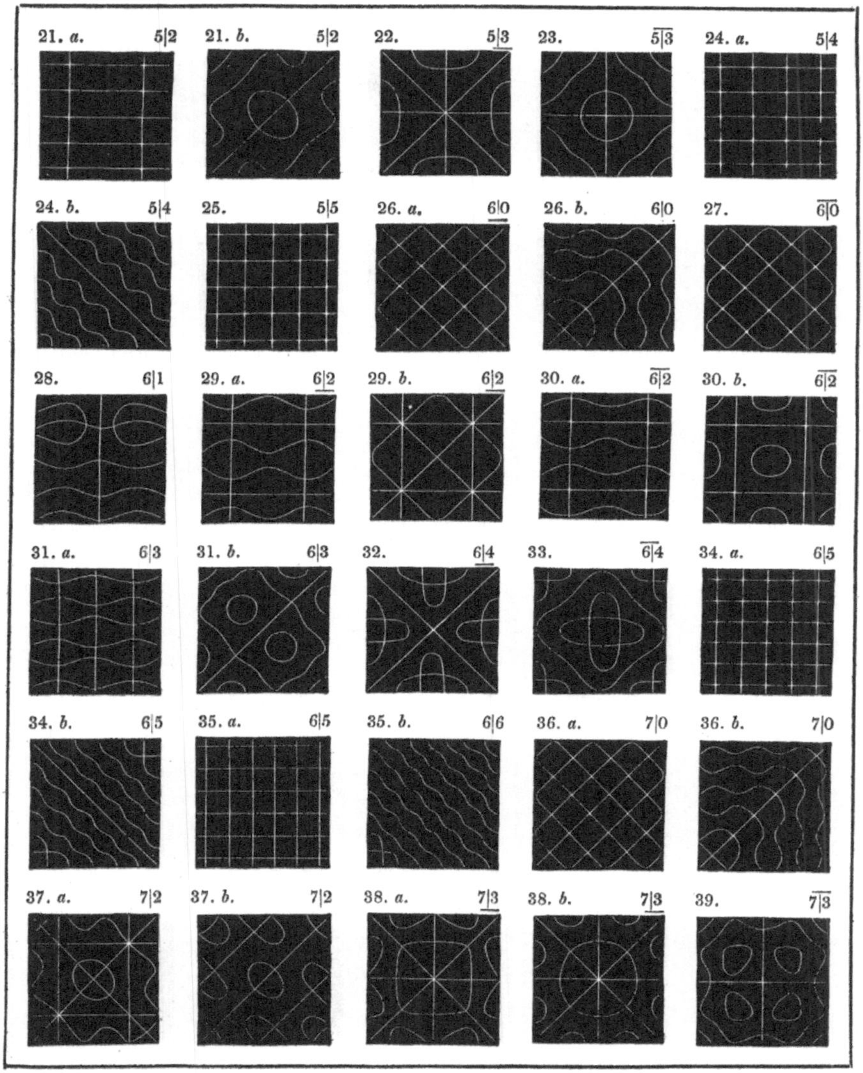

(Continuation of CHLADNI'S TABLE)

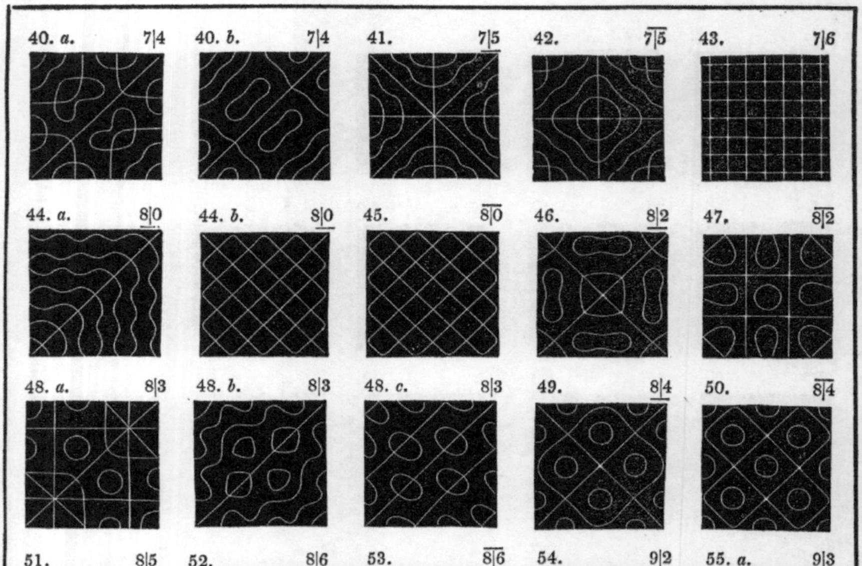

4 K 2

(Continuation of CHLADNI'S TABLE)

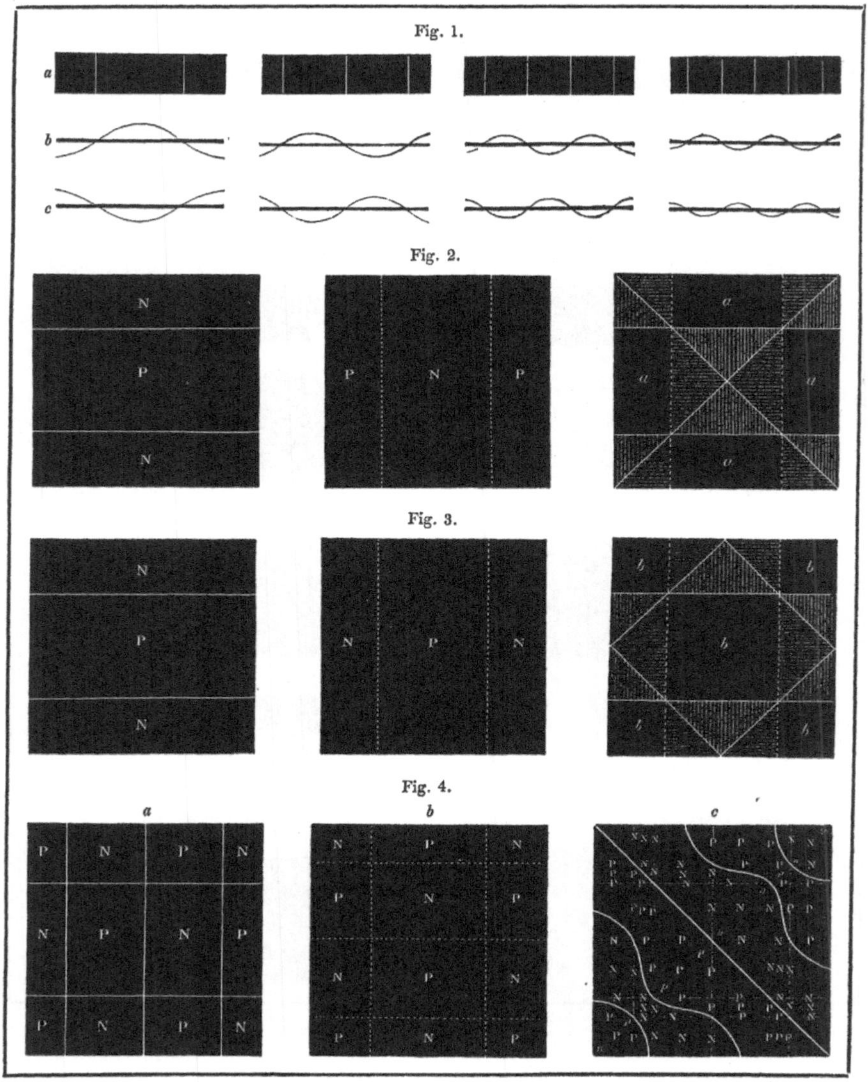

Tabular View

Of all the figures of a square plate, determined by calculation to result from two or four superpositions, when the quiescent lines of the primary figures do not exceed twelve in number.

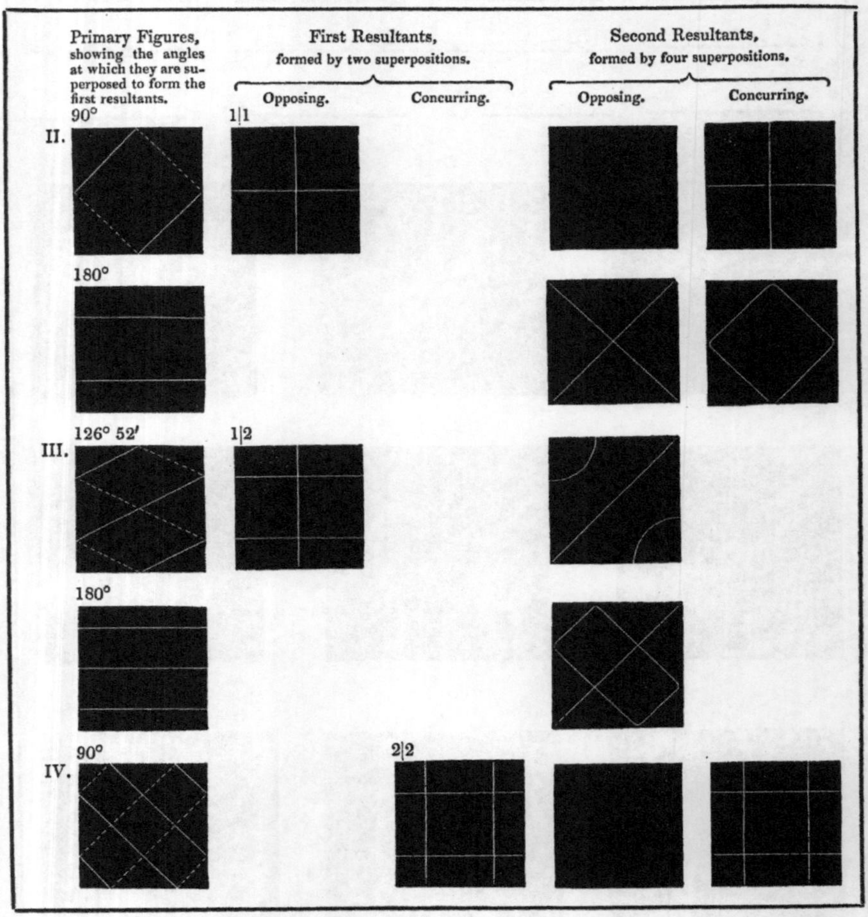

(Continuation of TABULAR VIEW)

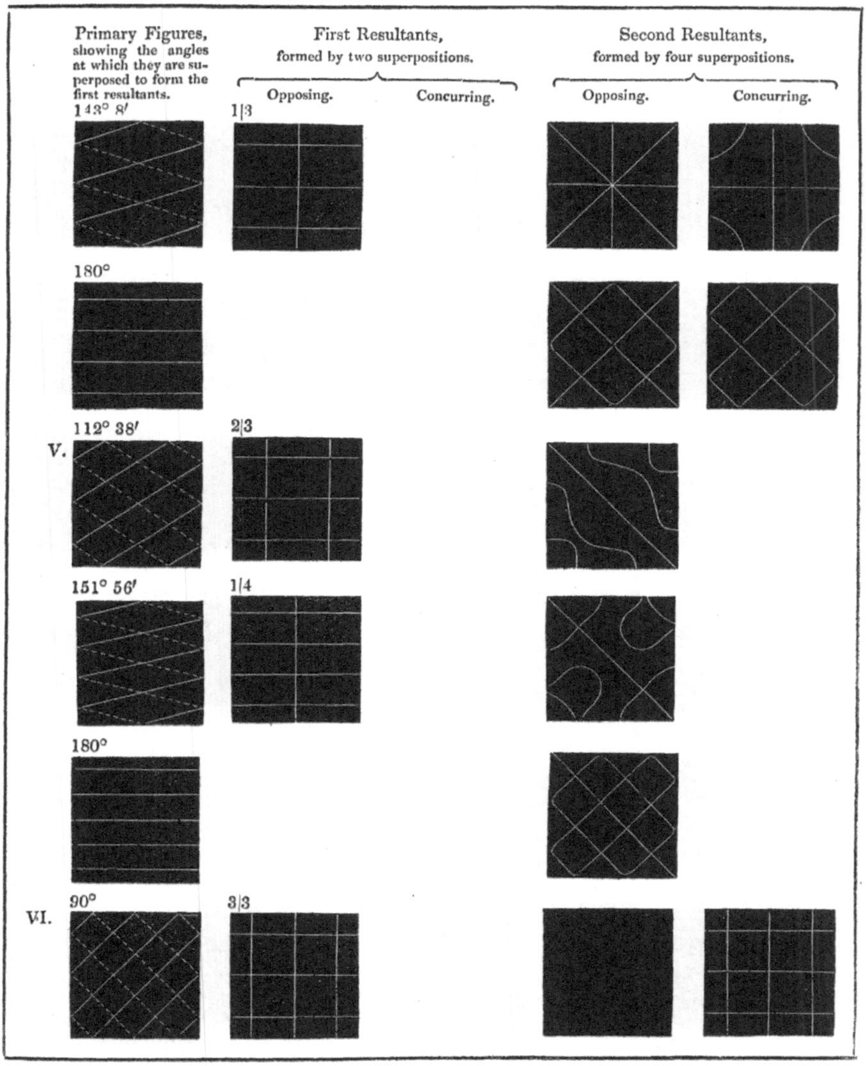

(Continuation of TABULAR VIEW)

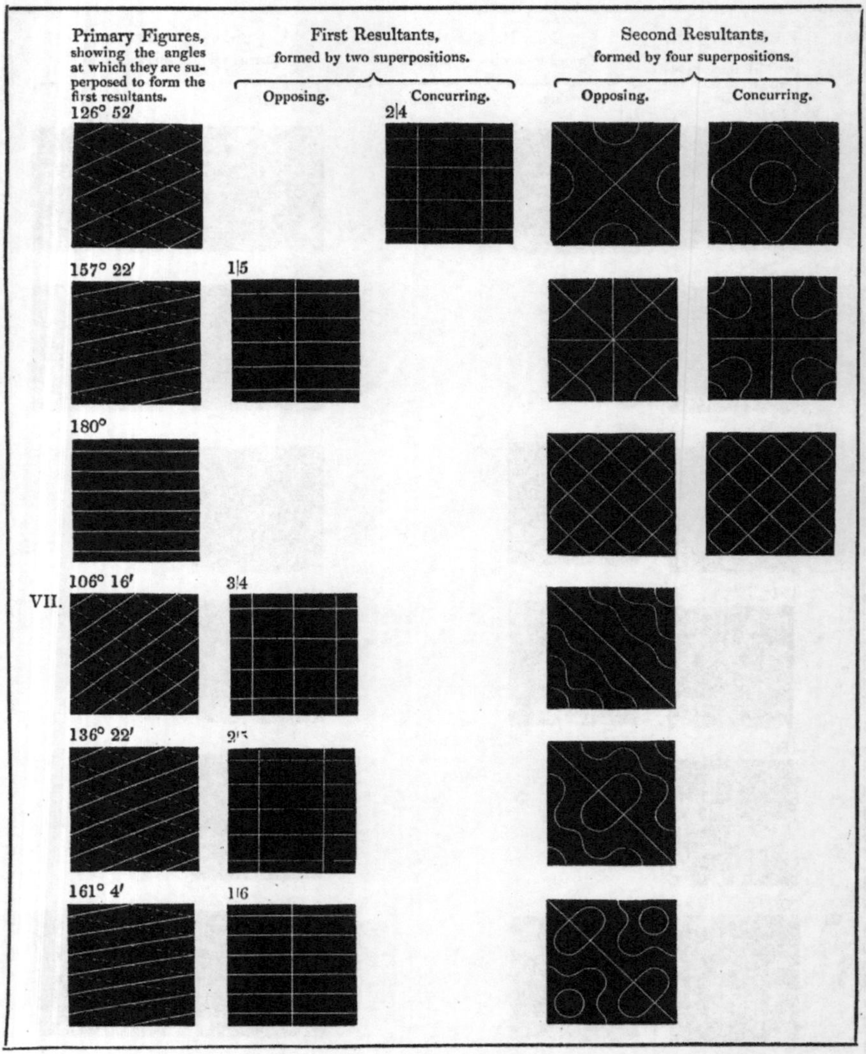

(Continuation of TABULAR VIEW)

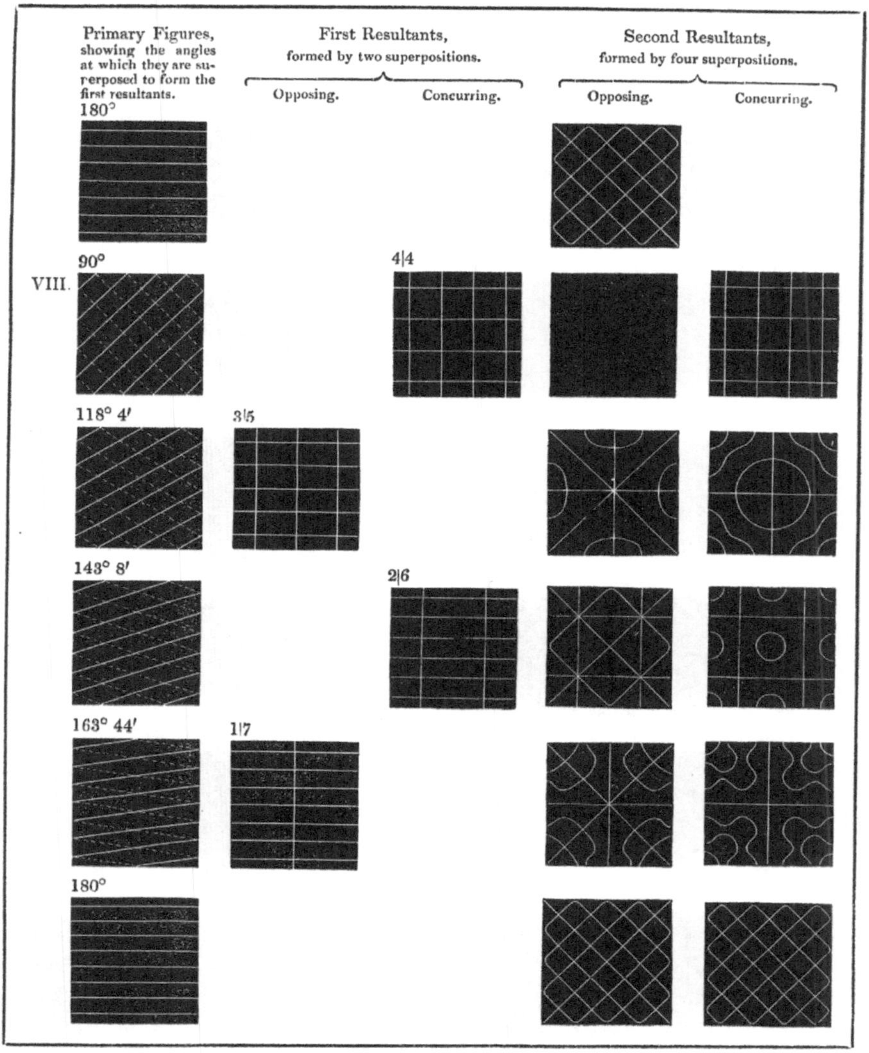

(Continuation of TABULAR VIEW)

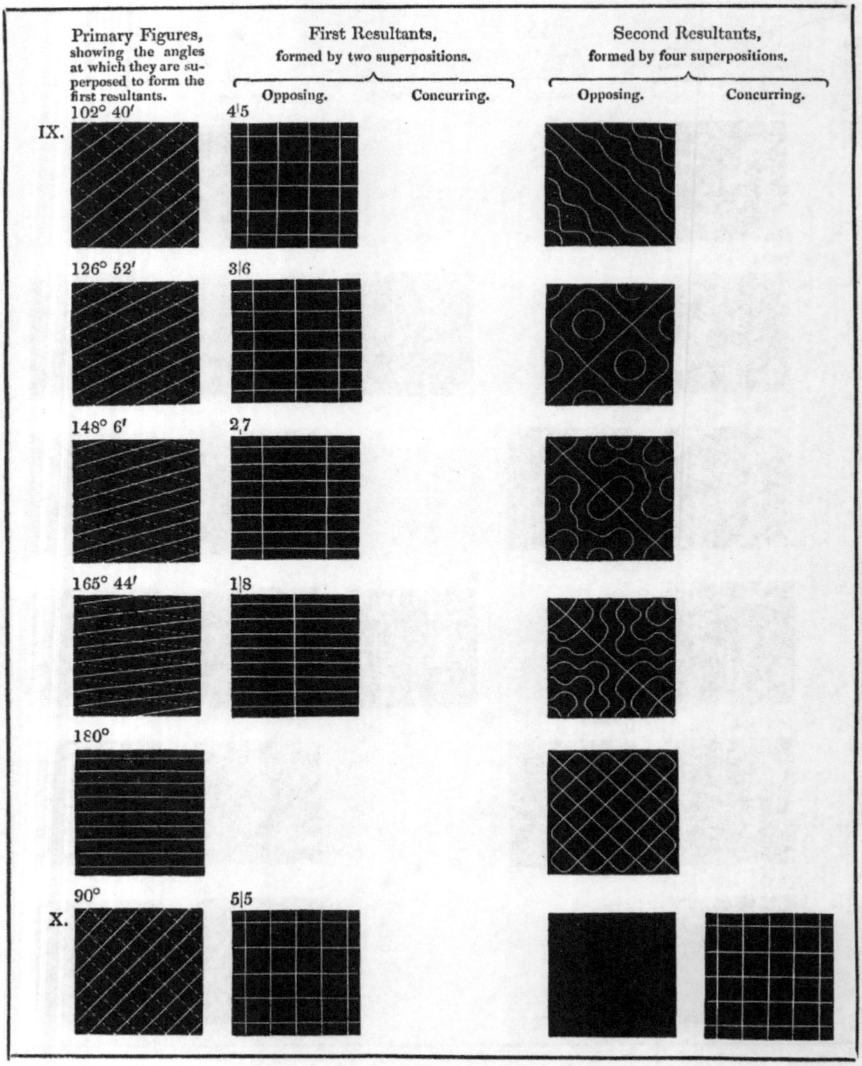

(Continuation of TABULAR VIEW)

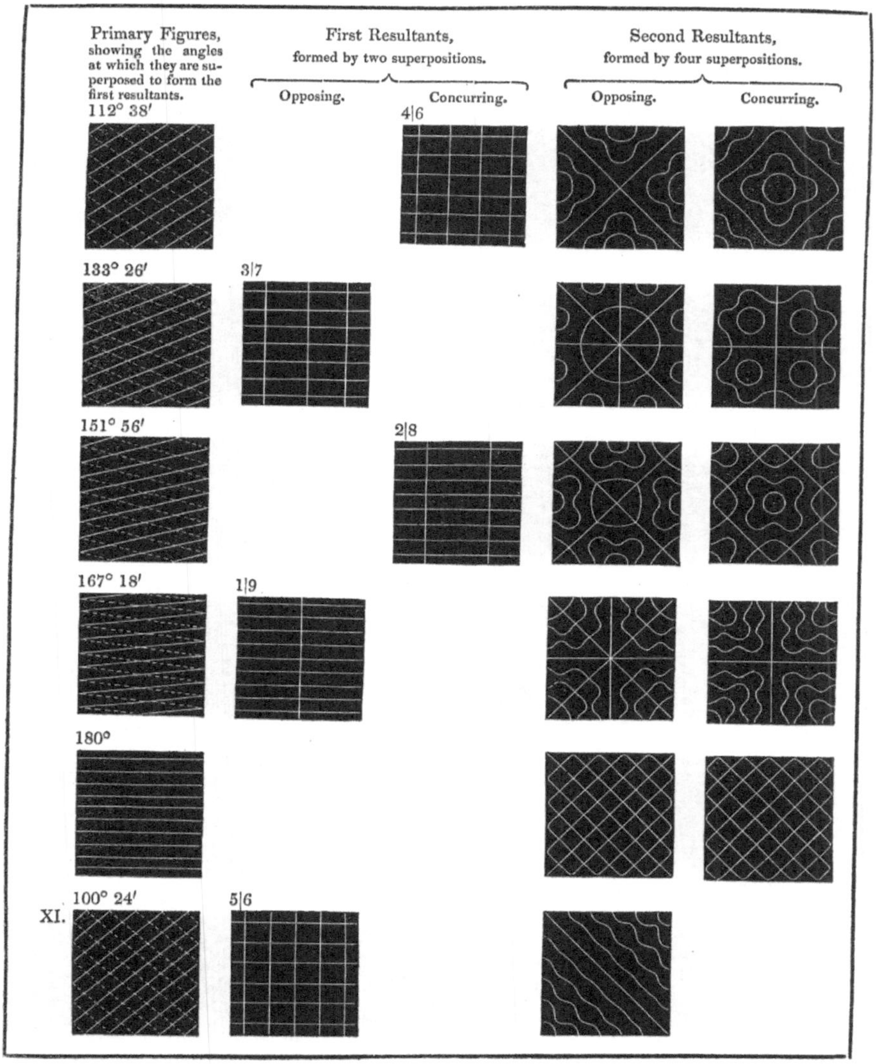

(Continuation of Tabular View)

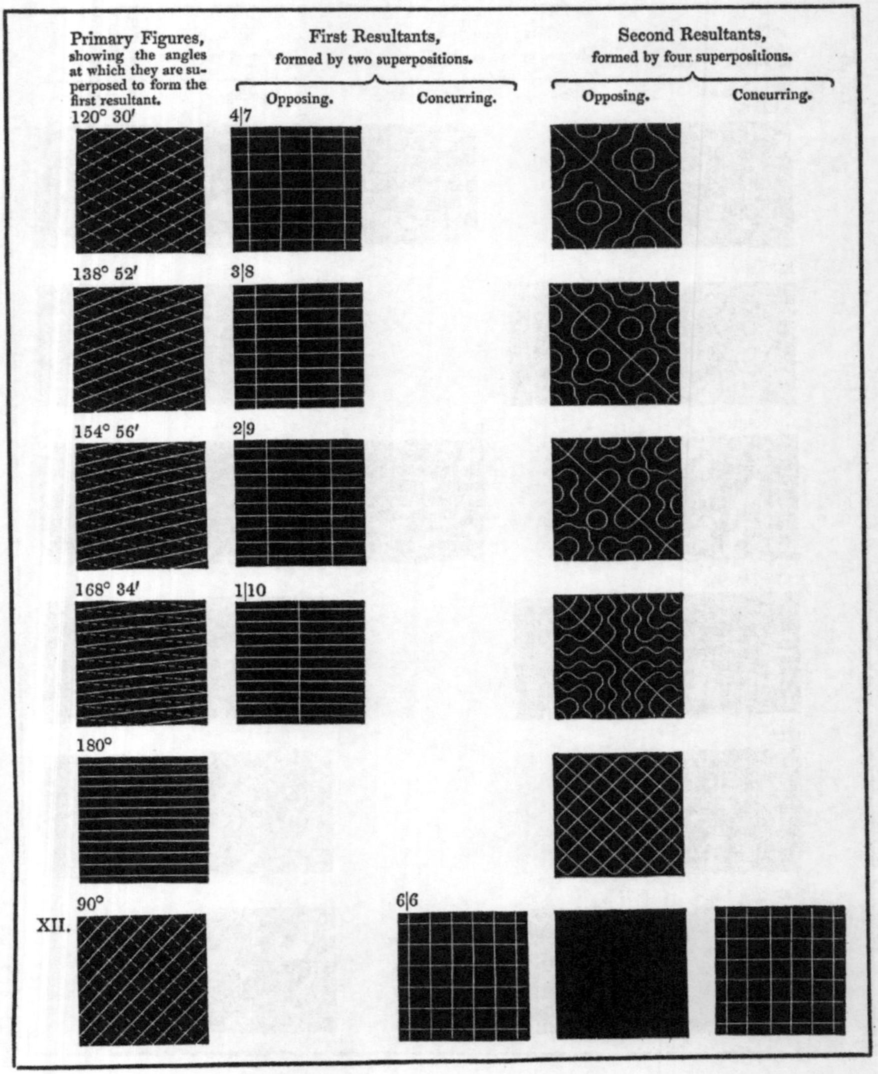

(Continuation of TABULAR VIEW)

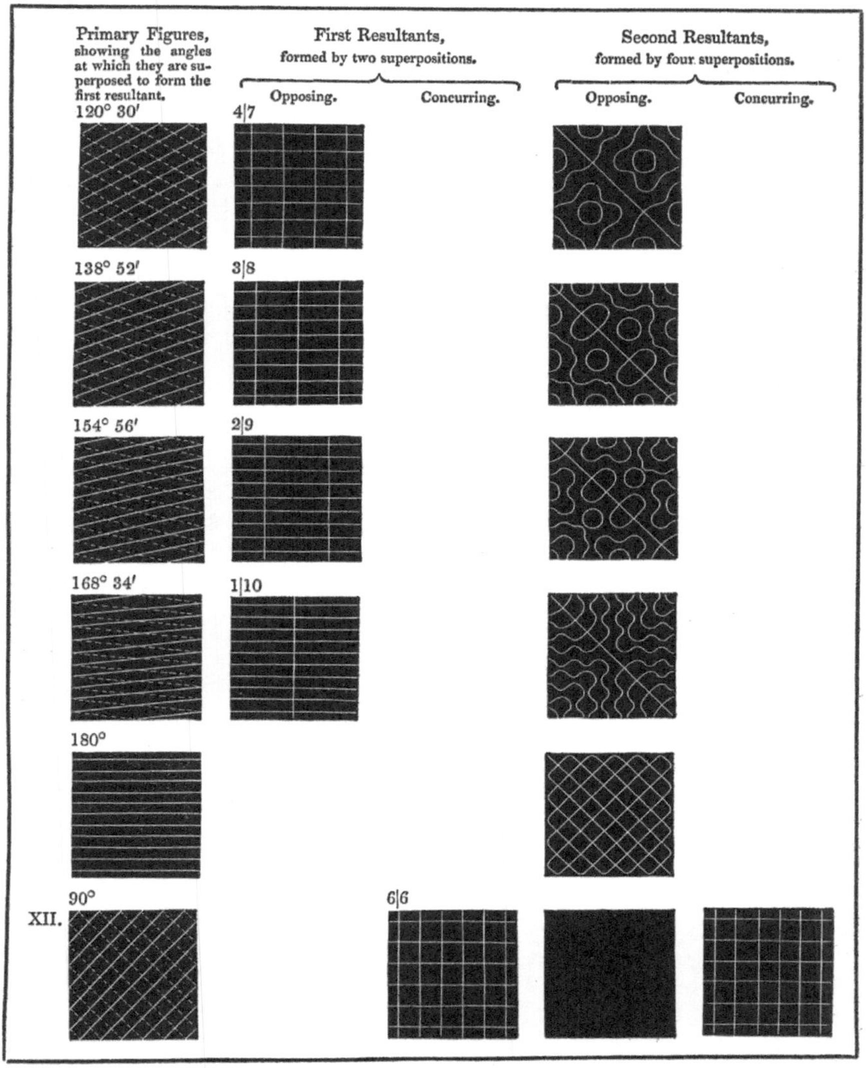

Editorial Headnote

15. Charles Wheatstone, "Lectures on Sound" (King's College London Library, 1835)

Having become the professor of experimental philosophy at King's College London (KCL) in 1834, one of Wheatstone's first tasks was to design a lecture series on sound. In this venture, he was determined to provide the most complete overview of sonorous phenomena to date. As he put it in a letter to John Herschel (1792–1871) in May 1835, Wheatstone had "just concluded a course on Sound, in which I endeavoured to include every experimental investigation which has hitherto been made in this department of physics".[1] This included accounts of his own Enchanted Lyre and Kaleidophone, Chladni's acoustic figures, and Kempelen's speaking machine. In total, he delivered some 140 experiments over the eight weeks. Such a demonstration of the beautiful order running through nature, and the communication of knowledge of nature to London audiences was consistent with the fledgling university's broader aspirations. Established in 1829 through the patronage of King George IV and the Duke of Wellington (1769–1852), KCL was very much a response to the formation of London University in 1826, a radical institution that sought to move away from the Anglican university model of Oxford and Cambridge. The overt secularization of London University, removing the commitment to the Church of England as a prerequisite for a university education, caused alarm among London's governing elites. KCL then, was to be a bastion of Anglican university education in the centre of industrial London. In providing a course of lectures on sound, Wheatstone was delivering sonorous knowledge through a distinctly Anglican framework. In particular, lecture five covered Chlandi's acoustic experiments, which Wheatstone was keen his audience try for themselves at home. After all, how better could they come to experience the order running through nature than to produce the visually satisfying patterns of acoustic figures? Wheatstone instructed his listeners on how to perform the experiments, ideally using a thick plate of glass, and achieve increasingly complex patterns through the use of sounds of higher frequency. In subsequent lectures, Wheatstone went on to examine the velocity of sound, how it could be communicated through solid conductors, and the acoustic principles of musical instruments. He would also draw attention to the potential of sound to help explain the operations of other natural phenomena, noting that there were undoubted connections "between the laws of these vibrations and those of other vibrating bodies".[2] From May until June 1837, he would revise his lecture series to expand his acoustic research into the study of light, heat, magnetism, and electricity. Wheatstone never published his 1835 lecture series on sound, but the script, although disjointed and not necessarily in the order intended to be read, provides the most extensive text on the contents of any of Wheatstone's lectures from the 1820s and 1830s. The original lecture notes are currently held in the library of KCL. Louise Spencer produced the transcript provided below.

NOTES

1 Royal Society Archives: HS 18.146, "Wheatstone to John Herschel" (13th May 1835).
2 King's College London Library: Wheatstone 5, "Lecture IV" in "Various: 8 lectures on sound" (ca. 1835).

15

LECTURES ON SOUND

Charles Wheatstone

Source: Charles Wheatstone, "Lectures on Sound" (King's College London Library, 1835)

LECTURE I

In the early stages of society men could not be long without using stretched chords or strings for various purposes, and the properties these possess of producing musical sounds, when they were touched by the hand or exposed to the wind, no doubt first suggested stringed instruments.

However this may be, we find in the early monuments of the Egyptians, the Assyrians, the Jews, the Greeks and the Chinese, which have transmitted to the present time, stringed instruments in which ingenuity seems to have been exhausted to contrive the varieties of form under which they have appeared. Several of the ancient forms are preserved in our early instruments, the Lyre for instance and the Lute, and the present form of the Harp is copied from a representation of an instrument brought by Mr Bruce from upper Egypt. It was long supposed that the ancients were unacquainted with finger boards, by which several sounds may be produced from the same strings; but on one of the obelisks [] at Rome, which was brought from Egypt in the reign of [], there is a representation of such an instrument.

The first discovery which was made

The way in which an impulse travels along a stretched cord, may be thus shown; the weight of the cord being considerable compared with its tension, the pulse travels with sufficient slowness to be followed by the eye. If the cord were of indefinite length the pulse would merely run along it, successively agitating every portion of the cord. But arriving at the fixed extremity of this limited and it is reflected in an inverted position, and continues to run backwards and forwards, 'till the rigidity of the rope, and the resistance of the air reduce it to rest. By continually renewing the impulses which I can do thus with the hand, progressive waves arise, which when they reach the fixed end and are reflected, will be observed to interfere with those still on their way, and as it were arrest them, producing a series of nodal points and ventral segments, the number of which depend on the tension of the cord, and the frequency of the impulses which I give to the moveable end.

DOI: 10.4324/b23406-15

The limited extent of the disturbance which I can thus communicate to a long rope, will not allow me to exhibit in it those modes of stationary undulation or vibration which correspond to the primary vibration of a string, and to its first harmonic subdivisions; but I can readily show these by employing a spiral wire along which an impulse travels more slowly than along the rope. If I disturb only a small portion of the wire, the single pulse which arises will travel to and fro as we saw in the rope; but if I renew the impulse every time the wave returns to its point of departure, which I can do with a periodical motion of the hand thus, the first mode of vibration arises; if I double the rapidity of the impulses communicated to the wire, the first harmonic subdivision is shown; and in a similar manner by increasing the frequency of the impulses, I can originate stationary undulations corresponding to the higher harmonic subdivisions of a vibrating string.

There is a complete analogy between the progressive waves produced in a narrow prismatic canal, and those propagated along a tended cord. The velocity with which a wave travels along such a canal is equal to that of a heavy body falling through half the depth of the fluid. The velocity therefore in this case is very small, and the crest of the wave may be attentively observed as it travels slowly along the surface, and is reflected when it arrives at the extremity. By timing the impulses as before, the stationary waves will be seen to arise, from the interference of the direct and reflected series. These stationary undulations on the surface of the water, are precisely similar to the vibration of an elastic cord, and originate in the same manner.

One of the most obvious phenomena we observe in a vibrating string, is that the pitch of the sound varies with every alteration of its length. The sounds become graver or lower as the length of the string increases, until they become inaudible; and if we then continue to lengthen the string, the vibrations become sufficiently slow to be counted. When the lengths of the string are such that their vibrations may be counted, it can be ascertained by direct experiment that the number of vibrations in a given time is inversely as the length of the string; thus if the length of the string be doubled the number of vibrations in the same time will be reduced one half. What is true of those slow vibrations of the string which we can follow with the eye, is also true of those rapid vibrations the differences of which are only perceptible to the ear; and this simple law holds both regard to the visible and the audible oscillations; that the number of vibrations is inversely as the length of the string.

If we take a string of sufficient length to allow us to count its vibrations, a very simple operation will enable us to determine of vibrations of any sound which corresponds to any smaller portion of the length. Thus if a string 40 feet long performs 4 vibrations in a second; when reduced to 20 feet it will make 8 vibrations, and when reduced to 1¼ ft it will make 128 vibrations in a second, and the note C will be heard. This experiment was first suggested and roughly made by Mersenne; Dr Young repeated it with every attention to accuracy and found that the result accorded with that obtained from other experiments, which I shall afterwards have occasion to mention, and with theory.

Interferences but of a different kind take place where two notes, nearly but not quite, in unison are sounded together. Let two organ pipes, or two strings commence their vibrations precisely at the same instant, and suppose one to make 100 vibrations in a second, the other 101. Their first vibrations will conspire to produce a sonorous wave of double force, and the impression on the ear will be double. But at the 100th vibration one has gained a single vibration on the other, so that the motions of the aerial particles in the two coexisting waves emanating from either string, are now not in the same, but in opposite directions; and the two waves being of equal intensity, they will instead of conspiring exactly destroy each other, and this will be very nearly the case for several vibrations on either side of the 100th. Consequently in approaching the 100th vibration, the joint sound will be enfeebled; there will be a moment of perfect silence, and then the sound will again increase till the 200th; when the one string having gained two vibrations on the other, the motions of the particles of air in the two waves will again completely conspire, and the sound will attain its maximum. The effect on the ear will therefore be that of an intermitting sound alternately loud and faint. These alternate reinforcements and subsidences of sound are called beats. The nearer the sounds approach to unison the slower are the beats, and when the unison is perfect the beats entirely disappear.

Sauveur employed these beats to ascertain the absolute number of vibrations belonging to different sounds. According to his ingenious idea, knowing the interval between two sounds, which may be ascertained in various ways, and the time which elapses between two beats, the absolute number of vibrations of the two sounds during this time are respectively the terms of the ratio between the numbers of the vibrations, this ratio being reduced to its simplest expression. Thus if the ratio between the two sounds be 80:81, and the beats 4 in a second, multiplying the former numbers by the latter, the sounds will be ascertained to consist of 320 and 324 vibrations respectively in a second, but as these are double vibrations, that the numbers may represent single vibrations they must be multiplied by two, and they will then be 640 and 648. Accurate experiments of this kind have been made by Sarti at St Petersburg, and by Dr Smith of Cambridge, the author of the well known work on Harmonics published in the last century.

When these beats, arising from coincident vibrations, are of very frequent recurrence they are heard as a musical sound. In all simple consonances the coincidences are very frequent, and a third sound is thus generated which has been called the grave harmonic of the two consonant sounds. If the two sounds really produced be expressed by the lowest whole numbers, the grave sound caused by their coincidences is always equal to unity.

*

The first mention of these grave resultant sounds, appears to have been made by Sorge in an instruction book for tuning organs and piano-fortes published at Hamburg in 1744. Romieu observed this phenomenon and communicated it to the academy of Montpellier in 1753. Tartini to whom this discovery has been almost universally attributed did not publish his *Trattato di Musica* in which it is mentioned until 1754. The Abbe Vogler employed in the organ the grave harmonic obtained from two smaller pipes, as a substitute for the great organ pipe which produced the same sound in the ordinary way.

There is no way of rendering evident to the eye what takes place when a mechanical impulse is propagated along the air, but we may form a general notion of the way in which the communication is made by considering how a tremor runs along a stretched rope, or a wave along the surface of water in a canal. In the first case, if the cord be struck at its middle, it will not, on account of its inertia, be put in motion at the same instant throughout its entire length, but a protuberance will be raised extending only to a small distance on each side of the point where the blow has been given. The point B which in the quiescent stake of the cord was in equilibrium between the two opposite tensions, is now solicited by equal tensions in the directions EB and DB, which may each be resolved into two others; those resolved parts which act in directions parallel to AD, EC respectively, counteract each other, while those which are perpendicular to them conspire to impel the point B downwards with an accelerated velocity. But as action and reaction are equal and contrary the same effort which E employs to drag B down will be exerted by B to drag E upwards, and by the line B has passed over half its course towards the quiescent position, the point E will have been raised above the line, and will have acquired a velocity capable of carrying it further in the same direction. The cord will at this instant have assumed this form. In the next moment the forces are reversed, B tends to drag D and E towards the line, and its own acquired momentum being expended in the effort, when it arrives at the quiescent positions it will have no tendency to go beyond it on the other side, its inertia being destroyed. D and E have however at this instant attained their greatest elevation, and the original protuberance is thus resolved into two others, but of less height, one on each side. In like manner the particles D and E in returning to their places drag up the next adjoining E and G, these the next and so on. The summits of the protuberances thus advance along the line, and correspond in succession to all its points; and the visible effect is an undulation which travels along the cord with greater velocity the greater is the tension of the cord, as it obviously ought to be since the rapidity with which each point returns from its displaced situation is greater as the former urging it is so; and this force is nothing more that the [].

We cannot see what takes place when a mechanical impulse is propagated along the air, but we may form a general notion of the way in which the communication takes place by observing this apparatus. Here are a number of short pendulums connected together by springs, and if I put the extreme one in motion, the impulse will travel along the whole row and with sufficiently slowness to be easily followed by the eye. Here is nothing you see but the motion which travels, the pendulums themselves cannot move beyond the limit of their vibrations. The tremor which runs along a stretched rope when it is struck, is another instance of the propagation of motion. And the same phenomenon is seen when a wave is excited in the surface of water in a canal. In the two last cases the small motions of the particles are transverse to the direction in which the wave is advancing, and the waves are forms or outlines travelling along the surface and must not be supposed to be things.

Thus isochronous waves originating at different points, when they meet, mutually act on each other or interfere, in such a manner as to reduce themselves to a complete synchronism. These considerations also enable us to account for the small augmentation in the intensity of sound when a number of unisons are emitted together; a hundred voices singing the same note does not produce by their joint affect or sound a hundred times as strong. The final intensity of the sound is the resultant of all the concurring and opposing waves which at the same time reach the ear, and a hundred unisons have not the intensity of a single note sounded with twice the force. The employment of unisons has however this advantage, that the air is agitated more completely and uniformly by them than by a single sonorous body oscillating only in one direction, and the ear placed in any direction will hear the sound with equal distinctness. This observation must be considered as applicable only to unisonant sounds, when different tones are simultaneously heard the augmentation of intensity is more considerable.

There is a quality in continuous or musical sound which is called pitch, and by which it may be distinguished as acute or grave. A sound may be very grave in the beginning, and during its continuance it may become more and more acute by imperceptible gradations. In this case we are sensible of a progressive alteration from the one state of sound to the other. Thus while I draw the bow across the string of a monochord, if I at the same time slide the bridge so as gradually to shorten the string, the sound from being grave, becomes gradually acute or high. (Ex1)

Grave and acute therefore are not different qualities, although they have different names, but are relative states or degrees of the same quality, like cold and heat, soft and loud, or great and small. A certain state of the atmosphere is considered neither hot nor cold. All states on one side of this are called warm or hot; and all on the other are cold. In like manner, a certain sound, for which however there is no established criterion, is the boundary between those that are called grave and those called acute. The chemist is accustomed to say, that the temperature of a body is highest when it is warmer, and lower when colder. In like manner we are accustomed to say, that the pitch of a sound is raised or depressed when it becomes more acute or more grave. There is in nature no necessary connexion between acute and high, grave and low, and accordingly as Dr Gregory has observed in his preface to Euclid's works, the more antient of the Greek writers regarded grave sounds as high, and acute ones as low, but this connexion was afterwards changed by the less antient Greeks, and has since prevailed universally. In the experiment I just now performed the alteration of pitch was continuous, but the experiment may be made in another way. After sounding the open strong, and while the bow is moving across it, if I put down my finger about 1 2/3 inches from the nut, the sound will change to one which is sensibly more acute than the former, and there is one abrupt alteration from one to the other. If I put down the finger 2½ inches from the nut; the sound will be still more acute, and it is evident that this alteration is greater than the former. Moreover if while drawing the bow across the string, I put down one finger at 1 2/3 inches, and immediately after, I put down another finger at 2½ inches from the nut, three sounds in succession will be produced, each more acute than the preceding, with two manifest degrees, or subsultory changes of pitch. Now since the last sound is the same as if the second had not been sounded, the mean of the two successive changes must be considered as equivalent or equal to the change from the first to the third. This change seems somehow to include the other two, and to be compounded of them, as a whole is composed of its parts, or as 2½ inches are compounded of 1 2/3 and 5/6 of an inch, or as the sum 15 is made up of 10 and 5. Thus persons who have thought on the subject have been led to conceive something like or analogous to a distance or interval, between these sounds. It is evident however that there is no real distance or space interposed between them; nor is it easy to acquire a distinct notion of the magnitude of these intervals. This conception is purely figurative and analogical; but the analogy is good, and the observation of it or conjecture about it, has been of great service in the science of music, by encouraging the search for some precise measure of those manifest intervals of musical sounds.

It is in the quality of pitch that sounds are rendered amenable to the laws of melody and harmony. Those not possessed of this quality such as the explosion of a musket, the smack of a whip, the voice of some animals (and many other sounds) both transient or continuous, are unmusical; and can neither be called grave or acute. But on the other hand many sounds which differ in a thousand circumstances of loudness, smoothness, mellowness and which make them pleasant or disagreeable, have this quality of musical pitch and thus may be compared. The sound of a Flute, a trumpet, a bell, a tuning fork, a string, or of the human voice, may all have the same pitch or we may be sensible of the interval between them. We can in all cases, alter the string of a monochord, till the most uninformed hearer can pronounce with certainty that the pitch is the same. We are indebted to the celebrated Galileo for the discovery of that physical circumstance in all those sounds which communicates this remarkable quality to them, and even enables us to induce it on any noise whatsoever, and to determine, with the utmost precision, the musical pitch of the sound, and the interval between any two such sounds.

Let us now consider more nearly the embouchure of wind instruments – first that of the human voice. In what way is it made to sound, or in what way does it produce the sound. Anatomy explains its tones by the vibrations of the ligaments of the glottis, and the ingenious Majendie saw this with his own eyes, as any one can see who will lay open the glottis of a living animal. Savart does not agree with this, and considers the human organ of voice as a pipe, similar to the bird call of the hunter. The great philosopher Biot explains the tone which arises in the human voice equally by the vibrations of the ligaments of the glottis, but the question, in what way do these ligaments vibrate? How are they brought to vibrate? These are yet nowhere explained, or therefore are only attempted to be explained.

We will towards the explanation of these simple phenomena, with which our great work is occupied in detail state only the following facts.

Whenever a tone arises or shall arise, only one ligament of the glottis vibrates, and when this happens, the other must be more stretched than the vibrating one. So soon as both ligaments assume the same degree of tension, all vibration must cease.

I can convince anyone of this truth by a simple experiment. Let us form a tube of very thin brass plate, which represents the larynx, about half an inch in diameter. At a rounded end of this let us form an artificial glottis and truly, as it always appears best to me from many other grounds, of the still moist skin of a pullet's egg. Let us cover the one mouth of the tube with two such skins, that in the middle of one of the entire diameters of the opening corresponding ligaments occur, as similar as possible to the human vocal ligaments in their ordinary state, and let us endeavour by means of the mouth, or an artificial bellows, to force a tone from the instruments: so it happens, in the case when both membranes are equally stretched, both through the pressure tension of the volume of air within the tube bent outwards, but thence no tone arises. If one membrane be pressed by means of a finger or an instrument somewhat further inwards in the axis of the tube than the other, so will the second, free, being equally to vibrate and sound. A like thing happens, when one membrane is more tightly stretched than the other, or also if one be replaced by a solid immoveable, to itself similar plate.

Fill a cylindrical bellows, or a gasometer with smoke, and pace on the opening thereof the artificial larynx; so long as both membranes have equal tension, the smoke streams in a diverging straight ray from the opening; but as soon as one ligament begins to vibrate, the ray of smoke immediately bends itself whirling towards the side of the resting plate.

The cause of all these phenomena is very easy to explain. The velocity of the current of air in the tube naturally depends on the pressure height of the column of water in the gasometer; this changed to a corresponding column of air, gives the velocity at the issuing orifice = Va. But since through the aperture narrowed by the membranes so much air is not allowed to issue, as the whole transverse section of the tube contains: the air in the wider tube must move slower, and this retarded motion can to the velocity at the issuing orifice only in the proportion of both openings stand, hence, if we assume this proportion as n:1, the expression Va/n arises, upon which the pressure of the membrane depends.

The great Daniel Bernoulli has in his known, greatest, ingenious, hydraulic work: *Hydrodynamica* S.260 more closely elucidates this object, and developed for the pressure of water upon a part chosen at pleasure of the sides of the tube n2–1/n2 a, which here also will for gas can be taken.

p.1

Every affection of the sense of hearing is called a sound, and Acoustics is the science which treats of the modifications and properties of sounds, and of their physical causes. The experience of sense teaches us that sounds differ, in duration being abrupt or continuous, in pitch being grave or acute, in intensity being loud or soft, and in timbre, a word for which we have none analogous in the English language, but which denotes that quality by which musical sounds differ which are the same in duration pitch and intensity.

Sounds, considered merely as affections of the sense of hearing, belong properly to that branch of physiological enquiry which relates to the nature and operations of the senses; but as all the sounds we can hear and all the differences we can distinguish among them are found to correspond to physical modifications in the medium which conveys these impressions to us, and in the sonorous body from which that medium has received them, the science which explains this connection, and the laws of the mechanical courses of sound properly belongs to general physics, and particularly so as it has enabled the mechanician to study, in a way that no other means would allow various circumstances respecting the elasticity and molecular constitutions of bodies.

Most sonorous bodies are in a state of agitation during the whole time they render sounds. This phenomenon particularly striking in the strings of the violin, harp, guitar and other instruments of this kind; but in all cases where sound is given these oscillations are too rapid to be counted; but the eye perceives them, it seizes the limits of the excursions of the cord, and it imagines that it sees it at the same time in all the intermediate positions, nearly as it sees a circle of fire when a glowing coal is whirled round with a sufficient velocity. CC' being for example, the primitive position of the cord, CLC' the position which is given to it by deflecting it with the finger or otherwise, when it is let go, it comes to CLC', returns to CLC' and performs completes thus some thousands of oscillations which gradually diminish in amplitude and which die away by themselves after some moments. The sound ceases with the motion and recommences with it. It is these oscillations, or these motions to and fro are called in acoustics vibrations.

In bells and harmonica glasses, these vibrations are less apparent; but they exist the same as in strings; we may assure ourselves of this by taking a glass bell near which a small wooden ball is suspended; striking it in order to produce a sound, and afterwards inclining it so that the ball may track touch the side, the ball is driven away with a rapid motion, and the repeated impulses which is produces in falling by its own weight are heard.

Lastly, it is sufficient to place the finger lightly on any sonorous body whatever, to affect, in every part of the body, an agitation which always accompanies the production of sound, but if rather a strong pressure be exerted at a single point, the motion is stopped in the entire mass and the sound is extinguished.

There are instruments, such as the flute and flageolet which seem to form an exception to the general principle we have stated; for nothing appears in these sonorous bodies to be in vibration; but we shall soon see that if the solid matter of these instruments does not vibrate, or only vibrates in an insensible manner, there is nevertheless a vibrating matter, and that matter is the mass of air which they contain. Thus, the principle is true in all is generality, when it is applied on to the sonorous body itself, and we shall in time endeavour to show that it is equally true when it is applied to the air or other medium which surrounds the sonorous body, and which receives the sound to transmit it to indefinite distances.

These vibratory motions may be produced by various means.

Having proved that sound arises from the vibrating motions of bodies, I will now endeavour to explain the cause of one of most important differences of sounds, viz its pitch. Sonorous bodies of the same nature yield different sounds when their dimensions are varied; thus the sound of a string varies in pitch on every attraction of its length, getting more acute or sharper as it is shortened and flatter or graver as it is lengthened; and the same changes are observed when water is poured into a musical glass, and when a pipe is shortened or lengthened either by means of a portion, or lateral aperture. Again sounds, produced from different sonorous bodies, may have the same pitch or be unison to each other, however they may vary in other particulars of quality, intensity &c from each other. The physical cause of this important difference of similar sounds, and sameness of dissimilar sounds was first discovered by Galileo, and was the first step towards establishing the science of acoustics. This great philosopher having demonstrated mathematically with respect to elastic cords, that the times of their vibrations were proportional to their lengths, and informed them that the musical pitch of a sound produced by a vibrating string depended solely on the frequency of the vibrations. And not being able to discover any other circumstance in which those sounds physically resembled each other, he concluded all sounds which were unison with any sound of a string made the same member of vibration in the same time, and therefore that the pitch of any sound whatever depended on the frequency of the vibration.

By a remarkable experiment Galileo illustrated this argument which seems to have been entirely overlooked in later times; he observed that in scraping a piece of brass with a chisel musical sounds were produced and on examining the plate of brass he found it marked with equidistant lines, as if the chisel had passed over the intervening spaces, which lines were closer together as the sound heard was sharper; when the chisel passed over the plate without producing a sound no lines were visible. Moving the chisel with different velocities he found that the spaces were two thirds where the sound was fifth sharper, and half when it was an octave higher. Had Galileo ascertained the time the chisel moved over a certain space experiment would have afforded the first direct sound method of ascertaining the number of vibrations in a given sound. As I shall not perhaps have occasion to revert to this subject, I may state en passant, that when a door or cartwheel creates, or when the finger is rubbed against a board, the sounds thereby produced, which have a definite musical pitch are the consequence of similar rapid periodical jumps.

Dr Hooke a celebrated contemporary of Newton, devised a very simple and ingenious mode of determining the number of vibrations for any pitch.

The sonorous vibrations of elastic bodies are in all cases too rapid to be immediately counted, therefore the ascertain the number of vibrations corresponding to any given sound recourse must be had to indirect methods. This information may be obtained in a variety of ways; those depending on a knowledge of the laws of vibrating bodies have been already mentioned in their proper place in the present lectures. I shall here explain only those modes which are independent of these laws and applicable at the same time a great variety of sonorous bodies through elastic media.

Dr Young has given a representation of an instrument for measuring minute portions of time, and counting the vibrations of sounding bodies, but it does not appear that it was ever actually constructed. He states that by means of this instrument the thousandth part of a second might possibly be rendered sensible. Two revolving pendulums are connected with a vertical axis in such a manner as to move two weights backwards and forwards accordingly as they fly off to a greater or smaller distance, the weights sliding during their revolution on a fixed surface; a small increase of velocity will immediately increase the distance of the weights from the axis and consequently the effect of their friction, so that the machine will be immediately retarded, and it may be thus made extremely regular. It is turned by a string coiled round the upper-part, and this string serves as a support to a barrel, sliding on a square part of the axis, which consequently descends as it revolves. Its surface being smooth may be covered with paper or with wax, and a pencil, or a point of metal may be pressed against it by a fine spring so as to describe always a spiral line on the barrel, except when the spring is forced a little on one side by touching it slightly, either with the hand, or by means of any body of which the motion is to be examined, whether it be a falling weight, a vibrating chord or rod, or any other moving body. In this manner, supposing a barrel a foot in circumference to revolve in two seconds, each 100th of an inch would correspond to the 600th part of a second; and the scale might be still further enlarged. The frequency of the vibrations of sounding bodies by connecting them with a point which will describe an undulated path on the roller.

I have imagined a mode of counting the oscillations of vibrating strings, rods and surfaces, depending on the well known phenomenon of the duration of luminous impressions on the retina. I take for instance, a glass plate, and cement to a part of it which I know to have the greatest motion a small glass bead silvered on the inside; the light of a single candle or lamp reflected from this small spherical surface will form a brilliant luminous point, and when the plate is caused to vibrate this point will describe a line equal to the magnitude of the vibration, and all the points of the path, will by reason of the law of vision I have referred to appear simultaneously visible. If I now move the surface in its own plane I shall compound this vibrating motion with a continued rectilinear motion perpendicular to it; and an undulatory or indented line will be visible. The continued motion may be circular instead of rectilinear, but in either case of it passes uniformly over a certain known space in a second, the number of undulations in that space will be equal to the number of vibrations made by the glass plate in a second and may thus be readily estimated. The same experiment may be performed with a tuning fork, a string, a rod or a harmonica glass.

I will now mention an improve.

If a mirror be caused to revolve slowly on an axis in its own plane, the image of a luminous point presented before it will continually change its place, keeping however always in a circle the radius of which is equal to the distance between the luminous point and the mirror. If the motion be made more rapid, instead of a travelling image we see a luminous path, which is straight if the point be stationary, and indented or waved if the point has a vibratory motion in a direction parallel to the axis of the mirror. The motion of the image in the mirror has therefore the same effect, as a continuous motion given to the vibratory body itself; and we have a mirror put in motion with a known uniform velocity is therefore an instrument by which the oscillatory motions of a great variety of bodies may be readily compared and counted.

The oscillations of a flame of hydrogen gas when producing sound in a glass tube, and those of a vibrating tongue admitting and excluding the light through it, may also thus be rendered visible and be counted.

For some experiments it is more convenient to revolve the mirror on an axis nearly but not quite perpendicular to its surface; for this purpose this mirror is so contrived that it may be fixed to revolve in any position with respect to the axis of rotation. The revolving mirror may be employed with equal success to measure the velocity of such motions as occur in ordinary machine; I have by its means approximated to the velocity of electricity in its passage through a conducting wire, and I do not despair of being able to measure by some modification of the apparatus the rapidity of light as it travels through the terrestrial atmosphere; but this is not the place for these discussions.

It was formerly thought that no other agitation of the air than that produced by the vibration of elastic bodies was fitted for affecting the ear as a musical sound. But this is not the case, any snap or noise whatever, if periodically repeated with sufficient rapidity is perceived as a sound having a definite pitch or note. Nothing can be more unmusical than the snaps of a quill when drawn along a comb, yet if a wheel be rapidly whirled and a quill or spring held against the teeth on its circumference, a very appreciable note will be heard varying with every different velocity of the wheel, and if the velocity be kept constant a definite musical sound will be maintained, which when high has some resemblance to the chirp of a bird. This curious experiment was first made by Dr Hooke, a celebrated contemporary of Newton, and he found by this means that the note C was produced by about 250 snaps in a second.

A very ingenious mode of producing musical sound by a series of periodical impulses and one which I have no doubt will be hereafter susceptible of many important applications, was proposed by the late Dr Robison of Edinburgh. He constructed a stop cock in such a manner that it opened and that the operation of a tube 720 times in a second, and on passing a stream of air through the tube a musical sound was produced corresponding to the G of the female voice; when the pulsations were reduced to 360 per second the sound he observes was like that of a rather harsh man's voice. The sound was particularly musical and clear, he states, when one third of the apertures of the pipe was left constantly open.

This is another form of the Syren, and an extremely elegant one. In the former case the apertures were perpendicular to the surface, and the plate was moved by wheelwork; in this the holes are pierced obliquely, and the action of the current of air will alone set the plate in motion. A register is likewise added by which the number of revolutions of the disc, when uniform, may be ascertained. On the upper part of the axis is an endless screw acting upon a wheel with 100 teeth, which at every revolution acts upon another wheel, so as to make it move one tooth; the axis of each wheel carries a hand which passes over a divided dial. The axis on which the plate revolves may be geared to the register or disengaged from it instantaneously, by pressing the face in one direction by or the other by pressing it in the opposite direction.

A pendulum vibrating seconds or a stop watch is a necessary appendage to this instrument; the register being put in gear for any number of seconds, the number of impulses in this time is shown by the dial.

The number, form and magnitude, of the apertures appear to have a very considerable influence on the quality of the sound, but no accurate experiments respecting this subject have yet been made. M. Cagniard de la Tour states that different effects are obtained by altering the length of the spaces between the apertures; he thinks that when the intervening spaces are small the sound approaches to that of the human voice, and that when they are very great they more resemble that of the trumpet.

The Syren derives its name from its property of sounding when plunged in water and fed by a current of that fluid. If the plate move with the same velocity precisely the same sound is produced as in air, thus proving that it is the number of impulses alone, and nothing depending on the medium in which the sound is excited, that influences our appreciation of its pitch.

We have seen that the pitch of a musical sound depends on the frequency of its vibrations, and I have brought before you a variety of means by which the absolute number of vibrations in any given sound may be determined. To facilitate the comparison of the absolute number of vibrations with the relative numbers, we may regard with Chladni each C, that is the musical note so called, as a power of 2 assuming the fundamental C to consist of two vibrations in a second. The lowest C of the Piano-forte or Violoncello will therefore be considered as consisting of 128 vibrations per second, and to ascertain the absolute number of vibrations for any other sound the relative number must be multiplied, for the first octave of the base by 128, for the second by 256, for the third by 512, &c.

No standard of pitch having been established before this admirable and simple one was proposed by Chladni, different numbers of vibration were attributed by different experimentalists to the same note or musical character.

We shall now consider the various experiments which have been made with the view to determine the limit beyond which sounds, whether acute or grave cease to be perceptible to the human ear. The range of audibility is stated by Chladni, Biot and other writers to be between 30, and 8000 or 12000 simple vibrations per second, which comprizes an extent of about 9 octaves.

The late Dr Wollaston made a series of experiments to ascertain the limit of hearing with regard to acute sounds, and concluded from them that it varied in different individuals. "The suddenness of the transition, says Dr Wollaston, from perfect hearing to total want of perception, occasions a degree of surprize, which renders an experiment on this subject with a series of small pipes among several persons rather amusing. It is curious to observe the change of feeling manifested by various individuals of a party in succession, as the sounds approach and pass the limits of their hearing. Those who enjoy a temporary triumph, are often compelled in their turn to acknowledge to how short a distance their little superiority extends."

The pipe which Dr W. states produced a sound at the limit of his own hearing was one fourth of an inch in length. The number of vibrations of this pipe could not have exceeded 18 or 21000 vibrations in a second.

The chirping of the gryllus campestris and the house cricket, and the note of the bat are sounds to which Dr W. found many persons insensible; but whether this resulted, as he supposed, from an incapacity to hear sounds of so high a pitch, or whether they were mere instances of deafness to these sounds which might be more audible by increasing their intensity, he did not satisfactorily determine. In a paper which I published about 6 years ago in the Quarterly Journal of Science, I led to the latter conclusion from observing that I could render myself insensible to the sounds of the gryllus campestris, by merely increasing the tension of the membrane of the tympanum by exhausting the Eustachian tube, whilst ordinary sounds remained nearly unaffected.

That this is the case has been fully proved by the more recent experiments of Savart. He found that longitudinal vibrations of a glass rod nearly 6 inches in length, which were 31000 in a second could be distinctly heard; and employing a wheel two feet in diameter: with 720 teeth on its circumference, by putting it in motion with proper rapidity, he obtained a sound resulting from 48000 simple vibrations in a second. The acute limit of hearing has therefore not yet been ascertained.

With regard to the opposite limit, that of grave sounds Savart has also made some interesting experiments. Vibratory motions slower than 32 in a second do not in general produce a continuous sound, but the separate impulses are distinguishable from each other by the ear. As this is undoubtedly owing to the component impulses being so short as to leave between them intervals of silence, Savart imagined that if the separate impulses were prolonged so as to fill up the entire intervals, graver sounds than those stated to be the limits of hearing might be heard.

To put this to the best of experiment he employed a new mode of producing sounds. A bar two feet and a half in length is put in motion by a rotatory machine, and at each side of the circular plane described by the bar, and corresponding to one of its diameters are placed two boards close to the edges of the bar which rotates between them. It is evident that in this arrangement, the number of the impulses is double that of the turns, since a blow is given at each semi revolution of the bar.

When the bar begins to revolve slowly, separate blows are heard which resemble extremely feeble explosions; but when the velocity becomes greater, the blows become more and more intense so as to induce the belief that the bar strikes against a solid body and at the same time an extremely grave continued sound is heard, but which acquires great intensity when the blows are sufficiently near each other to allow the impressions produced by each of them to last long enough to superpose themselves in a proper manner. Savart states that having tried the experiment with a great number of persons, they all agreed in fixing the limit at which the continued sound commenced, at 7 or 8 blows in a second.

Sounds differing in pitch are easily distinguished from each other whether sounded in succession or in combination. The ratio of the number of vibrations in one sound to that in the other is called a musical interval, and any two sounds however they may differ from two others in the absolute number of their vibrations, provided their ratios be the same are said to have the same interval between them.

Two notes sounded simultaneously or in harmony as it is called, produce very different effects according to the ratio or interval between them. Some are consonant or agreeable to the ear, others dissonant or disagreeable. Now it is universally found that in those combinations which please the ear or which affect it with a sensation of concord, the sounds have a very simple numerical relation to each as 1:2, 1:3, 2:3, &c, which on the contrary those combinations which produce a discordant effect are represented by more complex ratios such as 8:15, 24:25 &c. The concord is more satisfactory and pleasing as the terms of the proportion are lower and differ less from each other, and the discord is more unpleasant as the terms of the ratio are higher. The greater or less simplicity of the musical ratios of the vibrations is therefore the sole foundation of harmony. The ear, without counting the numbers themselves, perceives the effect of the ratios, and of the concurrence of the simultaneous vibrations when they arrive at the ear. The ear does for time what the eye does for space, when it is affected by the just proportions of form, without measuring or calculating the proportion themselves.

Next to the unison, which is the coincidence of two sounds having the same frequency of vibration, the most simple consonance is that whose ratio is 1:2, which is called an octave, on account of its interval being the same as that subsisting between the first and eighth notes of the musical scale, the formation of which will be hereafter considered. In this consonance one note performs two vibrations to each single one of the other, and the effect approaches to that of the unison, and so nearly are they allied, that in music notes differing by an octave are considered as the same. In like manner the double octave which consists of notes whose vibrations are as 1:4, is a perfect consonance and almost identical with the octave.

The next consonances in order of simplicity are those in which the prime 3 enters. When the ratio is 1:3 the vibrations of the graver sound are trisected by those of the acuter; this is a very agreeable consonance and is called the twelfth. The concord 2:3 differs from the one just mentioned in the lowest sound being taken an octave higher, and it has a still more agreeable effect, for though one of the numbers constituting the ratio is greater, yet as the distance between the two sounds is less they coalesce more perfectly and are less distinguishable from each other. The concord represented by 3:4 is also a very agreeable one.

The prime number 5 also enters into several very pleasing consonances. 1:5, 2:5, and 4:5 are all very agreeable; the last is called a major third, and the others are the octave and double octave to it. 5:8 is the minor sixth. 3:5 the major sixth, and 5:6 the minor third. The coincidence of the vibrations of the sounds forming the limits of these intervals are here represented.

No ratios are employed in music in which other prime numbers than 2, 3 and 5 occur. All musical intervals are expressed by ratios the terms of which are these numbers or multiples of them by each other. The primes 7, 11, 13 &c are entirely excluded, though it is has been suspected that the somewhat agreeable effect of several of the discords represented by a complex ratio, is owing to their being so near a ratio consisting of lower terms in which 7 enters are not to be distinguishable from it. Thus the flat seventh whose ratio in the musical scale is 9:16; is exceedingly near 4:7 and undoubtedly owes its pleasing character to this circumstance, though its use as a note of the scale does not depend thereon.

The ancient Greeks were much more confined in this respect that we are, for they admitted only the primes 1, 2 & 3 to constitute their intervals. We have no proof that they were acquainted with harmony, and this circumstance alone enables us to account for their rejection of some of our most agreeable consonances.

All the consonant intervals less than that of an octave are here arranged in the order of their distances from unity. If between the two notes an octave another note which forms a consonant interval with the lowest be interpolated, the interval which it forms with the highest will also be consonant; and one interval is said to be an inversion of the other. The fourth is thus an inversion of the fifth, the minor sixth of the major third; & the minor third of the major sixth.

We have hitherto considered those consonances only which result from the combination of <u>two</u> sounds, let us now enquire what are the concords produced by the unison of three sounds. For this purpose we may take any of the binary consonances we have already noticed and while <u>it</u> is heard change the pitch of another note through all the graduations between the two sounds forming the consonance, and observe what concordant and what discordant effects are produced during the transition. Operating thus with the consonance whose interval is 2:3 we find that the combination, or chord as it is called, is agreeable only when a note is interpolated which is either a major or minor third to the lowest sound, that is whose ratio to it is either 5:4 or 6:5. In the first case the combination is called a perfect major chord, in the other a perfect minor chord. It would occupy too much time to repeat this process with all the other consonances, it is sufficient to state that the result of every such experiment is one of the chords just mentioned or an inversion of it, an inversion of a chord being merely the transposition of one or more of its notes to an octave higher or lower. It is the remarkable peculiarity of a perfect chord whether major or minor that every consonant interval is included between the notes which compose it.

No other note can be added to the three of a perfect chord without rendering the combination dissonant, unless it be an octave to one of the sounds belonging to the chord.

A musical scale is the arrangement of the notes of a musical composition according to their pitch. The simplest scale of sounds that can be conceived is that formed of the notes of a perfect chord; every note succeeding another has an agreeable relation to it; but a scale thus formed has little variety, bugle calls and some national chords of very rude nations can alone be performed upon it. To avoid monotony it is necessary to increase the number of notes in the octave, and the most obvious and simple way in which this can be effected is to add the perfect chord of a sound which has the nearest relation to the fundamental sound of the first chord, which is either the fifth above or the fifth below it. A scale of five notes is thus formed, and as a great number of the original melodies of Scotland are confined to this series of notes it has been called the Scotch scale, but as Oriental and other national airs are founded upon it, it has been more appropriately termed the national scale.

By the addition of another perfect chord, the choice of which is determined by the same rule, a scale of seven notes is formed. This is the usual or diatonic scale. The ratios which the different sounds forming it bear to the fundamental or key note are here represented. It must not be supposed that in a musical performance the effect of each note of the scale depends on the reference which we make of it to the key note, the agreeable effects of the 2nd & 7th notes of the scale of which the ratios with the key note are completed are owing to the simpler ratios which they bear to other notes in the scale.

The names given to all musical intervals have reference on the diatonic scale. Thus we have the second, third, fourth, fifth, sixth, seventh, and octave or eighth. The thirds, sixths and sevenths are major in this form of the scale, but they may be minor; the major and minor intervals differ from each other by the intervals 24:25 or a minor semitone. The second, fourth, and fifth are called perfect intervals. Augmented intervals are the perfect and major intervals increased by a minor semitone, and diminished intervals are the perfect and minor intervals diminished by the same ratio.

If the diatonic scale be made to commence with the note C, that is with a sound the number of whose vibration is any power of 2, as 256, 512 &c the successive notes of the scale are named after the seven first letters of the alphabet. The words sharp and flat after the latter denote that the sound is higher or lower by a semitone than the natural note.

To enable us to form tables of the various sounds which the same elastic body is capable of producing in its different modes of vibration it is necessary to have characters to denote the sounds in different octaves. The ordinary musical characters are the cumbrous for this purpose. Chladni has adopted the following notation. He considers the lowest C of the Piano-forte or Violoncello as the base, and represents each sound in the first octave of the base by adding 1 after the letter, thus []. The sounds of the second octave of the base will be expressed by adding 2, those of the third and fourth octaves by 3 and 4, and so on. When it is required to refer to a sound graver than the lowest C of the Piano-forte, he denotes it be a line drawn under the character of the sound. To signify that the sound is a little graver or a little more acute he adds the algebraic signs minus and plus.

There would be but little variety in music were all compositions confined to a single diatonic scale. To avoid monotony it is necessary to change the key; this operation is most simply and agreeably effected by adding to the scale, according to the rule already laid down, another perfect chord, and at the same time withdrawing that which is the most remotely related to the new chord. Another scale is thus formed, which has two perfect chords in common with the first; and is therefore more nearly related to it than any other. Continuing this process, successively adding one chord and subtracting another, and number of diatonic scales may be obtained with perfect harmony and requiring no temperament whatever.

There are 15 keys employed in music and to have the harmony in all these perfect, 35 notes on the octave would be required. Were this indispensable, instruments with fixed tones would be extremely expensive and difficult of construction. But it has been found that if additional notes be made to the diatonic scale, dividing every tone into two semitones, a scale of twelve notes is formed, which sufficiently approximate to the 35 notes to be substituted for those without any very great offence to the ear. This is called the chromatic scale.

The proper mode of tuning or tempering these twelve notes so as to produce the most agreeable effect in practice has been a subject much discussed. Some contend that it is better to leave some keys perfectly in tune than the rest; others maintain that the imperfections should be equally distributed through every key. To effect the last purpose the interval between every two successive notes is made rigorously the same.

Edict of the Emperor Shang-hi made in the 52nd year of his reign

The superintendent of the music of my Empire has represented to me that the instruments which I ordered to be constructed being finished, it becomes necessary to cause them to be inserted in my book of solemn customs.

The instruments which were in use in the times of my predecessors were, it is true, very excellently constructed, but they were old, and the sounds they produced were dull and deteriorated; this has induced me after having examined them myself with much attention, to cause new ones to be constructed on the model of those which already exist; for I am not able to give any thing better in this respect, than what had been done under the preceding dynasty, and all the praises which the Tay-Tchang-see bestows on me, by making me the author of a new system of music and the inventor of new musical instruments, must be regarded only as an effect of his zeal for my service and the glory of my reign.

After having communicated my project to my prime minister, to the chief judges of the nine principal tribunals of my court, and to the other officers of my empire, I ordained them to tell me their thoughts freely respecting it. They have, with one common voice, sent me the following answer.

The musical instruments made under the preceding dynasty, are very imperfect. They can express, neither the delicacies, the embellishments, nor even the true tones of music, according to the principles of which we clearly see that they have not been constructed. But your Majesty has discovered by profound reflexion, the means of correcting their defects, and of making them capable of producing perfect and truly harmonious tones. We therefore believe, and we are fully convinced that your Majesty will render an essential service to the Empire, by giving orders that all these instruments be engraven, and that they be inserted in the book of solemn customs of the Empire, with the method of constructing them, their dimensions, and all the means which have been employed to render them such as they are. It might be feared that without this precaution, the remembrance of it might be gradually lost, and that in the course of time our music might fall into the state of imperfection from which your Majesty has withdrawn it. We therefore believe that in inserting them in the book of solemn customs of the Empire it is advisable to record not only the mode, and the entire theory of their construction, but also the Moon in which by the order of your Majesty, they were first brought into use. &c &c.

LECTURE II

Echoes

Acoustic Buildings

When the same sound is heard more than once it is called echo. An echo is produced when the sound in the course of its propagation meets with an obstacle of sufficient extent and regularity, and at a sufficient distance to prevent the repetition from being confounded with the original sound. We cannot well distinguish more than eight or nine different sounds in a second, therefore in order that the repetition of the sound shall be heard as a true echo and not simply as a resonance, the repetition must at least be 1/9th of a second later than the primitive sound. The distance of an echoing surface is easily estimated by the following rule, multiplying the velocity of sound by the interval between the sound and its repetition and half this quantity will be the distance required. The time which elapses between a sound and its echo, and the distance of the reflecting obstacle being determined, the velocity of sound may be ascertained; and it was in this way that Walker in 1698 determined this velocity for the 1st time in England, but his experiments did not come to near the truth as those made by the Florentine Academicians 34 years before.

Echo frequently arise from another cause than from the reflection of the sonorous waves from a solid obstacle. In narrow roads between two walls of rock, in long galleries and tunnels, in high chimnies, & where there are no solid obstacles to impede the progress of the sound echoes are frequently heard when the length is sufficiently great. Prof. Airy has lately made some remarks at the CPS on a very distinct echo produced in a high chimney erected at the gas works near that university. In these cases the sonorous waves are reflected precisely as I have shown them to be in tubes open at both ends, when the fixed vibrations of columns of air are produced, while in the ordinary cases of echoes the reflections take place as in tubes closed at one end. The resonance of tubes whether open at both ends or closed at one end is in fact nothing but a series of echoes following each other so rapidly that no intermissions are observable.

Among the echoes remarkable for distinctiveness or frequency of repetition are the following.

An echo at Coblenz on the Rhine repeats a word 17 times; one on the banks of the Lago del Lupo, above the fall of Terni repeats 15 syllables. According to Plot in his Nat His of Oxford there is an echo at Woodstock Park which repeats 17 syllables by day and 20 by night; in the Simonetta palace near Milan there is an echo caused by two parallel walls, which repeats a word 40 times, and the report of a pistol 50 or 60 times. Gassendi mentions an echo near the tomb of Marbella, which repeated the first verse of the Aeneid. The echo at Rosneath in Scotland is said to repeat a melody three times. In many accounts of echoes given by travellers we read that each successive repetition is given in a lower; this is very improbable and no doubt arises from the writers mistaking lowness of intensity for lowness of pitch.

If the echoing surface be concave towards the auditor, the sound reflected from its several points will converge after reflection towards him precisely as the reflected rays of light do; and he will hear a more intense sound than if the surface were plane, and the more so as it approaches to a sphere concentric with himself. An echo exited at one station is heard most distinctly at another when the two stations are conjugate foci of the reflecting surface; ie such that if the reflecting surface were polished, rays of light diverging from the one would be made after reflection to converge to the other. Hence if a vault be in the form of a hollow ellipsoid of revolution, and a speaker be placed in one focus, his words will be heard by an auditor in the other as if his ear were close to the other's lips.

Chladni mentioned that he observed, at Mayden near Amsterdam, an echo well known in Holland, caused by a semi-elliptical wall, the two foci of which were several feet distant from each other; the sound produced in one of the foci produced a greatly reinforced echo in the other focus; it appeared to proceed from the earth, which Chladni attributed to a slight inclination of the wall inwards.

In the cathedral of Girgenti in Sicily there is a remarkable echo of this kind; the slightest whisper is distinctly conveyed from the great western door to the cornice behind the high alter. It chanced that the confessional was placed in the precise focus of divergence at the former station, and secrets never meant for the public ear, became by this means known to the curious who having accidently discovered the circumstance resorted to the opposite point; after scandal had been for some time busily employed the fact became known and the confessional was removed.

A similar experiment has been made in a quiet night in two opposite recesses on Westminster Bridge; the effect is however by no means so great as in the instances above recorded.

Manzoni Palace at Venice, square room 25 ft high, concave roof, parabolic reflectors have been employed, and with some success when it has been required that the sound should proceed in a particular direction with increased intensity. In Heligoland on the highest point of the island there is a light house furnished with a clock; but it is at such a distance from the town beneath, which consists of one long street sloping to the shore, that it was seldom heard; a few years ago it was suggested that the clock bell should be placed in the focus of a parabolic reflector and the line of reflection directed upon the town; this was done and the strokes of the bell are now distinctly heard at all times. An application of parabolic reflectors or sound bounds as they are then called, to church pulpits has been published in the Transactions of the Society of Arts; from the testimonials there given and from their introduction into several churches there can be no doubt of their efficiency to a certain extent.

Sounds are propagated with great distinctiveness along a smooth surface, as the surface of water or a wall; in the whispering gallery of St Paul's Cathedral, which is a gallery round the circumference of the dome, a person speaking even in the lightest whisper at one point is heard distinctly by another person placed at another point as if the voice was close to his ear. The whispering gallery of Gloucester Cathedral is a passage with polygonal sides open at both ends, but this effect is by no means equal to that heard at the dome of St Paul's.

During the building of the Colosseum in the Regent's Park I had an opportunity of trying a few experiments on the circulation of sound round a circular wall, which could not be performed at St Paul's Cathedral. The internal diameter of the circular building is 130 ft, the circumference on which the panorama is painted is therefore about 407 feet. A temporary gallery was suspended from the ceiling and led from the stage whence the panorama is viewed, to the circumference of the building, to enable the artists who were employed to work at the upper part of the picture. At the end of this gallery near the wall I observed the following effects. On pronouncing a word it was repeated a great many times, extremely loud at first, but successively weaker until it became no longer audible. A sentence being uttered the last syllable appeared only to be repeated and this was echoed as many times as before. A single exclamation appeared like a peal of laughter. The tearing of a piece of paper was like the patter of hail. A continuous sound appeared only as a prolonged resonance. A large partition perpendicular to the wall was placed between me and another person; the distance between us was 10 feet, the faintest whisper he made was circulated and the repetition was far stronger than the original sound. Other experiments of a like nature were made, but I have ever since regretted that I omitted the opportunity of ascertaining the interval between successive echoes. The original proprietor of this popular place of exhibition intended to have made arrangements so that these curious effects might be shown to the public; but it having fallen into other hands this intention was abandoned; and the very facts I have mentioned forgotten. The smoothness and continuity of the wall, uninterrupted by doors or cavities of any kind, occasioned the experiments to be more perfect than those which can be tried at St Paul's. The effects were scarcely distinguishable when the same experiments were tried on the floor of the building though the wall was still circular. To hear the echoes it was requisite that the hearer should be at the circumference, when I was at the centre or at any part of the gallery between the centre and the circumference, no clear echoes were audible and I heard the voice of the speaker as in an ordinary room.

A correct knowledge of the doctrine of echoes is of importance to the architect in the construction of buildings for public speaking or music, both for the purposes of employing them when they are advantageous, and of avoiding them when their effects are prejudicial. It would be very useful to know the best way of constructing rooms in which the sound should be everywhere distinctly heard, without sacrificing architectural rules, or the qualities necessary for other purposes. Unfortunately so uncertain are the rules we possess that in most buildings which have succeeded, the success appears rather to be the effect of chance than of any accurate theory. Few writers have paid any attention to this subject; the two works most in repute are Saunders' treatise on theatres, including some experiments on sound, London 4th; and Rhode's theory of the propagation of sound for architects published at Berlin in 1800.

A hall or room is favourable for sound, when it is well adapted to facilitate the natural propagation of sound, and when the intensity of the sound is augmented by the resonance of other bodies, or by proper reflexions.

It is much easier to adapt buildings so as to be favourable to the natural propagation of sound, than to obtain proper reinforcements by artificial means; the most essential rules for this end are to avoid all kinds of decorations which project too much from the sides of the room, and to seat the auditors in successive rows above each other so that nothing shall intervene between the hearers and the stage or orchestra where the speakers or performers are placed.

According to Saunders 70 feet is the greatest distance at which an ordinary voice is distinctly audible; a theatre constructed according to this rule might contain about 2000 persons. The semi-circular form of most theatres is well adapted to contain a great number of persons at a moderate distance from the place where the sound is produced; but the horseshoe form of the Italian opera house and the form of some of the ancient theatres with diverging sides are capable of containing a much greater number of persons at the same distance.

When the building is so large that the distance of some of the auditors is greater than 60 feet from the speaker, properly adapted reflexions are supposed to be the best means to augment the intensity of sound.

Some architects have followed principles directly opposed to those which they ought to have followed. They have imagined that the sound which proceeds onward, ought to have a retrograde action; but this only occasions a disagreeable resonance which becomes an echo, which is more apparent as the reflecting surface is further from the person who hears the sound. To obtain a useful augmentation of the intensity of sound by reflexion, all retrograde action must be avoided; but the sound spreading to the right, to the left, above and below should be diverted towards the audience by the proper disposition of the sides, and thy must be heard almost at the same instant as the principal sound.

Of all the possible forms that can be given to rooms intended for hearing music or speaking the most disadvantageous is that of an ellipse. The principal property of an ellipsoid is to untie in one focus all the rays which proceed from the other. If it were indeed possible to concentrate an orchestra in one of the foci, and the audience in the other, the secondary sound would be very powerful, but there would be no utility in the arrangement, for if the building were small, no reinforcement would be required, and if it were large the secondary sound would be heard later than the primitive sound unless the ellipse were greatly elongated. The theatre and concert room at berlin was a regular ellipsoid, and had according to Chladni all the disadvantages of this form.

II

The theory by which Laplace accounts for the discrepancy between theory and observation with regard to the velocity of sound in the atmosphere has been so generally received by philosophers, and seems to be confirmed by such a variety of accurate and apparently satisfactory experiments, that very strong reasons ought to be adduced to justify us in rejecting it.

The uncorrected formula for the velocity of sound in air, is founded on the supposition that the elasticity is proportional to the pressure, and this is known to be the case when air is compressed or expands so slowly that the volume has had time to radiate or absorb sufficient heat to place it in equilibrium with the surrounding medium, but if the observation be made before this time has elapsed it will found that the elasticity has increased in a greater proportion that the pressure and this elasticity is obviously due to the heat cooled by the change of capacity, for if the pressure be continued and the heat be allowed to escape the elasticity will diminish so as to be in accordance with the law of Mariotte. Now the vibrations which produce sound are so rapid that the radiation and absorption which takes place during the compression and dilatation of the air may be disregarded, the heat therefore increasing at the same time with the pressure the elasticity must be increased in a greater proportion than if as supposed in the law of Mariotte the heat were to remain constant.

III

Each vibration of the sonorous body excites in the air an undulation of a determined length.

This proposition is one of the most important and difficult in acoustics; but we ought now to take it into consideration and endeavour with greater care to make it be understood, because it will serve us as a point of departure in explaining the theories of optics.

Let us imagine a horizontal tube TT', having, for example, 10,000 feet in length and 1 foot in diameter; the air which fills it is throughout at the same temperature and under the same pressure; a piston PP' fitting well to the sides, may in 1" of time complete an oscillation between the two positions PP" and SS' which are 1 ft distant. Nothing is beforehand determined respecting the increasing or decreasing velocities which it may assume in this interval.

All being at rest, the piston starts from PP' to arrive at SS'; during this motion, the air within the tube is modified in a certain manner, and better to study the modification which it undergoes, we will fix ourselves to the precise instant at which the piston arrives at SS', and we will suppose that all the particles of air remain in the state in which they there happen to be, we will suppose that those which are compressed cannot spring back, that those which are dilated cannot approximate each other, and that those which are at rest preserve their state of rest.

If the column of air were a perfectly hard solid body, it is obvious that one of its extremities being impelled by the piston, an equal portion of the other extremity of the tube would be pushed out at the same instant, but there is no perfectly hard body; the air is very fluid and compressible, and when the piston impels before it one of the extremities of the column, the other extremity cannot be affected by it at the same instant; time is required for the impression to be communicated to it, and according to the length we have given to the tube, we may safely affirm that no particle of air is pushed out of the open extremity T which the piston is passing from PP' to SS'. Therefore, the air within the tube to the right of the piston is compressed, since it occupies 1ft of length less than it before occupied. Besides it is evident that it is not equally compressed throughout the extent of the tube; for, during the second which the piston employs to move from PP to SS', the compression has been only communicated and felt at a certain distance, such as AA' for example. This portion of the column of air, capable of being modified during the motion of the piston, in what is called a wave or an undulation, and the length of the wave is the distance of its two extremities SS' and AA'.

In my first lecture I endeavoured to explain the nature and origin of sound; I shall now proceed to the consideration of the various circumstances which attend its travelling through the medium which conveys it to our ears.

That sounds are conveyed through the air is a matter of the commonest observation, but it is not equally obvious that they would not reach us as well through a perfectly empty space. To prove that the presence of the air, or of some other medium, is *necessary* for the transmission of sound we must have recourse to experiment, and the air-pump affords as a ready means of doing so. If a bell be placed in the receiver of an air-pump; while the air is being gradually exhausted the sound becomes weaker and weaker, and is at last, when the vacuum is complete, totally inaudible; it increases on the readmission of the air, and we should find, if we performed a similar experiment in a receiver containing condensed air, that the intensity of the sound would increase with the degree of condensation.

The rarefaction of the air, it is well known, increases with its distance from the earth's surface; consequently, the intensity of sound is greatly diminished in the higher regions of the atmosphere. Saussure in his "Travels in the Alps" relates that having fired a pistol on the summit of Mont Blanc it produced no greater report than a cracker would have done in a room. And many travellers have remarked the small distance to which sounds can be heard in similar elevated situations.

The increased intensity of sound in condensed air has been observed by persons who have descended in diving bells, and by others who have been shut up in reservoirs of air condensed by the action of bellows, such as are employed in many iron foundries.

It is therefore fully proved that a material medium is necessary for the propagation of sound, and that the intensity of the effect is sensibly proportionate to the density of the medium. This is true not only with regard to the atmospheric air but it holds also with all homogeneous aeriform fluids. Sounds are therefore transmitted with less intensity through hydrogen than through any other gas.

The vapours of liquids are no exception to the general rule; if I admit a small quantity of Ether into an exhausted receiver, the sound of a bell which was before imperceptible becomes distinctly audible.

The distance to which sounds are transmitted through the air depends on several circumstances; but principally on their original intensity, the hygrometric state of the atmosphere, and the direction of the wind. Sounds have sometimes been heard at very great distances. Chladni relates that he heard the guns fired at the Battle of Jena, at Wittenberg a distance of about 80 miles. Guns fired at Stockholm on the occasion of the death of one of the royal family in 1685 were heard by Dr Kearne at the distance of 180 miles; and it is related that in a naval action which took place in 1672 on the eastern coast of this island between the English and the Dutch, the cannonade was heard at a distance exceeding 200 miles from the place of conflict, having been heard at Shrewsbury and even in Wales.

Among feeble sounds the music of the Russian horns is said to be audible at the great distance of nearly 5 miles; and there are many instances of the human voice having been distinctly heard at an astonishing distance from the speaker. In Captain Parry's third Polar expedition, Lieutenant Foster found that he could maintain a conversation with a man across the harbour of Port Bowen, a distance of a mile and a quarter.

When a gun is fired at a stance, the report is always heard a sensible time after the flash is seen, and in like manner lightening precedes the thunder by a very observable interval. These are proofs that the transmission of sound through the air is not instantaneous, but that it requires time for its passage. Numerous experiments have been made in different parts of the world to determine the precise velocity of sound. Those of Mersenne, and of the Florentine Academicians in 1660, are the earliest on record. It was observed by the latter, that at a distance equal to 5739 English feet, the sound of a harquebass arrived five seconds after the flash; and repeating the experiment at half the distance, they found exactly half the time to be required. From these experiments the velocity of sound appeared to be 1148 feet per second.

The lecture of last week was devoted exclusively to the consideration of the various means we are able to employ to measure the pitch of sounds, and to ascertain the number of vibrations corresponding to any given degree in the sense of audibility. I endeavoured to show that though we find great difficulty in directly counting so many as 8 or 10 impulses made within a second, yet that we have means of counting in an indirect but in a certain manner 200 or 2000 or 40,000 within the same mark instant of time, and this within very narrow limits of error. And we can establish the number of undulations which the chirping of a cricket impresses on the air, with nearly equal certainty that we can reckon the vibrations of a pendulum.

In the present lecture we shall consider how other effects of sound are measured; its intensity, its timbre, and the velocity of its transmission through various media.

Air first with respect to its intensity.

LECTURE III

A Lecture on the Vibrations of Columns of Air in Cylindrical and Conical Tubes. Delivered at the Royal Institution. 15th March, 1832

Wind instruments of music are usually composed of cylindrical or conical tubes, which may be either straight or curved, and of an apparatus by means of which the air within them is put into vibration in the direction of their length. The modes by which the air contained in tubes may be continued in sonorous vibration are very various, and constitute so many classes of musical instruments; in the Flute, Flageolet, and some kinds of Organ pipes, the vibrations are occasioned by a current of air passing through a narrow orifice and impinging on a cutting edge; in the Clarionet and in Reed Organ pipes the sound is produced by a vibrating tongue; in the Trumpet, Horn, Serpent &c the original impulses are given by the vibrations of the performers lips, and in the Hautbois and Bassoon a reed is employed for this purpose which very nearly resembles in its action that of the lips. These means of producing sound are entirely independent of the columns of air with which they are usually connected; the embouchure of an organ pipe will sound when isolated from its tube, and the frequency of its vibrations is found to depend on the velocity of the current and on the distance of the cutting edge from the aperture or wind channel; the number of the vibrations of an elastic tongue, of which the Aeolina is the most well known example varied with its length, thickness and the substance of which it is made; and the vibrations of the lips are regulated by the muscular effort exerted to close them and by the velocity of the current of air which periodically forces them open; instances are by no means infrequent of individuals being able to imitate the effects of the Trumpet and other instruments of the same class by the lips alone. The combination of any of these means of obtaining sound with columns of air contained in tubes produces two results, firstly, those vibrations only, can be excited which the air in the tube is capable of maintaining, and secondly those sounds which are capable of being thus maintained are considerable modified and increased in intensity. All the modes by which the air is originally impelled into vibration, in any of the wind instruments I have mentioned are analogous in their action on the column of air to a mode of excitation, which from its greater simplicity I shall have occasion to employ, namely the vibrations of an elastic plate, or of a tuning fork. When either of these is brought near the tube, a tuning fork for instance, if the tube be of such a length as to be capable of vibrating in the same time with the tuning fork then then a strong reinforcement of the sound takes place; in this case there are no disturbing effect arising from a current of air passing through the tube, neither is there any complication arising from the reaction of the columns of air upon the vibrating body as frequently happens in the ordinary modes of exciting sounds.

One of the most remarkable properties of columns of air is their capability of rendering different sounds when their length is constant, by merely altering the force of the breath. This, so obvious to every performer on a wind instrument, could not have escaped the earliest writers on the philosophy of sound, and we accordingly find it the subject of much ingenious but erroneous speculation by them. It was Sauveur who first correctly explained this phenomenon, and showed that when a tube sounded any other than its fundamental note, it did so in consequence of a division of the column of air into portions separately vibrating, in the same manner as the harmonic sounds of a vibrating string had been previously explained. A few years after, the celebrated mathematician Daniel Bernouilli took up the subject and investigated it analytically; the results to which he was lead are of great importance and I shall endeavour briefly to lay them before you.

Let us at first suppose a cylindrical tube open at both its ends; by causing it to sound as an organ pipe and impelling the wind with more or less force, it may be made to render different sounds; and if the fundamental sound, ie the gravest which it can give, be denoted by unity, the other sounds will correspond with the series of natural numbers 2, 3, 4, 5, 6 &c; and whatever means be tried, it can never be made to render any sound intermediate between these. The sounds corresponding to this series of numbers are represented in this diagram by musical notes.

In the most simple mode of vibration of a cylindrical tube open at both ends; the particles of air, according to theory, alternately approach to and recede from the centre of the tube; and the motions of the air on each side of this point which is called a node, are made in opposite directions, indicated by arrows in the figure. At the node the condensations and rarefactions of the air are the greatest (because the actions of all the other parts of the air concur the actions of all the other parts of the air concur to produce this effect) but the motions of the particles are the least there: the rarefactions and condensations diminish as the distance from the node increases, whilst the excursions of the molecules become greater; and at the greatest distance from the node or at the open end, the excursions are the greatest; but the condensations and rarefactions are evanescent, and the density of the air remains the same as that of the free air which surrounds the tube. All these conditions of the air are well represented by this simple and ingenious apparatus constructed for the purpose by Mr Willis of Cambridge.

*

That the portions of air on opposite sides of the centre of the tube move in contrary directions to each other, has never yet I believe been experimentally proved. Now if this position be true, we should expect that the oscillations of two vibrating surfaces or tuning-forks isochronous with those of the column of air, and made simultaneously to and from both open ends would put the air into powerful vibration; and that, on the contrary, if the impulses were made at the same instant of time towards one and from the other end of the tube, their effects should neutralize each other and the air should remain at rest. But it is obviously impossible to make two independent vibrating bodies oscillate in such perfect synchronism as to produce these effects. If, however, the tube be bent nearly into a circle, so that its two ends may be brought close to the opposite sides of the same vibrating plate; the anticipated result may be perfectly obtained; the plate advancing in its vibration towards one end of the tube must at the same instant recede from the other, and these effects neutralizing each other, no augmentation of the original sound of the plate takes place. But if two halves of this circularly bent tube be jointed so as to move independently round the axis of the tube, the two ends may be brought to the opposite sides of portions of the plate vibrating on contrary sides of the neutral plane; in this case the impulses are made at the same instant towards both the ends of the tube, and the augmentation of the sound is very considerable. Thus we find an important point of Bernouillis' theory fully confirmed by experimental evidence.

In the second sound of an open tube there are two nodes, distant from the ends by a quarter of the length; and the column of air divides itself into a double part at the middle, and two single parts at the extremities; the division and the reciprocal motions are represented in these two figures. The sound is to the first of the same tube as 2 to 1, ie it is an octave sharper. At the centre of the tube, and at equal distances from the two nodes, is a point at which there are no variations of density, and at which the air is exactly in the same state as at the extremities of the tube; if at this point an aperture be made, or if the tube be even entirely divided the vibrations will remain the same; the part where the column of air may be this divided is called a ventral section.

In the third mode of vibration of an open tube, two double parts are formed in the middle, and two simple parts at the extremities; the ratio and sound is as here represented. In this, as in also every other higher subdivision, there is a ventral section between each two adjacent nodes.

The air at a node being in equilibrium between two equal and opposite forces; this equilibrium will not be disturbed if a solid partition be placed between the two portions of the air vibrating in opposite directions. The tube open at both ends is thus divided into two equal tubes, open at one end and stopped at the other, each giving the same sound as the entire tube. Those modes of division in which there is a node in the middle of the column of air are alone capable of being thus divided; when a ventral portion occupies the centre of a tube, as it does in each alternate mode of vibration, their division cannot be made. The progression of the sounds is therefore not according to the series of natural numbers as in an open tube, but follows the order of the uneven numbers 1, 3, 5, 7 &c, each alternate number of the natural series being intermitted.

These figures represent the three first modes of vibration of a column of air in a tube closed at one end when the length of the tube, instead of being half the length of an open tube, is equal to it. In the first mode of vibration of the air contained in a tube, one extremity of which is closed, the air alternately approaches to and recedes from the closed end, which performs the same office as a node in the other kinds of vibrations. This motion produces the lowest sound of which a cylindrical tube of the same length is susceptible; it is an octave below the gravest sound of an open tube of equal length. In the second mode of vibration a node is formed at one third from the open end, and two thirds from the closed end, and its sound is to that of the first as 3 to 1, ie it is an octave and a fifth above the former. In the third mode the situations of the nodes, and the motions of the air are as represented in the figure, and its sound is two octaves and a third above that of the first mode.

It is stated by Chladni, and I believe by every other author who notices the subject, that the end at which a tube is excited into vibration, or at which the wind enters, must always be considered as an open and; even if it be placed immediately to the mouth as in the horn and trumpet. Now I think I can prove that this statement is erroneous; that it is not in accordance with experiment, and that its general admission and adoption has prevented the proper application of the theory to the greater part of the wind instruments in use. If I take a cylindrical tube which open at both its ends gives a certain note, in this instance C, by blowing as in a Flute; on closing one of its ends it gives, for reasons I have already explained, a sound which is an octave deeper. If now I excite it as a horn, it gives a sound which is unison to that of the closed tube, though it is very different in quality; were the ordinary opinion correct it would be unison with the open tube. Moreover the series of harmonic sounds produced by different forces in blowing, are the same as those of a closed pipe excited at the open end; that is they correspond with the series of uneven numbers, and not with the more regular harmonic series of an open pipe. We find also the situations of the ventral sections to correspond with those of a closed tube excited in the ordinary way; thus in the first harmonic subdivision the ventral section according to theory, exists at a distance from the closed end which is one third of the length of the tube, and at this place an aperture may be made without injury to the tone; when the sound is excited in the manner of a Flute the aperture must be placed at two thirds of the length of the tube from the embouchure, and at one third when it is excited in the manner of a horn; in both these cases the aperture is as it theoretically should be, at one third of the tubes length from the closed end. These are unequivocal proofs that when sound is produced in a cylindrical tube, in the manner of a horn or trumpet; the tube must be considered as closed at one end, and the air as impelled into vibration at the node. The same observation extends to reeds, and to every case of excitation in which the communication with the eternal air is cut off.

I have here an organ pipe which consists of a reed and of cylindrical tube, it belongs to what is called the Cremona stop of the organ; comparing it with an open diapazon pipe which gives the same sound, it will be found to be only one half its length; it is consequently a tube closed at one end. But the most instructive instance I can adduce is the Clarionet, a well known musical instrument; it is in its internal bore a perfectly cylindrical tube, the bell at its lower extremity, and which gives it somewhat the appearance of a conical tube, being an appendage intended merely for ornament; now the fundamental sound of a Clarionet is an octave below that of a Flute of the same length, the series of its harmonic notes is that of a closed tube and do not resemble that of the Flute; in the Flute the first harmonic is an octave above the fundamental sound and all the intermediate notes are made by shortening the virtual length of the tube by uncovering apertures on its side. In the Clarionet, on the contrary, the first harmonic is a twelfth, ie an octave and a fifth above the first sound; and as the intermediate notes are also made by lateral fingerings, in the case of the instrument a much greater number of finger holes and keys are required than in the Flute.

It is a striking circumstance that in all cases of the production of sound at the closed end of a tube, the tone is invariably more powerful than when the sound is produced at the open end of the same tube. The cause of this is obvious, in the one case the sonorous impulses are made at that part of the tube where the alternate condensations and dilatations of the air are the greatest; and in the other case they are made where these condensations and dilatations are the least. It would follow, therefore, that if we could excite a tube at any intermediate point; the intensity of the tone would be intermediate between that obtained at the open end, and that obtained at the closed end; whilst the pitch of the sound would remain the same invariably. Now this is easily proved by changing the position of the embouchure of a reed or lip instrument from the closed end, or node, to different points of the side of the tube. I shall not detain you by making these experiments, as I shall be able to prove same thing in a more elegant and instructive manner. There is a mode of putting columns of air into vibration, to which I have not yet alluded, because it has not been applied in any instrument of music; if a flame of hydrogen gas be introduced within a tube, it has been observed that under certain circumstances, which have not been attentively investigated, a sound is produced which is either the fundamental sound of the tube or one of its harmonics; if the tube be open at both its ends any sound of the series competent to such a tube may be obtained, if it be closed at one end, the series is that of a closed tube excited in any of the usual ways. To examine the causes of the alterations of the intensity of the sound produced by the flame of hydrogen gas under the simplest circumstances, I take a tube sufficiently short so that I may be certain always of producing the fundamental sound. I observe that at the orifice of the tube the smallest possible flame is sufficient to excite the sound, which however ceases if the flame be made to move towards the node, which is the same thing in this case, the centre of the tube; but if at the same time that the flame is advanced in the tube, it be also enlarged in volume the sound continues, and with increased intensity; by continuing to move the flame towards the node, and at the same time proportionally enlarging its volume, the sound progressively increases in loudness, until it attains its maximum at the node, which in this case is at the centre of the tube; but in a tube closed at one end it would be at the stopped end. On moving the flame back again towards the orifice of the tube, it will be found that the sound will not be maintained unless the flame be gradually diminished. Some alterations of intensity and quality are obtained by changing the volume of the flame whilst it remains at the same point of the tube, but these variations are very limited. Thus by means of a flame of hydrogen gas the air in a tube may be put into vibration, either at a node, a ventral section or at any point intermediate between them; and the variations of the intensity in the sounds in each of these cases may be made the subject of accurate observation. It is probable that hereafter organs may be constructed in which columns of air excited by flames of hydrogen gas may be substituted for the organ pipes now employed; if this should be done such instruments might be made to possess a greater scale in the crescendos and diminuendos, than any instrument we are at present acquainted with.

Before I leave the subject of cylindrical tubes, I will say a few words on the influence of their diameter. It is a result both of theory and experiment that the diameter of a cylindrical tube does not affect the number of its vibrations; it however materially influences the quality and intensity of the tone, but in a manner just the reverse of that which is ordinarily supposed in the established works on Acoustics. I have here six tubes all of the same length, six inches namely, and varying in diameter from ¾ of an inch to two inches. What takes place in one of these tubes of a small diameter when excited by a flame at different parts of its length, I have already shown. As the diameter of the tube increases, smaller flames are required to excite them at corresponding points of their length, and the tones are much less intense than in tubes of less diameter; and as sufficiently small flames cannot be obtained to excite the tubes of larger diameter at their orifices; the sounds commence at a point so much nearer the node as the diameter is wider. In the largest tube no sound can be obtained even when the flame is at the node, in consequence of our not being able to procure a flame small enough; but if this tube be closed at one end the sound corresponding to that of a stopped tube, an octave lower namely, is heard; this results because lower sounds require a larger flame than higher sounds. These experiments serve to explain the cause of the loud shrillness of the trumpet, hautbois and other narrow bored instruments, and the mellow softness, as come with them of the Horn, Clarionet and other instruments of greater diameter than the former. The cause of the greater intensity of sound in narrow tubes is probably this, that the alternate condensations and dilatations of the air are more energetic when there is less lateral expression at the nodes. The general opinion is that the sound may be produced with greater force in a tube as the diameter is larger; the experiments I have adduced prove this to be erroneous.

The general results which I have obtained from a numerous set of experiments on the sounds of columns of air excited by flame, are these. The flame is required to be larger as the length of the tube is greater, as its diameter is less, and as the point of excitation is nearer the node; the same mode of vibration of harmonic subdivision always being understood.

I shall now proceed to the consideration of the vibration of the air in conical tubes. Bernouilli has successfully made this subject of analytical investigation, and I can do little more than to state the general results he has obtained and to show their accordance with experiment. But from the general misapprehension which has prevailed on the point which I have particularly insisted upon in this lecture, the supposed necessity of an embouchure being situated at the open end of a tube namely; the theory complete as it is has not yet been properly applied to explain the vibrations of the air in any of the musical instruments where the air vibrates in conical tubes.

"Let us at first consider a conical tube, open at its base and closed at its apex. Here as in cylindrical tubes, vibrations of different orders may be formed, according to the number of parts into which the column of air divides itself; and the more numerous these divisions are, the sharper the sound will be. In the first order of vibrations, which gives the fundamental sound, the entire column of air moves in the same direction, but alternately from the base towards the apex, and from the apex to the base. The variations of density are nul at the open end, and thence they progressively increase towards the closed point. On the contrary the excursions of the particles are nul at the apex, and go on increasing towards the open base

In the second order of vibrations, the mass of air divides itself into two parts separated by a ventral section, which remains stationary, and divides the tube into two equal parts. In the third order, these equal divisions are formed separated from each other by similar ventral sections; and so on indefinitely as we have seen to take place in cylindrical tubes. The distance between two consecutive ventral sections in a conical tube is therefore constant, and equal to the distance between two consecutive ventral sections in a cylindrical tube vibrating with the same number of subdivisions. With regard to the progression of the sounds and their absolute value, it is the same as in a cylindrical tube, open at its two ends and equal in length to the conical tube. In the latter the summit of the cone, distant from the first ventral section by a quantity equal to the distance between two adjacent ventral sections, is equivalent to an open orifice communicating with the atmosphere.

From these data it is obvious that a conical tube excited at its apex in the manner of a horn, or by a reed should exactly resemble in the progression of its sounds and in their absolute value, a cylindrical tube open at both ends and excited in the manner of a Flute. The Hautbois is such a conical tube, and we consequently find that this instrument is the same length as the flute, that it gives the same fundamental sound and series of harmonic sounds, and that the lateral apertures are similarly placed as in that instrument; whilst a Clarionet of the same length which is a cylindrical tube excited in a similar manner, gives as we have already seen a fundamental sound an octave below and a different series of harmonics.

The Hautbois, Trumpet and French Horn pipes of the Organ are also conical tubes, and they are, as we should expect them to be, double in length to the Cremona pipe giving the same sound, and equal in length to the open diapazon pipe. These pipes differ only in the divergence of the cone, which has the same influence as a variation of diameter in cylindrical tubes, ie they occasion no alteration in the pitch but considerably modify the quality of the sounds. The Bugle, Trumpet and French Horn pipes are also conical tubes excited at the closed apex, and their different sounds result from numerous harmonic subdivisions of the column of air; from the employment of these higher subdivisions these instruments are necessarily much longer than those as the Flute and Hautbois in which the fundamental vibrations of air in the tube are principally employed.

The correspondence between a conical and a cylindrical tube extends also to equal portions of the cone and cylinder, separated from the entire tube by planes drawn through two ventral sections. For as these portions before they were cut were consonant one to the cone, the other to the cylinder, and were therefore consonant to each other; they will be so after separation, since in this state they may continue their vibrations as before. These results may be easily verified by experiment; on comparing the sounds produced in the same way, of a cylindrical tube open at both ends, and any portion of the same length of a cone also open at both ends, it will be found that the sound obtained in both cases will be the same whether the conical segment be excited at its larger or its smaller orifice.

But though there is this perfect correspondence between the situations of the ventral sections in equal conical and cylindrical tubes; the same conformity does not obtain with respect to the distances of the nodes. The nodes are those points in the vibrating column of air where the variations of density are greatest, and where the motions of the air on each side of it are always directly opposed to each other. In a cylindrical tube we have seen that the nodes are invariably in the middle between two ventral sections; in a conical tube they are on the contrary, unequally distant from them; and there is this peculiarity in the variation of their places; that in the portion between two ventral sections which lies nearest the base the node most nearly corresponds with the centre, it deviates from it and moves towards the summit of the cone as the vibrating part lies nearer the apex, and at the first vibrating part the node actually corresponds with the ventral section. In the diagram the ventral sections are represented by the dotted lines, and the nodes by the thicker continuous lines.

In a cylindrical tube open at both ends and in the first mode of vibration; if a solid partition be placed at its node the tube will be separated into two equal portions each giving the same sound as the entire tube did when the partition was away. In a similar manner the segment of a conical tube may be divided by a solid partition at its node into two parts, each giving the same sound as the entire tube; but these parts will be of unequal length. There is no simple mathematical rule by which the situation of the node in a conical segment can be ascertained, and consequently we are unable to give a rule for separating such a tube into two unisonant portions; but I have found a very ready mode of obtaining the information by direct experiment. I have already shown that if a flame of hydrogen gas be introduced in a cylindrical tube the vibrations of the contained air acquire the greatest intensity when the flame is precisely at the nodal point; the same thing is true of truncated conical tubes, and it is found, but repeating the experiment with truncated cones of equal length but differently diverging, that the node is nearer the small end as the ratio between the diameters of the two orifices is greater.

If therefore, we take a series of tubes all of equal length and open at both ends, but differently diverging, divide them at their nodes and close the orifices presented by the divisions; we shall obtain a series of closed tubes varying in length from that of the entire tube to zero, and all giving the same sound. To verify these results the sound may be produced either by blowing across the edge of the open orifice, or by exciting the sound with the lips in the manner of a horn at the closed extremity.

The last observation I shall bring forward respecting conical tubes is that tubes of the same length but differently converging or diverging from the nodal point or closed end produce sounds differing in pitch. The sharpest note that can be obtained is equal to the sound of an open tube of the same length, and this is procured from a cone nearly closed at the summit, the sound becomes flatter as the ratio diminishes between the smaller and the greater orifice, and is an octave below where the converging sides become parallel and the tube is cylindrical; as the sides become more diverging from the node the sound becomes still flatter and is two octaves graver than the first when the proportion of the open to the closed orifice is as 1 to 4. The sound may be indefinitely flattened beyond this but they do not preserve much of the musical character. We have thus the means of making every note within the compass of two octaves from tubes of the same length; and as such tubes may be either excited as the stopped diapazon, or as the reed pipes of the organ; they might be advantageously employed when portability is required. A very simple empirical rule may be given for determining relative sounds of a conical segment when it is closed at its larger or at its smaller end; the number of vibrations are simply as the diameter of the open end. Thus when the diameters are as 1 to 2 there is an octave difference, when as 2:3 there is the interval of a fifth between them, and when as 1:4 they are at the distance of two octaves from each other. One sound is always sharper and the other flatter than the sound of a cylindrical tube closed at one end.

To conclude this part of the subject I will enumerate the principal wind instruments in use, and indicate to what classes they belong, in accordance with the principles I have endeavoured to develop.

The Flute, the Flageolet, and the open diapazon Organ pipe ends, and excited at the open orifice.

The Syrinx or Pandean Pipes, the stopped Diapazon organ pipe, and the Pitch pipe are tubes closed at one end, and excited also at the open extremity.

The Clarionet is a closed cylindrical tube excited at its closed end. This method of exciting a closed tube has the advantage of not preventing the employment of lateral fingerings, as the former method does. The crumhorn or Cremona pipe of the organ is an instrument of the same class.

The Hautbois, the Serpent, the Bassoon, the Horn, the Bugle, the Trumpet &c are all conical tubes closed at or near the apex, and excited at this closed point.

Theory of Wind Musical Instruments (this appears with Lecture 3)

LECTURE IV

So far as I have yet proceeded in explaining the laws of vibrating columns of air, I have confined myself to those results which experiment brings directly to our notice; I have now to undertake a more difficult task, to show how the motions of the air confined in a tube originate and are maintained, and why they differ so much from those of a wave advancing along the open air.

If we compare together the motions of the aerial particles in a progressive wave, of which this is a representation, with those in a stationary wave, which is here represented, and which is the same as the vibration of a column of air, these differences are observed: 1st in the progressive wave the condensations and rarefactions travel onward and each point of the path is successively affected by every vicissitude of density; in the stationary wave, on the contrary, the greatest compressions and dilatations occur always at the same point, and different fixed points are affected by different changes of density. 2ndly the amplitudes of the molecular excursions in the progressive wave are all equal, and the oscillatory motions of the particles commence one after the other, on account of the progressive communication of the motion; in the stationary undulation the oscillations of the particles all commence simultaneously, and the extent of the excursions differ at each point of the wave.

Now, though there exists so great a difference between the states of motion in these two kinds of waves, it can be shown that one arises from the other; that the stationary undulation is produced by the combinations or influence of two progressive waves advancing in opposite directions, or, which comes to the same thing, of two waves one of which is direct and the other reflected.

Before I proceed to trace the gradual transformation of progressive into stationary undulations, in the case of the sonorous waves in the air, I will endeavor to illustrate the subject by reference to what takes place under similar circumstances with regard to the undulations excited on the surface of water. The steps being much easier to follow that in the other intricate case.

In the last lecture I showed in this canal how a progressive undulation advances along the surface of water. I shall now attempt to show the stationary undulations which can be formed in this. For this purpose I have reduced the length of the canal by a partition.

If I agitate the water at one end of the canal, a wave is produced which travels to the opposite extremity and is thence reflected, and if the disturbance be periodically renewed at every instant the reflected wave reaches the original point of departure, the direct and reflected waves will unite and by their conjoined action give rise to a stationary undulation, or vibration, of the surface of the water. This is the simplest mode of oscillation of the surface of water and consists of one elevated and one depressed wave. If the agitation is repeated twice during the interval the wave takes in advancing and returning, another mode of oscillation is obtained, consisting of two elevated and two depressed waves. And by increasing the frequency of the agitation, other modes of oscillation may be obtained in a similar manner. The successive steps by which the conversion of the progressive wave into the oscillatory state is effected, are represented in the diagram with reference to the most simple mode.

A B represents the horizontal surface of the water in the canal when at rest. We will suppose the time required for the production of a complete wave to be divided into four equal intervals, and we will examine the changes which the [] undergoes in each of these intervals.

In the 1st interval half of the elevated wave is produced at a b; in the 2nd interval it proceeds onward to c, and the continued action of the rod gives rise to the second half of the elevated wave, so that now the entire elevated wave is formed. For the 3rd interval the elevated wave proceeds towards d, and behind it half a depressed wave is formed at a b. In the 4th interval the elevated wave advances and occupies the space edc; the half a b of the depressed wave proceeds to c, and its second half is formed at a b by the continued receding of the rod. Thus after the lapse of these four intervals a complete wave, consisting of one elevation and one depression is formed, which occupies the whole length of the canal. In the 5th interval, the elevated wave is reflected from the end A, and the direct and reflected halves being now superposed, the wave occupies only half the space it did before, and the surface of the water is raised to double its former height; the following depressed wave advances to bcd, and behind it at a b is a the half of a new elevated wave formed. In the 6th interval an elevated wave is completely reflected and now occupies the space <u>cde</u>; the depressed wave bcd having advanced at the same time occupies the same place, and the mutual interference of these opposite waves reduces the surface of water in that extent to a perfect level; in the meanwhile the elevated wave <u>abc</u> is entirely formed. In the 7th interval, the 1st half of the depressed wave is reflected from A and conspires with its unreflected half to depress the surface to a double depth; the reflected elevated wave cde proceeds to <u>dcb</u> and conspiring with the direct elevated wave which has proceeded from abc, forms a wave of double height <u>bcd</u>; ab is the newly formed half of a depressed wave. In the 8th interval, the depressed wave, the superposed halves of which before occupied the space ed, now occupies the space <u>edc</u>, and as the direct elevated wave dcb advancing occupies the same space, their mutual interference reduces the water to a level; the reflected elevated wave dcb proceeding in the contrary direction to the direct wave which just before conspired with it, in a similar manner interferes with the depressed wave <u>ab</u> which has just advanced to <u>abc</u>, and reduces this space also to a level; thus for an instant the whole surface of the water in the canal will be in the same level plane. In the 9th interval the direct and reflected waves conspire, so as to produce a depressed wave of double depth, and halves of elevated waves of double height. In the 10th interval the interference of the waves again produces a level surface. And in the 11th interval the waves conspire as they did in the 9th interval, but in such a manner as to produce conspiring waves in an opposite direction. If the motion be continued beyond this, the waves will alternately conspire and interfere, and give rise to that oscillatory motion which was visible in the canal, and the successive periods of which are represented in 9, 10 & 11.

This other diagram represents the successive steps in the formation of a stationary undulation, when three single waves occupy the length of the canal. It is unnecessary to go over them.

I will now apply the same considerations to the aerial waves, and endeavour to explain the way in which the various modes of oscillation in a cylindrical column of air are produced.

Let us take a tube open at one end and closed at the other, and let us suppose that a vibrating plate produces in the air which is in contact with it the regular alternations of condensations and rarefactions necessary for the production of sound. These alternate motions of the section of air at the orifice of the tube will give rise to a series of waves, alternately condensing and rarefying, which travel with uniform velocity towards the bottom of the tube, are thence reflected and return upon their former path with the same velocity. Let us consider the series of sonorous waves when the middle of one of them is in contact with the bottom of the tube: the commencement and termination of this wave coincide, and so also do all the extremities of the incident and reflected waves. These coincidences take place here, here, here and here, which are distant from each other by the length of a wave. In these pictorial representations the densities of the elementary parts of the wave are represented by the ordinates of x curve and also by the various shading of the length; the condensed waves are represented by the elevated parts of the curve and by the deepest shade, and the rarefied waves by the depressed parts of the curve and the lightest shade. The motions of the elementary particles of the wave are represented by the o curve, those to the right are indicated by the portions of the curve above the neutral line, and those to the left by the parts below it. At the moment we are considering the curves which represent the densities of the direct and reflected waves correspond, and each point of the column of air is activated by conspiring effects; the points where the curves cross the neutral line also coinciding in both waves, the natural state of density will still be retained at this point in the combined wave. Let us now examine what happens when the waves continue their motion of translation through this point; if a condensed wave passes in one direction a rarefied wave always passes in the other; and as the parts of these waves which pass simultaneously through this point are at the same distance from their origin, it follows that the compressions and dilatations are equal, and consequently that they mutual destroy each other. It is easy to see that the same holds good for these other similar points. But through all the intermediate points there passes always a direct and reflected wave of the same nature, the effects of which conspire. Thus, these points are the only ones which remain in a state of uniform density, and the condensations and rarefactions manifest themselves only in the intermediate points.

Independently of the condensations and rarefactions, displacements of the particles are also produced which it is of importance to examine. It is easy to see that at the bottom of the tube and at these points, the motions of translation of the direct and reflected waves are equal and contrary, and consequently, that they destroy each other, so that these sections which successively experience every degree of compression and dilatation, remain motionless. But this is not the case with these sections; the parts of the waves that traverse them at the same time are always of a contrary nature, and as their motions of translation take place in opposite directions, their effects conspire; so that these sections which remain in a permanent state of density, undergo periodic oscillatory motions.

Thus that state of motion is ultimately produced which is here represented.

Attempts have been made to fix these Aeolian sounds and to introduce them into musical instruments to be played upon like the Organ and Piano-Forte. In a report made some years ago by the French Academy on the discoveries of Chladni, one of the reporters states that about 50 years since he heard at Paris an instrument with keys like the Piano-Forte, called the Aeroelavichord, in which the sounds were produced by directing currents of air on metal strings. The sounds are described as being of great beauty, but deficient in expression and answering very slowly to the touch. A similar instrument, probably the same, is also slightly mentioned by Chladni, in his classification of musical instruments, under the name of Anemochord.

The construction of these instruments was however never made public; but the late Professor Robison of Edinburgh has described in the *Encyclopedia Britannica*, some experiments made by himself of a similar kind. "The writer of this article, says he, has attempted to regulate these sweet harmonic notes, and to introduce them into the organ. His success has been very encouraging, and the sounds far exceed in pathetic sweetness any that have yet been produced by that noble instrument. But he has not yet brought them fully under command, nor made them strong enough for any thing but the softest chamber music." His method was this. He made a wooden pipe the section of which was a double square. A partition in the middle divided it into two pipes, along side of each other. One of these pipes communicated with the wind chest, and was shit at the upper end. The other was open at the upper, and shut at the lower end. In the partition there was a slit almost the whole length, and the sides of this slit were brought to a very smooth chamfered or feather edge. A fine catgut string was strained in this slit so as almost to touch the sides. By this arrangement, when the wind entered one pipe by the foot, it passes through the slit into the other, and escapes at the top, which is open. In its passage it forces the string into motion, and produces a musical note, having the sweetness of the Aeolian Harp.

I have repeated this experiment, and finding its principal defect to consist in the uncertainty of the note produced, which is sometimes one harmonic subdivision, sometimes another, varying apparently in the most capricious manner; I attempted to remedy this inconvenience by limiting the portion of the string on which the wind acts; and to ascertain the most accurate adjustment of the aperture to the string, the sides of the aperture were made to approach to, and recede from each other at pleasure. By these means I have succeeded in obtaining very steady and powerful sounds, which seem to be perfectly manageable, and which there would not be much difficulty to introduce into musical instruments; I am now making some experiments having this end in view. The great augmentation of sound, beyond that in Dr Robison's experiment, and those of the Aeolian Harp, arises from this circumstance, that the wind escaping from a small aperture strikes the string with far greater force, than when allowed to rush with diminished velocity through an aperture equal in length to the string. There are some very curious phenomena attending vibrations of strings thus caused to sound; there is no need of sounding boards to reinforce the tone, as is necessary when continuous sounds are produced from a string by means of a bow. The tone of a string excited into vibration by a current of air forced through a narrow aperture, entirely depends on the undulations impressed by the vibrations of the strong on the issuing air; and if an efficient sounding board be added to the string we can distinctly hear two sounds of different qualities combining together in unison. Another peculiarity of this mode of exciting vibrations is that very good qualities of tone can be obtained from very short strings, which put in vibration by a bow would scarcely produce available sounds.

Some of the French Journals have recently stated that a M. Isouard has succeeded in producing sounds from strings by the action of wind, and that he is now attempting to construct a violin in which a bellows shall supply the place of a bow. But no account of his experiments have yet been published.

IV

In this lecture I shall first enter on the consideration of the principal phenomena of vibrating chords and wires, as ascertain by observation and experiment.

If a string or wire, stretched between two fixed supports be deflected from its straight position and then let go, it will vibrate to and fro on each side of its quiescent position, until its own rigidity and the resistance of the air reduce it to rest. In different musical instruments different modes of throwing the strings into vibration are employed, like the Harp, the Guitar, and other instruments of a similar kind the strings are disturbed from their quiescent positions by the finger and then suddenly let go; in some instruments now out of use, as the Spinnet, the Dulcimer &c a plectrum was employed for this purpose, in the Piano-Forte the strings are made to sound by blows given by hammers the striking ends of which are covered with leather; in the Violin, Violoncello &c a bow (which is an instrument consisting of a bundle of horse hair stretched and rendered adhesive by rubbing with rosin) is drawn along the strings, by this means these vibrations are continually renewed and may be maintained for any length of time.

This constant relation between the length of a string and the number of its vibrations, renders the monochord an exceedingly useful instrument the ratios and intervals of sounds.

There are the ratios of the numbers of vibrations corresponding to the successive sounds of the diatonic or usual musical scale; the number of vibrations of the second note compared with that of the first is as 9:8, of the third compared with the first as 5:4 &c. The lengths of string required to produce the two sounds represented by any of these ratios, must be inversely as these numbers; if the entire length produces the first note, the second note will be produced by 8/9ths of this length, the third note by 4/5 ths of the length &c. On this monochord there is a scale of all the perfect and tempered ratios which are employed in the construction of musical scales, and the relation of any sound to that given by the entire string may be ascertained by referring to it. But of these I shall not have occasion to speak in the present lecture.

The tone of a string also changes with every variation of its tension, and in all stringed instruments the chords are brought to their proper pitch by straining them more or less. If a weight be used as the tending force, this law will be found to hold; that the number of vibrations of a string is proportional to the square root of the weight which stretches it; that is to say, if we represent by 1 the number of vibrations a string stretched by a weight equal to 1, this number of vibrations in the same time will become 2, 3, 4, &c. when, without changing its length it is stretched by the weights 4, 9,16, &c.

The frequency of the vibrations of strings of the same matter, is in the inverse ratio of their thickness or of their diameter; if we take for example, two steel wires, one of which has a diameter double that of the other, stretch them by the same weight, and make them vibrate in equal lengths, the thinner string will make twice the number of vibrations which the thicker makes in the same time.

The frequency of the vibrations of strings of different substances is inversely as the square root of their densities; if we take, for example, a copper wire the density of which is nearly 9, and a catgut string, the density of which is nearly 1; if they have the same diameter, are stretched by equal weights, and are made to vibrate in equal lengths, the number of the vibrations of the copper wire will be three times less than that of the catgut string. It is obvious that the two last laws can only refer to cords which are homogeneous in their thickness and length, and that they cannot be applied, for example, to catgut or silk strings covered with wire, which are employed for the lowest strings of the Harp, and the fourth string of the violin. The enveloped metal wire is in these cases are an inert mass which has to be moved by the elasticity of the cord, and which consequently augments the duration of the vibrations.

A cord may vibrate either in its entire length, or divided into any number of equal vibrating parts separated from each other by nodes or quiescent points. When the cord divides itself into aliquot parts, each half, each third part &c vibrates as is if it were an entire chord itself; the sound produced in any mode of vibration is in consequence inversely as the length of a single vibrating part. The gravest sound is that obtained when the entire string vibrates, alternately forming the curves here represented. When it divides itself into two parts, one half is above the ordinary position, while the other is below it, and the curves are as here shown, the sound is higher by an octave than the first. If the cord divides itself into three parts, the curves are alternately as those which are here drawn in two different ways, and the sound is sharper by a fifth than the second; if it is divided into four parts* the sound is raised a fourth. In the more numerous subdivisions the sounds are as the numbers of the parts, or inversely as their lengths; the series therefore will be as the numbers 1, 2, 3,4, &c. When the gravest sound is the first C, the series of possible sounds will be as here shown; the figures indicating the number of parts into which the string divides itself, and the characters beneath them, the musical sounds corresponding to each mode of division.

The sounds of this series are called harmonic sounds from the agreeable harmonic relations which subsist between those sounds which form the laws of and terms of the series. A sound, the number of vibrations of which is any multiple in whole numbers of another sound is called a harmonic of that sound. The series of sounds is precisely the same as that produced by the different partial vibrations of a column of air contained within a tube open at both ends, the laws of which I dilated upon last week.

To produce the sounds for which the cord divides itself into aliquot parts, a finger must be very lightly placed on a nodal point, and a violin bow applied nearly at the middle of a vibrating part. Too much pressure must not be made on the nodal point, otherwise the communication of the motion would be impeded from one part to the other, and the pressure of the bow must be much less than for the fundamental sound. The modes of division may be rendered visible by placing small bent pieces of paper on different parts of the string; those which are upon vibrating parts are driven away by the vibrations and fall off, but those which happen to be on the nodal points are immovable.

If two strings are tuned to the same note, and one of them is caused to vibrate either in its entire length or in any of its harmonic subdivisions, the other will assume precisely the same state of vibration, as may be shown by placing the pieces of paper on different parts of the string which is not touched by the bow. In this case the vibrations are principally communicated from one string to another through the board to which they are fixed; but the vibrations may also be communicated through the air, and any powerful sound at a distance which is a harmonic of the fundamental sound of the string will immediately cause it to assume the mode of vibration corresponding to that harmonic sound.

Harmonic sounds are frequently employed by performers on the Violin, Violoncello, Harp and Guitar, and to produce them the same precautions are employed which I have already noticed, that is, the string must be lightly touched at a nodal point, and caused to vibrate by striking it with the finger, or drawing the bow at the midst of a vibrating part. The harmonic sound of strings have always a much sweeter quality of tone than sounds of the same pitch produced by the entire vibrations. The extraordinary effect of those beautiful sounds must be in the recollection every person who has heard Paganini.

There was formerly an instrument in use consisting of a single long string stretched on a sounding board, and various notes were obtained from it by exciting its harmonic subdivision by means of a violin bow, and beneath the extreme nodal points of each mode of vibration frets or bridges were placed to indicate where the finger should lightly touch the string. This instrument of the 16th century was called the Trumpet Marigni from the name of its Italian inventor, and because it produced the same series of notes as the Trumpet with all the omissions and imperfects of its scale.

A much better known instrument the Harp of Aeolus, or as it is more commonly called, the Aeolian Harp, owes all its agreeable effects to the harmonic sounds of strings excited into action by the wind. This instrument consists of a rectangular sounding board on which are stretched from six to twelve strings, all of the same length and thickness and tuned precisely to the same note. When the wind acts upon the strings of this instrument is placed in the open air or in the window of a room, it excites them differently; one it causes to vibrate in its entire length, others in various modes of subdivision; and as the strings are seldom impelled to vibrate in a greater number of parts that 6, the different co-existing sounds are always related to each other as consonances; sometimes the 7th harmonic is faintly heard, and sounding with the other notes gives rise to a combination resembling the discord of the flat seventh, but more agreeable to the ear, because the ratio 4/7 is more simple than that of 9/16 for which it is the substitute.

The first popular mention of the Aeolian Harp was by Thomson in his Castle of Indolence, about the time of the publication of this poem it was first manufactured in England, and acquired a great celebrity; it was however long before described by Athanasius Kircher in his Musurgia. The sounds of strings produced by wind is mentioned in the poems of Osscair, and by Eustathe an early commentator of Homer.

Chladni states that Captain Haas of Basle exposed in the open air very long and thick wires, which gave different sounds, particularly on changes of the weather; and in Brugnatelli's Annali di Chimica for 1800 similar observations by Berrettari are recorded.

String sounded by air.

Several, or all the modes of vibration which may be separately obtained from a string may coexist. If one of the lowest strings of a Piano-Forte be violently struck, on listening attentively we hear a great number of simultaneous sounds corresponding to the different divisions of the string. But the most obvious way I know of hearing these coexisting sounds is to stretch a first violin string at the bottom of a door which does not closely fit to the ground; when the door is shut the current of air entering beneath sets the string in vibration, and if a fire be in the room, with such intensity that a great variety of sounds are simultaneously produced; by this means chords are produced as perfectly as on an Aeolian Harp with a number of strings. There can be no doubt however that in the Aeolian Harp itself chords of several notes are occasionally produced from the same string.

The observation of constant coexistence of harmonies with the fundamental sound in some cases, induced Rameau and other musical theorists to regard this coexistence as necessary not accidental, and they went so far as to affirm that a sound is musical only because it contains in itself all its harmonics; when this theory was very in vogue Jean Jacques Rousseau attacked it with this reasoning: if a sound is musical because it contains within itself all its harmonics, these harmonics being musical sounds must also contain their harmonics, then again theirs, and so on ad infinitum which is an evident absurdity. Erroneous as Rameau's theory obviously is it is still occasionally maintained in modern works on the theory of music.

That the coexistence of harmonic sounds is not necessary is evident, because in all modes of vibrations where there are nodes, we can by touching them, exclude every kind of vibration in which these parts should be in motion, and produce the mode of vibration required and the sound which corresponds with it, without mixture of any other.

When two sounds are at once produced from the same string, it vibrates in a compound curve the form of which may be calculated by adding together for every point of the string, the ordinates representing the motions of the elementary curves. The are some the compound curves thus formed calculated by Count Riccati. This is the compound figure resulting from coexistence of the two simplest modes of vibration, in the one the string vibrates entire, in the other divided into two parts. When the entire vibration of the cord coexists with its divisions into three parts, under different circumstances two different curves are formed which are here represented. This figure shows the combination of the two harmonic sounds, for which the string subdivides itself into three and four parts.

The vibrations of a string are not always confined to one plane, but a rotatory motion frequently co-exists with the vibrations, without at all affecting them or the corresponding sound. The coexistence of the rotatory motion of a string, with its already compounded vibrations arising from the simultaneous production of different sounds, gives rise to a very singular class of phenomena which were first observed by Dr Young. He took one of the lowest strings of a Piano-Forte round which a fine silvered wire was wound in a spiral form and contracted the light of a window, so that when the eye was placed in a proper position, the image of the light appeared small, bright, and well defined, on each of the convolutions of the wire. The chord was then made to vibrate and the luminous point delineated its path, which presented to the eye a line of light which was accurately observed by a microscope. According to the different ways by which the wire was put in motion the form of the path was diversified as representing in these figures.

All the properties of vibrating chords to which I have alluded, and which I have hitherto considered only as experimental facts, may be shown to be the necessary consequence of the following simple proposition, a result of the mathematical theory, which has been confirmed by the most satisfactory experimental evidence. The proposition is this: The Velocity of a pulse or undulation, propagated along a tended chord, is equal to that which a heavy body would acquire by falling through a height equal to a portion of the length of the cord the weight of which length is equal to its tension, that is its extending weight. Thus, if the weight or force causing the tension if the cord be equivalent to the weight of 200 feet of the same cord, the velocity which a falling body acted on by gravity would acquire by falling this height would be 80 feet in a second, and every impulse would be conveyed with this velocity from one end of the cord to the other.

The brothers Weber have put this theorem to the test of experiment by employing cotton threads about 50 feet in length, the weight length and tension of which were carefully ascertained, and measuring with a chronoscope the time occupied by an impulse in travelling from one end of the thread to the other and thence back again to the original point of departure. The results were in accurate accordance with the theorem.

The same is true with regard to the vibrations of strings, their modes of oscillations whether vibrating in their entire length or in their harmonic subdivisions, are the resultant effects of the superposition of two series of waves proceeding from opposite directions or of a direct and a reflected series. Hence, knowing the velocity with which an impulse travels along a tended chord, which the theorem I have mentioned gives us, we are able to determine the time of vibration for any given length of that chord; for a vibration is performed in the time occupied by any impulse travelling through twice its length, or which is the same thing which it occupies in its direct and reflected path along the string; and this holds good whether we consider the entire vibrations of a string or its harmonic subdivisions. Hence, it is obvious, that the time of vibrations of a string depends on two circumstances, 1st it is directly as the length of the string, or part of the string, that vibrates, and 2nd it is inversely as the velocity with which an impulse is transmitted along that same string. The tension and weight of the string influence the time of its vibration, and consequently the pitch of its sound, by affecting the velocity with which the impulse is transmitted along the cord; thus, the tension being increased by any given quantity, the thickness of the cord remaining constant, the modules of tension, that is the length of cord the weight of which is equal to the tension, must increase in the same proportion as the tension itself; but the velocity of a falling body is as the square root of the space through which it falls, and it being asserted in the theorem that the length of the modulus of tension corresponds with this space, it follows that the time of vibration of a cord is inversely as the square root of the tension. Again, if the weight of the cord be increased by any given quantity, the tension remaining the same, the modulus of tension must in order to preserve the same weight decrease by the same quantity; and the velocity of transmission being as the square root of the length of the modulus, the time of vibration is as the square root of the weight of the cord, the length and tension remaining constant.

The sounds produced by the transversal vibrations of elastic rods must now, for a few moments occupy our attention.

Vibrating rods may be either of metal, glass or wood, and they may be made to vibrate under three different circumstances. 1st when one end is fixed, 2nd when both ends are fixed, and 3rd where both ends are free. As in the case of strings the sounds may be excited by a blow, or by the action of a violin bow.

The laws of the vibration of rods were theoretically determined by Daniel Bernoulli; and Chladni has confirmed his results by numerous experiments. Bernoulli demonstrated that for a same rod to which different lengths are successively given, the numbers of vibrations performed in the same time are inversely as the square of these lengths. This law is equally applicable to cylindrical and prismatic rods, and to laminae, whatever be the nature of their substance; provided only they be homogeneous, and equal in breadth and thickness throughout their whole extent.

This law may be verified by experiment; for this purpose it is sufficient to fix a rod in a vice, and to give to it successively lengths which shall be as the square roots of the lengths producing the required sounds on a monochord. The same sounds will then be produced in both cases from the rod and wire.

The number of vibrations is directly as the thickness of a rod or lamina, doubling its thickness therefore sharpens the sound are octave; but the breadth of a lamina has no influence in changing the sound.

A single rod, may like a single string, or a single column of air produce a series of different sounds, corresponding to different modes of vibration. These are not, however, as in the former sounds, the relations they bear to each other are as the squares of the uneven numbers. The same absolute and relative sounds are obtained, whether the rod is fixed at one end, or both ends, or is free at both ends, as may be seen in this table. The lowest sound capable of being obtained from a rod fixed at one end does not belong to this series, and this sound is to the first of the series as the squares of 2 is to the square of 5.

Chladni employed the vibrations of a lamina fixed at one end, for the purpose of determining both by the eye and the ear the number of vibrations constituting each sound. He took a plate of brass about the 20th of an inch thick, and half as much in breadth, preferably equal throughout. The frequency of the vibrations of such a plate is inversely as the square of the length, other circumstances remaining the same. He fixed one end of the plate in a vice, and allowed the length to be such as to make four vibrations in a second; these variations, visible to the eye, were compared with the oscillations of a pendulum vibrating seconds, and the length of the projecting apart was then accurately determined and divide into 2, 4, 8 and other parts. When the plate was fixed in the vice so as to project only half the former length it made 16 vibrations in a second; these vibrations are too slow to be herd and too rapid to be counted, but when the projecting part of the plate was reduced to one fourth of the original length, it made 64 vibrations, and an octave below the lowest C of the Piano-Forte was heard. Causing the plate to vibrate with one nodal point the same lengths gave respectively 25, 100 and 400 vibrations in a second. Chladni states that when the experiments were well made the results were always conformable to theory.

(Coexisting vib. Kaleidophone)

The vibrations of rods and laminae have been employed in a great variety of musical instruments. In musical boxes, rods fixed at one end are caused to sound by the action of pins fixed on a revolving cylinder; the Violon de fer consists of a circular sounding board on which and near the circumference are inserted a series of iron rods the sounds of which are elicited by the action of a bow; In Chladni's clavicylinder, the Edephone, the Melodion, the Terpodion, and other musical instruments which originated from the same source, the sounds are produced from rods fixed at one end just in vibration by the friction of a revolving cylinder, the compass in all these instruments is that of the Piano-Forte, and they are played upon in a similar manner by keys. Rods and laminae free at both ends have also been employed for musical purposes; the Staccato and the Harmonicon or glass staccato are instances of this application, and similar instruments are very prevalent in oriental countries; in these cases the rods vibrate in their simplest mode and are supported at their two nodes.

Aeolina

I shall now proceed to give some account of longitudinal vibrations, which are excided in rods, tubes, and cords. If I take a long glass rod, and holding it with one hand exactly in the middle, slightly rub one of its halves with a piece of wet cloth, a sound is produced of considerable clearness and brilliancy. The vibrations which are thus produced are evidently longitudinal vibrations. Continuing to rub the rod in the same way backwards and forwards, but with different velocities and pressures, a series of different sounds is obtained, and if the first or lowest sound of the series be represented by 1, it is easy to find that the others are represented by the successive natural numbers 2, 3, 4, 5, &c.

Thus a rod or tube free at both ends and longitudinally excited produces the same series of sounds as a strong does. A rod fixed at one end, in its lowest mode of vibration gives a sound an octave deeper than the same rod at both ends, and its series of subordinate sounds correspond with the series of uneven numbers; in this resembling a column of air in a tube closed at one end.

We can assure ourselves by experiment that rods of the same length always give the same sounds whatever may be their breadth or thickness. Thus glass rods, tubes, and laminae 6 feet in length give the same sound and whether they are thin or thick.

Rods of equal length but of different substances give different sounds. Chladni found that rods two feet in length gave the following sounds. Whalebone, A^5, Tin B^5, Silver D^6, Brass F^6, Copper nearly G^6, Glass F, Iron $C\#^7$, and Deal wood nearly D^7; other woods varied between F^6 & C^7.

The longitudinal vibrations of rods affords a ready means of ascertaining the velocity with which sound is transmitted through different solid substances. For a sonorous impulse travels over any length of a solid in the time that length makes a simple vibration; therefore the length of a single undulation multiplied by the number of vibrations in a second will give the velocity required. For example, a rod of pine wood 8ft in length, vibrating longitudinally produces a sound corresponding to C^5 or 2048 vibrations in a second; their number multiplied by 8, the number of feet in the rod, gives 16,384 which is the velocity of sound in pine wood, or the number of feet which it travels through this substance in a second.

The longitudinal vibrations of solid bodies were discovered by Chladni, who also established their principal laws and their correspondence with the vibrations of columns of air. Savart has recently examined these vibrations more minutely, and has ascertained that besides the nodal points which are in accordance with theory, spiral nodal lines coil round the rod or tube and extend from one fixed node to the other, making one or more turns. Weber has also taken up this subject, and the more remarkable results obtained by these two philosophers I shall briefly state.

Let us at first experiment on a long glass rod, from which we obtain only the fundamental sound; holding the tube horizontally, I place a small paper ring on one of its halves, and rub the other with the moistened cloth. So soon as the sound is heard the ring briskly slides along the surface, and at last stops at a certain point to which it invariably returns when it is removed from it. This point may be marked, and it forms evidently a point of a nodal line. I now turn the rod to bring another edge on the upper side; on recommencing the vibrations, the ring again slides and stops; this second point may be marked like the first. Continuing to turn the rod gradually and in the same direction, all the points of the nodal line may be successfully marked, and it may thus be demonstrated that it forms a helix with threads very wide apart, and making several turns round the tube. The red line on this rod represents the nodal helix in the first mode of vibration. Reversing the tube to place the ring on its other half we find this remarkable peculiarity, that one of these curves is not the continuation of the other, but that both starting from the middle turn in the same direction.

If a tube be employed instead of a rod, its interior surface presents a nodal line exactly similar to that traced on its external surface; but these two lines are diametrically opposite. The spiral line of the internal surface is represented on this rod by the blue line, which is so placed that when a point of the red line is above a point of the blue line is directly below it on the opposite side.

To trace the internal nodal line Savart places within the tube a small quantity of very dry sand, or a small ball of cork or any other light material.

We find nodal lines analogous to these, when instead of the fundamental sounds, one of its harmonic notes is produced. For the sound 2 there are two inversions in the direction of the helix: thus commencing from one of the ends of the tube, the helix turns from right to left, until about the first quarter of the length; then it inverts itself to turn from left to right until the third quarter; and then finally it again inverts itself to proceed as it first did from right to left. The nodal lines and their inversions for the second sound of a tube are represented on this rod. For the fourth sound four inversions take place.

In repeating the experiments of Savart, Weber has been led to some new results which are worthy of notice. He found that when he took two glass tubes of the same length but of different thicknesses the spiral coils were more numerous in the thinner tube than on the thicker one, though they both gave the same sound, as has been already explained. The number of coils which the spiral line makes has therefore no influence on the sound. He also ascertained that in very short tubes the spiral lines were converted into equidistant rings, possessing different properties in their opposite halves; one half consisting, as it were, of attracting points towards which the paper rings or particles of sand moved; and the other, as it were, of repelling points, from which the rings of paper and grains of sand were driven.

There are few subjects relating to the vibrations of elastic bodies more obscure, than the origin of these secondary nodal lines in rods and tubes longitudinally excited; which exist simultaneously with the nodal lines which theory indicates and which occasion no change whatever in the number of vibrations determined by theory. Conjectures however have not been wanting to supply the deficiency; Weber supposes that they are the nodal lines of longitudinal modes of vibration which are excited at the same time, but are so acute as to be beyond the limits of audibility; Poisson with greater plausibility states that they are probably normal vibrations of the rod or tube, similar in frequency and consequently unisonant in pitch to the longitudinal vibrations which excite them. This would account for the coils becoming more numerous as the tube is thinner; but there are other circumstances which cannot be so easily explained.

These spiral nodal lines are also always present in the longitudinal vibrations of strings, even in the thinnest; and Savart has by a very delicate experiment detected them in a spiders line.

Commencement Lecture IV: Longitudinal vibrations of strings

In the last lecture I called your attention exclusively to the phenomena of vibrating strings, I showed that the sound of a string varies with every alternation of its length, of its tension, and of its weight, and I endeavoured to explain to you the simple laws, which theory has determined and which experiment has fully confirmed, by which these changes are governed. I then proceeded to prove that the sound or pitch of a string depends solely on the number of oscillations or vibrations which it makes in a given time, a second for instance, and I indicated the means by which the number of vibrations corresponds to the notes of the musical scale may be ascertained. (The musical notes, and the numbers of vibrations made by a string producing them, are represented in this table.)

Our attention was next occupied with those sounds of a vibrating string, which are produced by its spontaneous division into different numbers of aliquot parts, each part vibrating as a separate string; the series of sounds which thus arise was shown to correspond in the number of their vibrations with the series of natural numbers. The sounds of the Trumpet Marine, of the Aeolian Harp, and the Harmonica notes of stringed instruments, were all shown to depend on this cause. In the concluding part of the lecture I explained in what manner the velocity of an impulse travelling along a stretched cord has been ascertained, theoretically and experimentally, and I stated that the time of one complete vibration of a string is equal to the time of a pulse running over double the length of the cord. I refer to these general results of what I have already fully dilated upon, because it will be necessary to bear them in mind in order fully to understand the subject on which we are about to enter.

In all the phenomena of vibrating strings which have yet been offered to your consideration, the particles have been put in vibration perpendicularly to the axis of the string, thus causing the string during its motion to assume curved forms on each side of its position of equilibrium. These were formerly called the transversal vibrations of the string; Savart calls them normal vibrations, and this expression is now generally adopted, for reasons which will be evident hereafter.

The longitudinal vibrations of strings were discovered by Chladni about 50 years ago; they consist of alternate contractions and dilatations of the string or of its aliquot parts; the vibrating particles moving in the direction of the axis alternately to and from the nodal points and fixed extremities. The vibrations are not visible whilst the string is sounding as is the case when the normal vibrations are produced. In the most simple longitudinal motion, the entire cord alternatively moves towards each of the fixed points (D). The second mode of longitudinal motion is that in which the cord divides itself into two equal parts, which tend alternatively towards the nodal point at the middle and the fixed extremities (D); in the third mode of longitudinal vibration, the motions of the parts are alternatively as here represented (D). This series of sounds is exactly the same as the series of harmonic sounds of a string vibrating in the ordinary manner or normally; that is the numbers of vibrations are related to each other as the numbers 1,2,3, 4 &c.

There is another point of resemblance between the longitudinal and the normal vibrations of a string; in both cases the sound or number of vibrations is inversely as the length of the string; but in other respects they greatly differ; the tension and thickness of the string materially influence the normal vibrations, but the longitudinal vibrations are not affected by these circumstances, and the longitudinal vibrations are so greatly sharper than the normal vibrations produced by the same string, that to obtain appreciable sounds we are obliged to make our experiments with strings of very considerable length. To obtain the harmonic sounds it is requisite to use strings 40 or 50 feet in length.

To produce these sounds a vibrating part of the string must be rubbed longitudinally with a violin bow, held very obliquely or with a piece of cloth roughened with resin. To obtain the divisions of the cord into aliquot parts, a node of vibration must be slightly touched with the finger.

The length and the materials of which it is made are the only circumstances that affect the longitudinal vibrations of a string. A brass wire gives a sound sharper by a sixth than a catgut string, and the sound of a steel wire surpasses that of a brass wire nearly by a fifth.

Chladni concluded because the laws of these two kinds of vibration are very different from each other, that there exists no relation between the normal and the longitudinal vibrations of the same string. But Poisson has, by the application of that powerful of analysis of which he possesses such complete mastery, discovered a very simple and beautiful relation between these two orders of vibrations. When a weight acts on a string it lengthens it in a small degree and the elongation of the string is proportionate to the tension. Poisson's proposition is this: The square of the number of longitudinal vibrations, is to the square of the number of normal vibrations, as the length of the string is to the elongation which it has experienced in virtue of its tension. Now as the length of the string is very great compared with the increase of length occasioned by the tension, so the number of longitudinal vibrations is very great compared with the number of normal vibrations.

There is another way of putting this proposition equally simple and comprehensible; it is this: The square of the number of normal vibrations is to the square of the number of longitudinal vibrations as the tension which acts on the string, is to the tension which would be required to double the length of the string, supposing the law of its extension to be constant. This very simple relation between the number of longitudinal vibrations and the number of normal vibrations of the same string has been experimentally verified by M. Cagniard de la Tour on a very long string, the normal vibrations of which are visible and sufficiently slow to be counted.

Longitudinal Vibrations of Rods IV

These vibrations, whose existence, modes of division, and relations of tone were first discovered by me, follow the same laws as the vibrations of the air in wind instruments. According as the rod is free at both ends or at one end*, or is fixed at both ends, the tone agrees with the series of even or uneven numbers. They are very obviously higher than those of the transversal vibrations.

The tones of different substances appear to be proportionate to the square root of the longitudinal rigidity, and as the inverse square root of its gravity. In the same mode of vibration, and the same substance, the tone is inversely as the length, and depends neither on the thickness nor breadth of the rod. In the substance which gave the highest tone, tried by me, it was something more than four octaves higher, then the tone of a column of air of the same length in the same mode of vibration.

The Transversal Vibrations of Rods IV

On comparing together the results of Chladni's experiments on rods 1st fixed at both ends, 2nd fixed at one end and free at the other, and 3rd free at both ends, it will be observed that the same series of sounds is produced in all these cases, and if the same rod be employed identical sounds are obtained. The same sound in the different series is not accompanied by the same number of nodal points, and no analogy between the same sounds of the different series has yet been pointed out. There is however an obvious one viz: that whenever the same round is produced the rod assumes the same number of curvatures, or which is the same thing there are the same number of quiescent points in the axis (considering as quiescent points both the nodal and the fixed points).

Thus it appears that rods of the same length, and assuming the same number of curves, give the same sound, whether one or both ends be fixed or free, and consequently the form of the curve, other circumstances remaining the same, does not influence the sound.

The sounds of the above series are not as might at first be supposed inversely as the squares of the number of curved portions, but they increase in a more rapid progression nearly agreeing with the squares of the odd numbers, 3, 5, 7, 9 &c; indicating that besides the increase of elasticity from diminished lengths, it is also increased by the number of bends of the curve.

If instead of fixing a rod at both ends, these ends be made merely to, rest against, a solid body, the curves will be the same as in the former case, but the first sound will be lowered a major ninth and the sounds of the series will be merely inversely as the squares of the number of curved portions; this last circumstance indicates that the cause of increased elasticity from more numerous bends of curvature is here absent.

On comparing together these two series of sounds it will be observed that the difference between the corresponding sounds of a rod pressed at both ends, and one fixed at both ends is a constant number multiplied by the number of curved portions.

	P.P. F.F	
1^2	$4+5=9$	$F\#^1 + VIII + II = G\#^2$
2^2	$16+10=26$	$F\#^3 + 6 = D^4$
3^2	$36+15=51$	$G\#^4 + 5 = D^5$
4^2	$64+20=84$	$F\#^5 + 4 = Bb^5$
5^2	$100+25=125$	$D^6 + 3 = F^6+$
	$144+30=174$	$G\#^6 = B^6-$

The first sound of the series of a rod pressed at both ends being known, any sound of the series when both ends are fixed may be known by the following formula, in which S indicates the sound or number of vibrations, N the number of curves, and A as constant number which is to the first sound of P/P is 5:4.

$S = N^2 + NA.\ S = 4N^2 + 5N$

There are two other series of vibrations of a rod, 1st. one end fixed and the other and pressed, 2nd. one end free and the other pressed. The sounds of both these series are identical, the number of curves being the same, and are the means between the sounds of P/P and F/F

P/P ------ – F#^{1} F#^{3} G#^{4} F#^{5} D^{6} G#^{6}

F/P & U/P ------ – D^{2} Bb^{3}+ B^{4} G#^{5} D^{6}+ A^{6}

F/F ------- – G#^{2} D^{4} D^{5} Bb^{5} F^{6}+

It is worthy of remark that if a rod F/F vibrating, in 2,4, 6, 8 &c curves be divided at the central nodal point, and simple pressure be there applied the sound remain unchanged. A similar result is obtained by cutting a rod U/U at a central nodal line. Hence the progression of sounds of a rod P/F or U/P is the same as the modes of vibration with an even number of curves of F/F or U/U; but two octaves lower because the length of the rod is double the bisected portion.

A Table of all the sounds of a straight rod

Number of Curves	**U/F F/F U/U**	**U/P F/P**	**P/P**
1	G#^{2}	D^{2}	F#^{1}
2	D^{4}	Bb^{3}+	F#^{3}
3	D^{5}-	B^{4}-	G#^{4}
4	Bb^{5}-	G#^{5}-	F#^{5}
5	F^{6}+	D^{6}+	D^{6}
6	--	A^{6}	G#^{6}

There is one other sound, the lowest which the rod can produce, not included in the above table; it has one curve only and differs from all the others in having the concave part of the curves externally and having no point but the fixed point in the axis. Its sound is to the lowest of the regular series as 4:25 or $2^2:5^2$.

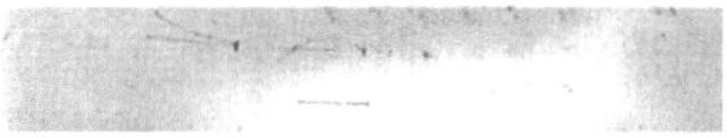

LECTURE V

The phenomena which, during the present hour, will occupy our attention are exceedingly curious and interesting. Few experiments, when seen for the first time, are calculated to excite greater admiration in the spectator than the elegant and symmetrical figures which start into existence, when sand is strewed on a plate of regular form, caused to vibrate, and which change with every different sound which is obtained from the plate.

We are indebted to Chladni for our first knowledge of thee phenomena; and the diagrams before you exhibit the numerous and diversified figures which he ascertained before experiment could be produced from plates of different regular forms. The history of the discovery is curious and I shall relate it in Chladni's own words.

"As an admirer of music, the elements of which I had begun to learn rather late, in my nineteenth year, I remarked that the theory of sound was more neglected than most other branches of natural philosophy, which give rise in me to the desire of supplying this defect, and of being useful to this part of science by some discoveries. Making in 1785 many very imperfect experiments, I had observed that a plate of glass or metal gave different sounds, when it was pinched and struck at different places; but I could nowhere find any information concerning these modes of vibrations. At this time some notices appeared in the Journals of a musical instrument, made in Italy by the Abbe Mazzocchi, consisting of bells, to which he applied one or two violin bows; this made me conceive the idea of using a violin bow to examine the vibrations of different sonorous bodies. Where I applied the bow to a round plate of brass, fixed at its middle, it gave different sounds, which compared with each other, were equal to the squares of 2, 3, 4, 5 &c; but the nature of these motions to which these sounds corresponded, and the means of producing each of them at will, were yet unknown to me. The experiments upon the electric figures which are formed upon a plate of resin, discovered and published by Lichtenberg in the Memoirs of the Royal Society of Gottingen, made me presume that the different vibratory motions of a sonorous plate might also present different appearances, if a little sand or some other similar substance were spread on the surface. On employing this means, the first figure which presented itself to my eyes, upon the circular plate of which I have spoken, resembled a star with 10 or 12 rays,* and the very acute sound was, in the series alluded to, that which agreed with the square of the number of diametral lines. Imagine my astonishment at seeing this phenomenon which no person had yet seen. Having reflected on the nature of these motions, I did not find it difficult to vary and multiply the experiments, the results of which followed with tolerable rapidity. My first memoir appeared at Leipzig in 1787."

I have quoted this account at length, because it affords an instructive example of the way in which important discoveries are frequently made. Two eminent philosophers had nearly anticipated Chladni in their researches, Galileo had noticed that light particles laid on the sounding board of a musical instrument were violently agitated at some parts, while at others parts they did not appear to move; and Dr Hooke had proposed to observe the vibrations of a ball by throwing flour upon it. Neither of these experiments were tried under advantageous circumstances, and it remained for Chladni to combine all the elements of success.

The following are a few of the general conclusions drawn by Chladni from the numerous experiments the results of which are represented in this series of diagrams.

Two vibrating parts separated by a nodal line always perform their motions in different directions, so that one is above the ordinary position while the other is below it. A vibrating part situated at the edge of the plate is always smaller than a part contained within nodal lines. These lines may cross the place in all directions straight or curved, or may return upon themselves; but they can never terminate except at the edge of the plate. The nodal lines may assume the forms of different curves according to circumstances. In general the curvatures of two serpentine lines, or of two similar lines separated by a straight line mutually approach to and recede from each other. Towards the places where this nodal lines interest each other, they always become thicker, so that the form of the vibrating parts is not angular but more or less rounded having frequently the appearance of a hyperbola.

In all the possible modes of vibration of a plate, the modal figures may be reduced to a certain number of lines which either cross the plates from one side to the other, or which are parallel to the circumference or to parts of the circumference; for example, all the figures of rectangular surfaces may be reduced to lines parallel to one or the other of the two dimensions; those of a circular plates to a certain number of diametrical and circular lines; those of a semi-circular plate to a certain number of semi-diametrical and semi-circular lines &c. All figures which do not correspond with these primitive types, Chladni considers as distortions of them, and capable of being reduced to them by the aid of certain rules.

I shall now proceed to give such practical directions as will enable any person to repeat the experiments and to render visible the various figures which are here represented; premising that the lines on which the sand accumulates are motionless, while all other parts of the plates are in vibration.

The plate must be held at a quiescent point or at several, and put in motion at a vibrating part by a violin bow, after having spread a little sand on the surface, which is repelled by the agitations of vibrating parts and accumulates upon the nodal lines.

The plates may be of glass, or of a sufficiently sonorous metal, of copper, or of brass for example, wooden plates may even be employed, but the figures will not be regular, because the elasticity is not the same in all directions. For ordinary purposes plates of glass are preferable, because it is easy to procure them of any size and thickness, and their transparency allows the places to be seen when they are touched on the under side. Very thin are preferable to thicker plates, being more easy to put in motion, and because they can assume a greater variety of forms. The size is arbitrary: for the most simple figures a diameter of from 3 to 6 inches will be sufficient; but to obtain more complicated figures, larger plates must be employed. In order that the figures shall be regular, the thickness must be everywhere the same. Plate glass is not preferable to window glass, because the surfaces are not always parallel, and it is in general too thick. It is not necessary to grind the edges of the glass plate as many persons recommend; with proper care a violin bow will last a very long time, when employed for these experiments.

If two nodal lines intersect each other in the figure, the plate must be held at the point of intersections, because if it be held at any other point of a nodal line, the vibration of the adjacent parts are too much impeded. On this account figures in which there are no intersecting nodal lines such as * are the most difficult to produce; those whose fingers are too large, or who have not sufficient strength will never succeed in obtaining them.

The plate must be held tightly by the ends of the thumb and a finger, so that the plate shall not move when the bow is applied to it. Those who have not sufficient strength, or who are unaccustomed to these experiments, may make use of a pair of nippers with pieces of cork or leather fastened to the nipping ends, or of a vice screwed to the table. An accurate eye is much more useful than any measure, to determine with exactness the most proper places at which the plate should be touched in order to produce each figure, because the thickness, form and material of a plate are scarcely ever sufficiently exact to agree precisely with the measure. The production of many very complicated figures will frequently depend on chance, but to produce less complicated figures, it is necessary to imagine that the required figure is previously visible. If the plate has not been touched exactly at a proper place, so that the figure appears less perfect than it should do, the position of the fingers must be slightly changed to place them at the exact point. When an interesting figure has been obtained by chance, which is required to be produced again, its re-production may be facilitated, by marking the places where the plate has been touched and the bow has been applied. It frequently happens that the quiescent point where the plate is held, and the vibrating part where it is put in motion, belong to more than one figure; in these cases, to exclude the other motions, it is necessary to touch at the same time, with the extremity of another finger, a place where there is a nodal line belonging to the mode of vibration required to be produced, but which ought to be in motion in the other modes of vibration. It is sometimes very convenient to employ small blocks covered at the top with cork or leather; if the plate be pressed on one of these with the extremity of the finger, and another nodal line be touched, if necessary with the end of another finger, it is the same thing as if the plate were held in the manner I have before recommended.

The bow must be held firmly in a vertical direction, and moved so as always to rub against the same place. The bow should always be applied at the middle of a vibrating part which is not to far distant from the place where the plate is held. In those cases where the same way of holding and exciting the plate may produce several sounds, care must be taken to move the bow continually with the velocity and pressure most proper to produce one of these sounds, excluding the others, for the co-existence of a different motion or sound destroys the figure which is required to be obtained. In general the more simple figures, which are those accompanied by the lower sounds, will appear more easily if the bow be moved with greater pressure and less velocity; but to facilitate the more complicated motions, the sounds of which are sharper, it will be proper to employ more velocity and less pressure. The figures will be more distinct, if (all the other precautions being attended to) the bow be drawn its full length, and if after having slightly reinforced the motion of the bow it be suddenly removed; to allow the plate to vibrate freely.

Dry sand is the best material to spread on the surface of the plate, but iron filings or any other similar material will answer the purpose. If the sand be too fine the particles adhere too much to the surface and the figures will not be distinctly delineated. Too much sand must not be placed on the surface, and it must be spread equally; if there is too great an accumulation at one part and too little at another, the sand may be transferred to the deficient places by holding the plates for a few moments inclined towards the side where the sand is wanting.

I have been rather minute in these directions because there may be many persons present who might wish to repeat these experiments. The materials are easily to be obtained, and the great variety of figures, which with a little practise may be obtained from plates of different forms, will amply compensate for the time employed in repeating the experiments. I will only add to these instructions that the easiest way of transferring a figure, which has been obtained, to paper, is to place this plate on a sheet of paper and draw its outline with a pencil, then removing it to a short distance parallel to the paper, trace with a pencil, the point of which is at right angles to the holder, the nodal lines of the figure. In this way the nodal figures may be rapidly and correctly delineated.

The unwearied industry of Chladni has then presented us with this immense mass of experiments. He produced all these figures on plates of different forms and sizes, he delineated them, and he ascertained the series of sounds which corresponded with the series of figures of each plate of different form. But the causes of these phenomena were still enveloped in obscurity, no reason was assigned why plates of different forms gave series of figures so essentially differing from each other, nor why plates of a given form presented a certain series of figures and no others. No attempts, in fact, had succeeded to reduce these complicated phenomena to simple principles, of which they should be shown to be the necessary consequences. Desirous of obtaining the key to these hieroglyphics I paid the subject considerable attention, and at last succeeded in deciphering them, and I hope that you will not consider a few minutes misemployed while I attempt to give a general notion of the mode of calculating these figures to which I was led, and the results of which are in such strict accordance with experiment that there is not the slightest bend in a line which it does not fully account for. The first part of these investigations was published in the *Philosophical Transactions* for 1833.

Let us at first attend to the modes of vibration of a plate having its length considerable in proportion to its breadth; a very little consideration will show us that they are precisely similar to those of a cylindrical or a prismatic rod, free at both ends. Increase of breadth makes no difference in the sound of a rod while the mode of vibration or the number of nodal points remains the same; and the only difference between a rod and a plate vibrating in the same way is that the nodal points of the former case become nodal lines, which may be rendered evident by strewing sand on the plate. Continuing to augment the breadth of the plate it may become ultimately a square, and we shall see that though the parallel nodal lines appear so long as the two dimensions are unequal, whatever the proportions may be, yet immediately the dimensions become equal in the square, no effort we can make will enable us to obtain these parallel lines, but instead of it this figure consists of two diagonal lines will appear, accompanied by the same sound. Why is this figure peculiar to a square plate; simple because the breadth and length of the plate being equal to each other, the same mode of vibration, accompanied by precisely the same sound, may exist in two different positions of the plate and as the middle of a side is a place of greatest vibration for each mode of division, there is no reason why, on the bow being applied there, one figure should appear in preference to the other; the plate therefore is simultaneously affected by both nodes of vibrations, one in the direction of its length the other in the direction of its breadth, and a new figure this arises totally different from those which concur to form it. It will not be difficult to show how this resultant figure is formed, by referring to these diagrams. These two figures represent the component modes of vibration, which resemble each other in every respect except position, the nodal lines in one being at right angles to those in the other. The nodal or quiescent lines separate parts vibrating in opposite directions. To fix our ideas we will consider the plates only during a single instant of its vibration, and we will suppose the parts marked P to be vibrating above the natural surface of the plate, while the parts marked N are vibrating below it. It is obvious that the results of the superposition of two similar modes of vibration will be the following.

1st. The points where the quiescent lines of each figure intersect each other remain quiescent points in the resultant figure; 2nd. The quiescent lines of one figure are obliterated when superposed by the vibrating parts of the other; 3rd. New quiescent points, which may be called points of compensation, are formed whenever the vibrations in opposite directions in each primary figure neutralize each other; and lastly, at all other points the motion is as the sum of the concurring, or the difference of the opposing motions.

Applying these considerations to the particular instances before us, viz when the primary modes of vibration present two parallel nodal lines, it is easily seen that lines of compensation can exist in the rectangular spaces marked a, as every point included within them is actuated by concurring motions; but in all the other rectangular spaces they are necessarily formed, since every point within them is affected by two opposing motions, and the compensations must occur at every point equally distant from the two rectangular quiescent lines, one belonging to each component figure. The resultant figure is thus found to consist of two diagonal lines, perpendicular to each other and passing through the center of the plate, as here represented.

In the case just considered, the two component modes of vibration, though similar, are opposed to each other, that is, while the middle vibrating part is above the surface in one it is below it in the other. A different resultant figure is obtained when both modes of vibration concur, that is when the corresponding vibrating parts of each figure move simultaneously in the same direction; the resultant figure is here represented, it is calculated in the same way as the other, and to obtain it experimentally the bow must be drawn at a corner, which is a place of greatest vibration in both figures.

In a similar manner are the resultant figures formed, when the primary figures consist of any number of nodal lines parallel to the sides; or when the primary figures consist of lines parallel to a diagonal of the square.

In these two cases; that is when the component figures consist of lines parallel to a side, or of lines parallel to a diagonal, there can be only two superpositions at right angles to each other; but if the lines of the primary figure are inclined in any other direction they may be superposed four times, because there are four positions in which such mode of vibration can be placed on a square surface. All the more elegant and complicated figures are the products of four superpositions. It would be a tedious and laborious process to ascertain a resultant figure by combining its four component modes of vibration; but the same purpose will be effectually answered by combining them first in pairs, and then combining two of these first resultants rectangularly together. Here is an example of the process; this figure has been obtained by combining together two primary modes of vibration in which the lines are neither parallel to a side nor to a diagonal; here is the same figure in a different position; combining these two figures is the same thing as combining the four primary figures; the result is here represented with the steps of the process, but time will not allow me to follow it out.

The table to my [word missing here] shows every resultant figure of a square surface, when the number of quiescent lines in the primary modes of vibration does not exceed twelve. I have carefully calculated each figure by the rules given in my memoir, and of which I have just endeavoured to give a general notion, and have shown in the table the successive processes of superposition. The first vertical row exhibits the two primary modes of vibration superposed at the required angles; one figure being represented by the black lines, the other by the red lines. The second row contains the first resultants which arise from the opposing superpositions of the preceding; and the third row those that result from their concurring superpositions; it will be seen that these all consist of straight lines crossing each other rectangularly, differing in number in the two directions, but equal when added together to the number of lines in the primary figure. The fourth and fifth rows exhibit the perfect result figures which are formed, <u>these</u> by the opposing, and <u>these</u> by the concurring superpositions of the first resultants which the plate has already been found competent to produce. These are the sheets on which the calculations were originally made.

On comparing these calculated figures with those obtained experimentally by Chladni, the greater number are found exactly to agree; and where differences exist, the causes of difference are accounted for by the theory. There is an obvious cause of error in delineating figures from experiment in this circumstance, – that the sand accumulates in the spaces where two convex curves are near and opposite to each other, the motion being there very small, so that it is difficult to ascertain whether the curves join or not; and secondly inequalities in the plate sometimes occasion lines which ought to intersect each other to appear as separate curves.

All that I have said has been in reference to square surfaces; the dissimilarity of the acoustic figures on plates of other forms arises from two causes; the superpositions differ in number being in some cases very numerous; and when the same number of superpositions occur in plates of different form, the angles at which the primary figures are inclined to each other vary in the different cases. But whatever be the form of the plate, it appears from these investigations, that every acoustic figure may be reduced to a primary figure, consisting of parallel lines only, giving the same sound. And in any mathematical investigation which may hereafter be made, these alone need claim the attention of the enquirer; and thus the analytical difficulties of the subject will be materially diminished.

I do not know whether I have succeeded in rendering this subject intelligible; but I will now leave theoretical views and return to experiment.

For the experiments I am now about to bring forward we are indebted to Savart, who has greatly extended the original discoveries of Chladni. If a sheet of any membranous substance as fine parchment or paper be carefully stretched on a frame of any regular form, and sand be strewn on its surface, any sound being produced near it will agitate the surface and cause the sand to accumulate into acoustic figures. These frames are covered with papier vegetal, a substance which Savart recommends as the best that can be employed; it is exceedingly thin and of very uniform texture. The experiments are very difficult to perform in a lecture room, because the temperature and moisture of the atmosphere continually varying, the tension of the membrane scarcely remains for a moment the same but changes ere the figure can be well formed. These are some of the most regular figures that can be obtained when the form of the frames is circular, square or triangular.

There are two ways of changing the acoustic figures of a stretched membrane, by altering its tension which the sound remains the same, which may readily be done by breathing on the surface; or by changing the sound while the tension remains the same. In a room where the state of the atmosphere is constant, the figures will change at every note, when a slow air is played on a Flute brought within seven or eight inches of the frame.

There is one important difference between the vibration of a plate of glass or metal; and those of a stretched membrane. In the former case a plate of a given form can only produce a certain series of figures; thus, between the two first figures of a square surface there is the interval of a fifth, and no figure or sound intermediate between these two modes of vibration can be produced. But, on the contrary all the principal modes of vibration of a stretched membrane may transform themselves gradually from one to the other through a great variety of intermediate figures, and a surface of this kind is susceptible of resonating to any note whatever; though with less intensity than if the figure be one of the natural modes of subdivision.

Savart has investigated a variety of these transformations as they take place on square surface. Some of them are here represented. It will be sufficient to follow one of these series.

Having succeeded in producing this mode of division, if the tension of the membranes is constant, and the sound becomes a little sharper, it may happen that the angles opposite to the summit separate as in the second figure, which if the sound continues to ascend gradually assumes the appearance of the next three figures, and finally that of No 6 consisting of four parallel lines. The same figures may change into each other by a different series of transformations as here shown. It results from these observations, that not only susceptible of producing all possible numbers of vibrations, and that each sound corresponds with a particular figure, but that the same sound or number of vibrations may be accompanied by several figures.

I shall conclude this lecture with a short notice of the peculiar phenomena attending the vibrations of curved surfaces.

When a plane surface is curved it becomes more and more unfitted to produce the acoustic figures, and when the curvature is increased beyond a certain degree, the only figures which can be obtained are those which consist of straight lines parallel to each other. It is obvious that the superpositions which have been shown to give rise to the figures of plane surface, can not take place on surfaces thus curved.

Bells and harmonica glasses are the curved surfaces most usually employed for the production of sonorous vibrations. These vibrating bodies may divide themselves into 2, 4, 6, 8 or in general into any number of even vibrating parts, separated by nodal lines which intersect at the middle where the stem or neck of the glass is inserted.

When a bell or a goblet is struck, the lowest sound is predominantly heard; but on listening attentively, it will be found to be frequently accompanied by a confused mixture of sharp in-harmonious sounds; each sound however of which the goblet is susceptible may be separately produced by touching one or more than one nodal line of the mode of vibration required, and applying the bow at the middle of a vibrating part. We cannot employ sand to render the nature of the vibration visible, because it is not a plane surface; but if water is poured into the goblet, according as the first, second or any other mode of vibration is produced, it is repelled by 4, 6 or more vibrating parts, so that the crispations which are here represented are seen on its surface. The same crispations are shown on the outside if the glass is surrounded with water. When a little dry lycopodium powder is spread on the surface of the water, the divisions into 4, 6 or a greater number of parts are rendered visible by more permanent figures.

The most simple mode of vibration, which gives the lowest sound, may be produced unmixed with any other sound, if the glass be touched, at two points distant from each other by a quarter of a circle, and the bow be applied at a distance of 45° from a nodal line, the position of which is determined by the touching finger. The four parts move in such a manner that those apposite to each other are bent inwards while, the two others are bent outwards, and vice versa, the points between the vibrating parts remaining motionless. The glass alternately assumes the curves here represented. If water is poured into the goblet the agitations upon the surface are as here shown.

The vibrating parts and nodal lines are very obviously shown in this experiment. Eight cork balls at equal distances from each other, rest against the external surface of the goblet; the alternate balls ae black and red, and when the glass is excited so that the middle of a vibrating part is in contact with one of the balls, that and the other three of the same colour are violently driven away; while the four which rest against the nodal lines are scarcely moved from their places.

When an harmonica glass is put into vibration by rubbing its edge, or a part of the circumference near the edge, with a wet finger, continuing to move in the same direction, or when the glass is made to revolve on its axis and the finger pressed upon it without moving, it also divides itself into four vibrating parts, but the position of these parts changes every instant. The mode of vibration and the sound are the same as if the glass were struck, or a violin bow were applied; but the place where the motion is produced has another relation to the position of the nodal lines and vibrating parts. When the motion is produced by striking, or by applying the bow in the direction of a diameter, this place is the middle of a vibrating part, and the nodal lines are 45° distant from it; but when the motion is produced by friction in the direction of the circumference, a nodal line passes through the place where the friction is applied.

In the second mode of vibration, a glass or bell divides itself into six vibrating parts. To produce this sound, the bow is applied at a distance of 90° from a place where a nodal line is determined by touching. The glass bends itself alternately in the curves here represented, and when water is put into the glass the crispations are as here shown. The third mode of vibration is that in which the glass is divided into eight vibrating parts. In other modes of vibration the glass may divide itself into 10, 12 or as many vibrating parts as its size and thickness will permit. When the same glass is employed the number of vibrations is as the square of the number of vibrating parts.

We have now to enter on another class of phenomena, relating to the sound of surfaces.

When, instead of sand, a light powder is strewed on a vibrating plate a different class of phenomena is presented. Chladni had early remarked that if light particles were mingled with the sand they would serve to show the centres of vibration in the vibrating parts which make the greatest excursions, and he remarked that in some figures the Lycopodium powder arranged itself into circular or straight lines, but he offered no explanation of these singular appearances. In 1827 Savart made a more extensive series of experiments on this subject, and observed the following facts. On a circular plate, when the sand showed one circular line, the powder collected at the centre, and also in a circle between the nodal lines and the edge of the plate; when the sand described more than one circle, in addition to the central point and the external circle the powder formed other circles intermediate between the nodal lines. In other modes of vibration, those of rectangular plates for instance, the lines formed by the light powder were less decided; but the general appearances led Savart to conclude that whenever an acoustic figure was formed, it was always accompanied by a secondary mode of vibration, so related to the primary one that it consisted of the same nodal lines, and of others intermediate to these. Thus the figure of a circular plate with two circular lines, is according to him, always accompanied by the figure with 4 circular lines; the figure of a circular plate with two diametrical lines is invariably accompanied by the figure consisting of one circular and two diametrical lines. When I was engaged in the investigation of which I have just given a very short outline I found it necessary to examine this question, and was compelled in the course of the examination to dissent from the conclusions of Savart; I found that if two different modes of vibrations coexisted in the same plate, the nodal lines of one superposed on vibrating parts of the other would be obliterated; that the sound supposed to coexist was imperceptible to the ear; and on applying the test I described in the first lecture, that is by fixing a polished bead reflecting a single point of light on the edge of the plate which was moved at right angles to the direction of the vibration, I found no indication of co-existing vibrations; besides Savart's theory did not account for the constant appearance of the dot of lycopodium powder in the centre of the plate whenever circular nodal lines are formed. The phenomena appeared to me more in accordance with the supposition that they arose from tangential vibrations occasioned by the contractions and dilatations in the direction of the surface which must necessarily accompany the motions of flexion of the ordinary vibrations; and this opinion seemed to be favoured by the sliding motion of the particles. The accumulations of the powder in all parts of the plate included within nodal lines, exactly accorded with this hypothesis, but there was one objection according to this supposition, on the vibrating parts at the edges of the plate the accumulations of the powder should occur at these edges, but experiment shows that they are much nearer the nodal lines. Whilst I was in this uncertainty, and delaying the publication of my paper for the want of a satisfactory solution of the difficulty, the real explanation occurred to Dr Faraday, and with that quickness of apprehension which distinguishes this eminent philosopher in all his researches, he rapidly devised a number of beautiful and conclusive proofs of the truth of his explanation. These phenomena, he has shown, depend on currents of air excited on the vibrating surface; the air is impelled upwards more strongly at those parts which vibrate the most, and as the plate recedes while the particles of air continue their upward motion by virtue of their acquired momentum, the air rushes into the partial vacuum thus formed and the powder slides along the surface from the nodal lines to the parts of greatest vibration. In the external vibrating parts, the currents sets in from the edges of the plate as well as from the nodal lines, and hence the accumulations occur between them. It would require an entire lecture to repeat all the proofs which Dr Faraday has given of this explanation, suffice it to say, that any obstacle, a piece of card for instance, intercepting the current, changes the position of the collected powder, which it would not do did the phenomena depend on the inherent motions of the plate; the lycopodium lines disappear in the vacuum of an air pump when exhausted to 28 inches; and heavy sand presents the same phenomena when plates are caused to vibrate under water.

There is still another class of phenomena depending on the vibrations of surfaces, which will require a few moments consideration. These are the beautiful reticulated appearances which are presented on the surface of a thin film of any liquid substance when spread on a vibrating plate. My first experiments on this subject were communicated to the Academy at Paris in 1823. To examine these phenomena.

Part (End) lecture V

Vibrations of Surfaces of Unequal Elasticity

There is not one of the physical sciences which is not capable of affording elucidation to almost every other. Of this I shall in the course of the present hour adduce give a remarkable example. No two sciences appear, on a casual glance, to have less mutual connexion, nay even to be more widely separated, than Acoustics and Crystallography; yet we shall shortly see that the knowledge derived from one has been employed to illustrate the other; and that a mode investigating the molecular structure of crystallized bodies has been suggested by the laws of vibrating surfaces, which if pursued with the industry and success with which the enquiry has been opened, must lend to the most valuable results.

M. Savart, to whose name is so intimately connected with the advancement of the science of sound, in almost every department is also the author of these instructive researches.

The only means hitherto employed in order to obtain accurate information respecting the intimate structure of crystals has been, in the first place, by cleavage for both opaque and transparent substances; and 2ndly by the phenomena exhibited by polarized light in its transmission through them; a most valuable mode of investigation but only applicable to transparent bodies.

The first of these means has informed us that crystallized bodies are aggregates of laminae parallel to certain faces of the crystal; but it does not indicate the elastic state of these laminae, nor the force with which they adhere together. The second, much more powerful as an instrument of experimental enquiry than the first, reveals to us actions which depend on the internal arrangement of the particles themselves, and enable us in some degree to scrutinize the delicate refined and invisible architecture of nature.

According to the results of Savart's experiments, crystallized substances resist compression with different degrees of elastic force, according to the direction in which it is attempting to compress them; and all the phenomena depending on the elasticity are affected by this cause. Thus the nodal figures of plates of exactly equal dimensions vary in form and position, and the sounds in pitch when they are cut from different directions of the crystal.

If a circular plate, of equal thickness throughout, and perfectly homogeneous be made to vibrate so as to produce the most simple figure, that consisting of two diameters intersecting each other rectangularly, it is obvious that the figure may place itself in any direction, its position being determined alone by the print at which the plate is excited; thus being always the middle of a vibrating part. If we now take another plate, still perfectly circular and of equal thickness, but possessing in its plans a different degree of elasticity in two perpendicular directions, a plate of wood for example cut parallel to the fibres, the symmetrical disposition round the centre no longer exists, the nodal lines place themselves invariably in the directions of the greatest and least resistance to flexion, and if the bow be afterwards drawn at the extremity of either of these lines the figure will present itself as a hyperbola having for its second axis that diameter which is in the direction of the greatest resistance to flexion. It results from this, that by forming, with different substances, circular plates of very equal thickness, we may by the fixed or indeterminate position of the acoustic figure ascertain whether the properties of the substance are the same in all directions.

To discover the laws of these phenomena Savart commenced with the most simple case, that in which the elastic state of the body is different only in two directions. He took a cylinder of beech wood with the annual layers concentric with the circumference, which is here represented. AY is the axis of the cylinder, BCDE a plane passing through the axis, and nn a line perpendicular to this plane. It is obvious that plates taken perpendicularly to BCDE, and according to the different directions 1, 2, 3, 4, 5 &c ought to present different phenomena, since they all contain in their plane the axis of least elasticity nn', and the resistance to flexion goes on increasing in the direction of the figures, until the plate coincides with the axis of greatest elasticity AY.

In the first plate, all being symmetrical round the centre, the figure places itself indifferently in every direction. But on examining the others it will be found that the perpendicular lines can place themselves only in one position; and as the differences of elasticity in the two directions become greater, the hyperbolic lines of the other figure become straight and the two sounds are separated by a wider interval.

After this more simple case he examined the changes of form and position in the figures, and the variations of sound, in circular plates taken from a block of wood in which there were three rectangular axis of elasticity. In order to represent more easily the relative positions of the lines round which the different series of plates were cut, and the relations they have to the planes of the ligneous layers, as well as with the direction of their fibres, he refers them to the edges of a cube AE, the face AXBZ of which is parallel to the ligneous layers, and the edge AX to the direction of the fibres; these considerations allow the edges AX, AY, AZ to be regarded as the three axis of elasticity. This series represents the successive changes in plates perpendicular to the face AXBZ, and taking round the axis of mean elasticity; this, those which take place when the plates are taken round the axis of least elasticity and perpendicular to the face AYCZ, and this series is taken round the axis of greatest elasticity and perpendicular to the face AYDZ. To ascertain the relative resistance to flexion in wood, according to them three axis of elasticity, Savart cut three small square prismatic rods of precisely equal dimensions, in these three different directions, and ascertaining the number of their vibrations when caused to sound to be respectively the first C, the first G and the third C, and guided by the known law that the resistance if to flexion is as the square of the number of oscillations, he found this resistance to be in the three different directions as 1, 2¼ and 16. It is evident that the elasticity in every other direction is intermediate between the directions just particularized.

The fourth series consists of sections made round the diagonal AD, and perpendicular to the plane BCYX; and the fifth row is a series cut, round the diagonal AE and perpendicular to the plane rst. Time will not allow me to particularize the many instructive changes which take place in these different series of plates. I must therefore confine myself to a few of the general results which have been obtained by a comparison of the various experiments.

1st. When one of the axis of elasticity happens to be in the plane of the plate, one of the nodal figures always consists of two straight lines intersecting each other rectangularly and one of these lines is invariable placed in the direction of this axis; the other figure is then formed by two curves which resemble the two branches of a hyperbola. This rule is exemplified in the first four series.

2nd. When the plate contains none of the axis in its plane, the two nodal figures are always hyperbolic curves; straight lines never enter into their composition. The truth of this rule is evident by reference to most of the figures of the 5th and sixth series.

3rd. The plate which gives the sharpest sound, or which is susceptible of producing the greatest number of vibrations, is that which contains in its plate the two axis of greatest and mean elasticity; and the plate which is perpendicular to the axis of greatest elasticity is that which gives the lowest sound or which makes the fewest vibrations.

4th. When one of the axis is in the plane of the plate, and the elasticity in the direction perpendicular to this axis is equal to that which itself possesses, the two nodal systems are similar; they are each composed of two straight lines intersecting each other rectangularly, and are situated 45° from each other. In a body possessing three unequal axis of elasticity there are only two sections which have this property (No 3–2nd row).

5th. The first axis of the nodal curves coincides always with the direction of the least resistance to flexion; it hence follows that, when this axis in a series of plates is placed in the direction previously occupied by the second axis, the elasticity has become relatively less in the former direction.

(2ᵈ series)

6. In a body which possesses three unequal axis of elasticity, there are four planes in which the elasticity is so distributed that the two sounds of plates parallel to these planes become equal, and the two modes of division transform themselves gradually from one to the other, turning round two fixed points which Savart calls nodal centres (No 3–1st series).

7th. The number of vibrations is only indirectly connected with the figure, since two similar nodal figures, as No 3 of 2ndsʼ, and No 3. of 5thsʼ are accompanied by very different sounds; whilst, on the other hand, the same sounds are produced attended by very different figures, as in the instance of No 5. 2dS.

A more general consequence which may be deduced from these different facts is that when a circular plate does not possess the same properties in every direction, or in other words when the parts which constitute it are not arranged symmetrically round its centre, the modes of division of which it is susceptible assume positions determined by the peculiar structure of the body; and that each mode of division considered separately, may always, undergoing however modifications more or less considerable, establish themselves in two determined positions, so that it may be said that in heterogeneous circular plates all the modes of division are double.

By the aid of these data, we may form some idea of the elastic state of crystallized bodies by submitting them to the same kind of research. Savart has done this for rock crystal; and I shall proceed to give a short notice of his experiments.

Rock crystal ordinarily occurs under the form of a hexahedral prism, terminated by pyramids with six faces. Although this substance is not capable of cleavage by ordinary means, it is admitted from analogy that its primitive form is a rhombohedrum such as would be obtained were the crystal susceptible of cleavage parallel to three non adjacent faces of the pyramid. Guided by these notions with which we are furnished by minerology, it is obvious that circular plates taken parallel or perpendicular to the axis, parallel to a cleavable or one-cleavable face of the pyramid &c should present different phenomena with regard to sonorous vibrations, since the cohesion and elasticity are not the same in these different directions.

After having proved that sections which are parallel to each other have precisely the same elastic state, Savart proceeds to examine the phenomena presented by plates cut in different directions with respect to the axis of the crystal.

The first series consists of plates cut parallel to the axis of the hexahedron. All these planes are far from possessing the same elastic state; but if we take any three of these plates; limited only by this condition that the angles they form with each other are equal then their elastic state is the same. Thus 1, 5 & 9 have the same figures and yield the same sounds. There is one very remarkable circumstance relating to this series; all the plates which compose it affect polarized light exactly in the same manner and no differences can be observed in the phenomena when examined by either of these plates; whence we must conclude the knowledge we acquire respecting the structure of crystals by means of polarized light, is of a different order to that are attain by sonorous vibrations. It appears, from this, that the last process more especially indicates the elastic state, and the force of cohesion of the constituent particles in different directions of all the planes, whilst the phenomena of polarized light, depending more particularly on the form of the particles, and on the positon they assume round their centre of gravity, are to a certain point, independent of the mode of junction of the different plates which compose the crystal.

The second series is composed of plates which are cut round a line joining two opposite edges of the prism, and pass through the axis of the crystal. The transformations of the nodal lines in this series are exactly analogous to those which are observed in a series of plates cut round an intermediate axis in bodies which possess three unequal and rectangular axis of elasticity.

The third series is formed of plates which are cut round a line perpendicular to two opposite faces.

It is impossible for me to enter into the minute details of the curious results which Savart has obtained with rock crystal, Iceland spar, sulphate of lime and other minerals/crystals; my intension has been merely to give such a general notice of the experiments, as may attract attention and stimulate curiosity. Those who wish to acquire more definite notions on the subject may consult the 40th vol of *Annals de Chimie*.

The relations which exist between the modes of division and the primitive form of crystals, allow us to presume that the primitive forms of certain substances, which do not submit to a simple mechanical division, may be determined by sonorous vibrations. It is equally natural to suppose, that less imperfect notions than those we now possess respecting the elastic state, and cohesive force of crystals, may elucidate many peculiarities of crystallization: for example, it may not be impossible that the degrees of elasticity of a given substance are not exactly the same for the same direction referred to the primitive form, when the secondary form is different; and if this should prove to be the case, as Savart has been induced from several facts to suspect, the determination of the elastic state of crystals will lead to the explanation of the most complicated phenomena of the structure of bodies. And finally, it seems that the comparison of the results furnished with regard to the constitution of bodies on the one hand, by the means of light and on the other, by the means of sonorous vibrations, must necessarily concur to the progress of the science of light itself, as well as to that of sound.

Though we discover a great analogy between the phenomena presented to us by rock crystal, and those which have been observed in bodies in which the elasticity is different in three perpendicular directions, yet the experiments before us sufficiently indicate that rock crystal cannot be included in the number of substances having three rectangular and unequal axis of elasticity, and still less in the number of those all whose parts are arranged symmetrically round a single right line. For the same phenomena are constantly produced in three different positions; and it appears that all may be referred to the different cleavage planes and to the faces and the edges of the primitive rhomboid.

Thus all the plates cut parallel to the natural faces of the hexahedral prism possess exactly the same properties and these properties are very different to those of the plates equally parallel to the axis, but which are perpendicular to two faces of the prism. Likewise, the plates parallel to the planes of cleavage of the pyramid give the same sounds and produce the same acoustic figures; whilst the planes parallel to the three other faces present different figures to those of the preceding plates. It appears therefore to result from this identity of phenomena for three distinct positons, that there are in rock crystal three systems of axis or principal lines of elasticity.

After comparing together all the phenomena presented both by rock crystal and wood, Savart infers that the directions of these axis of elasticity are as I am about to state. The intermediate axis of elasticity he considers to be the greater diagonal of the face of the primitive rhomboid, which is the same as the line perpendicular to the axis of the prism and joining its two opposite edges; and he is confirmed in this induction by the almost exact resemblance between the series of plates cut round this line, and the series of plates of wood cut round the axis of mean elasticity in that substance. The axis of greatest elasticity he states to be parallel to the plane of cleavage of the pyramid, forming the smaller diagonal of the face of the primitive rhomboid; and the axis of least elasticity to be included in the diagonal plane which contains the intermediate axis, and is perpendicular to it. Thus, to define their positions more concisely, the axis of greatest elasticity and the intermediate axis are both contained in the plane which forms the face of the rhomboid, and are perpendicular to each other, and secondly, the intermediate axis and the axis of least elasticity are contained in the diagonal plane and are also perpendicular to each other.

The only striking difference which there appear to be between the structure of calcareous spar and that of quartz is that in the first of these substances, the smaller diagonal of the rhomboidal face is the axis of least elasticity, whilst in the latter it is the axis of greatest elasticity. To prove the accuracy (Fig 6) of this assertion nothing more is requisite, than to cut in a rhomboid of calcareous spar, a plate parallel to one of its natural faces, and to examine the disposition of its two nodal systems, one of which consists of two lines crossed rectangularly which always place themselves on the diagonals of the losange, and the other is formed of two hyperbolic branches; but with this peculiarity that it is the smaller diagonal which becomes the first axis of the hyperbola, whilst it is the second in the corresponding plate of rock crystal. It is remarkable that this difference is found in two substances, one of which, quartz has positive double refraction, while in Iceland spar the double refraction is negative. How far this difference of structure influences the phenomena of light in these substances, and whether it affects other substances in a similar manner, remains to be determined by future experiments. I may however observe that a similar difference is remarked in the effects dilatation by heat; in carbonate of lime the dihedral angle of the faces of the summit diminishes; in quartz on the contrary this angle increases.

Such are the principal facts regarding the structure of crystals observed by M. Savart.

I shall conclude with a few words respecting the structure of metals and other substances, to which Savart has applied the same method of investigation; it is not improbable that some of these may suggest useful applications to the arts.

Melted metals have been hitherto considered as the solid substances which most nearly approach the conditions of homogeneity; they have been considered as aggregates of an infinite number of small crystals united together without any order. But if we examine, by means of vibrations, small circular plates either cast in moulds, or cut from large masses, of metal, or taken from rolled sheets, we shall find that they present the same phenomena with reference to sonorous vibrations, as if they had belonged to a fibrous or to a regular crystallized body. For if we endeavour to produce the mode of division consisting of two crossed lines, it will be seen that the figure can only place itself in two determinate positions, and will almost always be distorted into two hyperbolic curves which are accompanied by sounds, which sometimes differ only insensibly from each other, but sometimes have the interval of a third, a fourth or even a fifth between them. But though each of the plates cut from a mass of metal appears as if it belonged to a regularly crystallized body, we cannot consider the entire mass as a crystal. The distinctive character of crystallised bodies is that their structure is found to be exactly the same for all parts of the same plane, and for all parallel planes taken in any direction with reference to the faces of the crystal; but this symmetry does not hold in masses of metal, for if we take a cylinder of metal and cut a number of parallel planes perpendicular to the axis, we shall find that there is no correspondence between the nodal figures, and the sounds of the different plates, and the same want of uniformity is observed when a number of discs are cut out of the same flat plate of metal. These and many other experiments of the same kind clearly show that metals do not possess a homogeneous structure, but they also prove that they are not crystallized regularly. There only remains one supposition to be made, that they possess a semi regular structure, as if, at the moment of solidification, distinct crystals were formed within them, of very considerable size, but the similar faces of which are not turned in the same directions. According to this idea, metals should be regarded as groups of crystals, each of which consider separately, presents a regular structure, whilst the whole mass appears entirely confused. It is a natural consequence of this structure, that the differences of elasticity in the same substance will appear greater in proportion as the circular discs which manifest them are of smaller diameter, since the number of crystalline systems which these plates will contain will be fewer as their decimetre is less. Experiment perfectly confirms this, and these inequalities of structure are consequently observable in large plates.

It does not appear that there is any difference between the structure of metal discs cut from large masses, and those formed by melting the same substance in circular moulds; nor does hammering, or tempering, or sudden cooling, or passing an electric current through the metal in a state of fusion, exercise any appreciably influence on the general disposition of the phenomena. Among the numerous means which Savart has employed, one only has succeeded in impressing perfect uniformity of structure on plates of metal. By communicating a rapid succession of minute impulses to the mould while the metal is solidifying, the formation of the crystalline systems is disturbed, and so great an uniformity of elasticity is thereby determined, that the circular plates which have been so treated almost always produces only a single sound, and the nodal system of crossed lines places itself indifferently in any position. Savart remarks that it would be curious as well as important to examine whether the metals, the crystallization of which has been thus disturbed, are as tenacious as in the ordinary state, and to see whether they do not acquire by the process some new properties which would render them more fitter for certain operations in the arts. One application immediately suggests itself; it is at the present time considered so important an object to obtain for optical purposes, glass of perfectly uniform structure, that it would be worth while to try this process in addition to the ingenious ones which have been already employed, for effecting so desirable an end.

From numerous experiments it appears that the deviation from uniformity is greater as the metals are purer, and is much more considerable in tin, lead, and zinc metals, which readily melt, than in copper, bismuth, iron, antimony and silver. It is always very small in alloys; thus the two sounds of brass, and particularly of bell metal discs are so near each other that it is almost impossible to distinguish them. As bell metal and brass enter more readily with vibration, and continue their vibrations longer, than the pure metals, it is evident that uniformity of structure is a more important condition for a sonorous body that either density or specific elasticity. I have had, in my experiments on the communication of sound through solid conductors, frequent occasion to observe the want of uniformity of structure in metal wires; both theoretically and experimentally an iron wire and a deal wood rod cut in the direction of its fibres have the same specific elasticity, and both should transmit sound with the same velocity and the same intensity; but I have found that the transmitted sounds are very sensibly enfeebled after passing through wires of iron and other metals, while on the contrary a succession of deal rods joined together, end to end, will transmit sounds with almost unimpaired intensity to very considerable distances; the uniformity of elasticity in wood is the only cause of this superiority. Professor Weber of Gottingen has made a series of experiments which also obviously shows the absence of uniformity in metal wires; weighing equal lengths of wire drawn through the same hole in a draco plate he has rarely found that they are of equal weight.

Reverting to Savart's experiments, he observes that analogous phenomena are to be found in glass, sulphur, common rosin, copal, amber, slate &c, but the differences are much less than in metal, the sounds seldom differing by a semitone. Among all the bodies which he has examined, he has only found one in which the nodal system of crossed lines can place itself indifferently in any position; this is sealing wax; in this compound substance the particles of the colouring matter, prevents the particles of resin from arranging themselves regularly.

The last observation to which I shall call your attention is one which perhaps applies to all bodies which do not crystallize regularly. It is that immediately after they are solidified, they are much less readily put into vibration than they are several hours, several days, or even several months after. It frequently happens that a plate, when at first produced only vary dull sounds, obtained with great difficulty, ultimately vibrates with such facility and energy, that its particles disintegrate and are thrown to considerable distances. Savart addresses the following remarkable instance of this property. A circular plate of sulphur was cast in a proper mould, and on endeavouring to make it sound immediately after it had cooled, it was found impossible to succeed; but after a few days some dull sounds were produced from it. The number of vibrations for a determinate mode of division was then recorded, and the plate allowed to remain at rest for one or two months; after this time had elapsed it vibrated with extreme facility, and its sound for the same mode of division had become much sharper, being more than a tone above its original pitch. It seems to follow from this, and other experiments which might be brought forward, that during the act of solidification the particles are arrested in positions from which they afterwards attempt to escape, and that they do not arrive at a state of stable equilibrium until a certain time has elapsed which is sometimes of very long duration.

In concluding this lecture, I will merely again remark that the subject I have endeavoured to explain, affords another instance, in addition to the numerous ones we already possess of the advantage of studying the difference branches of physical science in connection with each other, even then apparently the most dissimilar.

LECTURE VI

All the phenomena which have occupied our attention for the last four lectures result from the interferences of sonorous undulations when reflected upon themselves in elastic bodies of linear or superficial dimensions. In the cases of columns of air, of extended wires, and of rigid rods, the interfering waves move in directions exactly opposite to each other; and I have already fully entered upon the consequences of these interferences, as giving rise to the various modes of vibration which characterize these bodies. With respect to surfaces; which present, as we have seen, an almost infinite number of figures, mostly symmetrical, showing to the eye the limits of those vibrating parts into which the surface divides itself; these also are the effects of interfering waves; not however of two waves, merely, meeting each other in appropriate directions as in the case of linear elastic bodies, but of a number of waves crossing each other in various directions. Though this has only yet been proved with regard to the normal vibrations of surfaces, there can be no doubt that more extended enquiry will enable us to apply the same considerations to the tangential and oblique vibrations of surfaces, and to the vibrations of bodies of three dimensions, a case of great complexity as the interfering waves then cross each other in different planes. It is not too much to expect that where analysis shall bring its powerful aid to assist the inductions of experiment, we may ultimately be able to determine the figures and the sounds of bodies of any dimensions whatever, when these dimensions, the nature of the primitive disturbance and the point at which the disturbance is made, are given.

There are, besides these, other phenomena of interference to which we have not yet directed our attention; these arise from the meeting or coincidence of waves in the open air which proceed in the same or in nearly the same direction. They may be divided into two classes; those which originate from isochronous vibrations and which consequently have the same musical pitch; and those which arise from non-isochronous vibrations and have different musical tones. To the first class I shall confine myself at present.

It may be a matter of surprise to persons who have not studied the laws of sound, how it is that when different musical instruments sound the same note in different places, the ear is affected precisely in the same way as it would be had the sound originated from one point only. For it is obvious, that though the originating vibrations at the different points are isochronous or performed in equal times, yet they cannot all be synchronous or commence and terminate exactly in the same time; some must commence after the others, and it might be imagined that in this case the pulses of the different systems of undulations would not be co-incident on the ear, and thereby occasion confusion. This modification of an ingenious contrivance of Dr Young's, to which he gave the name of the harmonic sliders will enable me to show what takes place under these circumstances and how all confusion is avoided.

The curve formed by the ends of these rods may represent a system of aerial waves; the different elevations and depressions representing the various states of condensation and rarefaction. This board has an exactly similar curve upon it, and if it be placed beneath the rods, they will be raised by it and a new curve will be formed at their extremities, which will be the resultant of the two curves before separately shown. If the two curves be made to coincide so that their elevations and depressions exactly agree, the elevations, the depressions and the intermediate points will be in the same positions as before but the ordinates of the compound curve will be double what they previously were. If, on the contrary, the curves be so superposed that the elevations of one shall corresponding with the depressions of the other, which agrees with the case when a wave of rarefaction of one system coincides with a wave of condensation of the other, a level surface is obtained which indicates that the opposing waves completely neutralize each other. In this case silence is the effect of the coincidence of two sounds, as I shall presently have occasion to show, and as in the undulatory theory of light darkness is the effect of two systems of luminous waves under similar circumstances.

When the superpositions of the two curves take place in such a manner that the elevations and depressions neither correspond nor are in direct opposition to each other; the depressions and elevations of the resultant curves remain at precisely the same distances from each other as before, that is the lengths of the waves remain unaltered, now commence, at points, and, consequently at times, intermediate to those at which the component waves commenced.

I shall now proceed to the consideration of one of the most important and difficult propositions in Acoustics, but of which I shall only attempt to give the most elementary notions. The proposition is that each vibration of the sonorous body excites in the air an undulation or wave of a determined length. Let us imagine a straight tube, of equal diameter throughout, and of indefinite length, filled with air of uniform density and elasticity; and let a piston fitting the sides of the tube perform in 1" an oscillation between the two positions PP' and SS' which are distant one foot from each other. We will say nothing at present with regard to the increasing and decreasing velocities which the piston may assume in this interval.

All being at rest, let the piston start from PP' and arrive at SS'; during this motion the air in the tube is modified in a certain manner, and to study better the modifications it experiences, let us take the precise instant when the piston arrives at SS', and we will suppose that all the particles of air remain as they then are, or rather we will suppose that those which are compressed cannot expand themselves, that those which are dilated cannot approach each other, and that those which are at rest preserve their quiescent state.

If the tube were filled with a perfectly hard body, it is manifest that one of its extremities being impelled by the piston, the other extremity would be driven out of the tube at the same instant and by the same quantity. But there is no body in nature which is perfectly hard. The air is a <u>very compressible</u> fluid, and when the piston impels one of the extremities, of the column before it, the other extremity does not obey the impulse at the same moment; time is required for the impression to be transmitted to it, and as the length of the tube is supposed to be very long, we may affirm that no particle of air has issued from the open extremity of the tube, which the piston moved from PP' to SS'. Therefore the air is compressed within the tube since it now occupies a foot in length less than it did before. But besides, it is evident that it is not compressed <u>equally</u> throughout the whole of the tube; for during the lapse of a second which the piston occupied in its passage from PP' to SS', the compression has travelled only to a certain distance AA for example. That portion of the column of air which is capable of modification during the motion of the piston is called a wave or undulation, and the breadth of a wave is the distance between its two extremities SS' and AA'.

Let us now examine how the air is modified in different parts of the wave; and for this purpose let us imagine transverse sections which divide the column of air into small portions or slices, such as abcd, cdef &c all having the same thickness; as it is probable that the motions of all the particles in each section are exactly similar, it is sufficient to know what happens to one particle of each section to understand in what manner the whole mass of air is affected. Now since the air contained between PP' and AA' has been compressed throughout and reduced to occupy only the space SS' and AA', it is clear that in each section the particles have experienced two effects: 1st they have been compressed' 2nd they have received a certain impulsive velocity which has caused them to recede from the centre of agitation, or from the piston which has impelled them.

The motions which characterize a wave from its origin to its termination, may be represented to the eye by a figure; for this purpose all that is required is to raise upon the line SA which indicate the breadth of the wave a series of perpendiculars the heights of which represent the degree of compression of the corresponding sections; the extremities of these perpendiculars will form a line the curvature or sinuosities of which will accurately represent the order in which they compressions of the successive sections succeed each other. At S the height of the perpendicular is nothing, because there is no compression, and at A the same thing takes place; at X the height of the perpendicular is XX', at Y it is YY', at M it is MM &c so that the curve representing the compressions may be the semi-circumference of a circle. It is easy to understand that upon this length SA, a great number of different continuous curves may be drawn passing through S and A, as here represented; and even, one of these curves being given, a motion may always be attributed to the piston such that it may give rise to a wave the successive compressions of which are presented by this curve.

Having analysed the different modifications which the piston, while it passes from PP' to SS' in the interval of a second, is capable of impressing on the column of air, let us endeavour to see what takes place in the following moments, the piston remaining at rest at SS'. The air momentarily compressed cannot remain in this state; for the tube being open at T', after a certain time the surplus air must pass out, and the entire column return to a state of rest. Now, it is demonstrated, in mechanics, that the compression and the velocity are successively communicated in the following manner; in the first instant of the second second, the velocity passed to the right of AA', reaches a first section and at the same time the section which touches the piston returns to rest; in the 2nd instant a second section to the right of AA' is attained, and a second section before the piston returns to rest; in the 3rd instant the motion reaches the 3rd section, and the 3rd section before the piston is reduced to rest &c; so that at the end of the second second the air is at rest from S to A, and it is agitated from A to B; the length AB is equal to SA, and besides, the compression and velocities from A to B are exactly what they were from S to A. Thus the undulation advances, preserving in its passage its breadth and all its characters; at the end of the 3rd second it will be at BC; at the end of the 4th at CD, &c.

The wave I have just described, in which all the sections are compressed, and all the velocities impulsive, is called a wave of condensation.

But it is easy to see that reverse phenomena are developed at the left of the piston PP' whilst it is moving towards SS'. In this case, a greater length is given to the column of air; the 1st section rushes after the piston and is rarefied, the second section rushes after the 1st and assumes its place, &c &c; and after the 1st second, when the piston stops at SS' the rarefaction has proceeded as far as a. The wave which is thus produced is called a wave of rarefaction; its breadth is exactly the same as that of the wave of condensation which is produced before the piston; the rarefactions are at SS' and aa', and in all the sections the velocities are apulsive, that is directed towards the centre of agitation.

This wave of rarefaction is also propagated successively throughout the whole extent of the column of air, always preserving the same length and the same succession of velocities and rarefactions.

We will now return to what takes place in the column of air to the right of the piston. If the piston, after it has stopped during an imperceptible instant, returns to its primitive position PP' passing through the same velocities, it is manifest that it will excite before it, to the right of SS' a rarefied wave exactly similar to that which it excited to the left during its advance, and that this wave will immediately follow the first condensed wave, so that at the end of the second second, the condensed will be between A & B, and the rarefied wave between A & S. On the other side, on the contrary, the rarefied wave will occupy from a to b, and the condensed wave from a to s, then another motion of the piston forwards and backwards will again excite similar waves, and similarly disposed, which will run after the first and so on. A series of waves thus following each other is called a system of waves. A continuous sound will thus be produced if the motion of the piston be sufficiently rapid.

Between the sound at the right side of the tube, and that at the left, there is only this difference, that in the former it is the condensed wave which first moves onwards, and in the latter the rarefied wave.

In this account of the origin and formation of aerial undulations I have followed Pouillet; and it is easy to see that the same reasoning would apply were any vibrating body whatever substituted for the piston, and were the waves, instead of being confined in a tube allowed to diffuse themselves circularly or spherically in every direction.

The breadth of a wave is equal to the space which it travels over during the time the sonorous body makes one vibration; therefore the breadths of waves must vary according as they are excited by sounds of different pitch, or which is the same thing, vibrations of different frequency. The breadths of the undulations in the air which correspond to audible sounds vary between 32 feet and a quarter of any inch.

The breadth of the wave corresponding to any given sound may be ascertained by dividing the space sound passes over in a second, by the number of vibrations which the sonorous body makes in the same time. Thus if we wish to know the breadth of the wave corresponding to the note C consisting of 256 vibrations a second, we must divide 1090 by 256, and the result will give us 4¼ feet as the breadth of the wave required.

However sounds may differ in other respects, if they are of the same pitch, that is if they are unisons of each other, they excite undulations of the same breadth in the air; and this whether the sounds are produced by the regular vibrations of elastic bodies, or by the rapid succession of mechanical impulses as in the case of the Syren which I alluded to in the last lecture.

Other characters of sound are also found to correspond with modifications in the sonorous wave. The intensity depends on the degree of compression and velocity of the particles of air in the wave; and the timbre or distinctive qualities of sound, in all probability, depend on the order in which the velocities and charges of density succeed each other in the different sections of the air comprized between the two limits of the wave.

Thus the different qualities of sound might be represented to the eye by the different forms given to <u>these</u> curves.

Waves of all kinds, however the sounds which produce may differ in pitch, intensity or quality, travel through the air with the same velocity. Were it otherwise a concert of music heard at a distance would from the acceleration of some sounds and the retardation of others be confined and discordant. But we find that at whatever distance music is heard, it still preserves its proper harmony and time.

In what I have hitherto stated regarding the velocity of sound, I have considered only the results given by experiment. Newton and his successors have investigated the subject theoretically, and the general results of their calculations are, that the velocity of sound is uniform – is independent of the nature, extent and intensity of the primitive disturbance, and is equal to the velocity which a heavy body would require in falling through half the height of atmosphere supposed to be homogeneous. (V = 1(2gH)

Reduced to numbers, and supposing the temperature to be at zero, as we have all along done, this expression is equal to 916 feet 322. This it is seen differs widely from the result of experiment; 174 feet is one sixth of the whole amount, and is far too great a difference to be attributed to errors in the experiments which were made with the utmost accuracy. There was something therefore radically deficient in the theory; Laplace supplied this omission by supposing that the heat disengaged by the compression of the molecules of the air modified the law of its elasticity, and accelerated the propagation of sound. This, as Sir J. Herschel observes, is a really existing course, which must act to increase the velocity, and is fully adequate to do so to the extent observed.

When sound is allowed to diffuse itself especially in every direction, its intensity diminishes as the square of the distance increases, so that at the distance of ten feet from the sounding body its intensity is only one hundredth as great. But if the diffusion of sound be prevented by causing it to pass along a pipe, it may be conveyed to considerable distances with very little diminution. The speaking pipes employed to transmit messages from one apartment to another are familiar instances of this. Biot made some experiments with the water pipes of Paris upwards of 3000 feet in length and found that the lowest whisper made at one extremely was distinctly heard at the other.

This is a model of Mr Charles celebrated acoustical illusion the Invisible Girl. The sound is conveyed by pipes artfully concealed within the frame, which open opposite to the mouths of the trumpets from which the sound appears to proceed.

With respect to the velocity of sound in the different gases I need only mention that both theory and experiments concur in proving that the velocities are inversely as the squares of their specific gravities.

The length of a column of air vibrating so as to produce its lowest sound in a tube open at both ends, is, equal to the length of a single progressive wave. Therefore, knowing the velocity of sound in the air, and the number of vibrations in a second of any given sound, it is easy to determine the length of an open pipe capable of maintaining this sound; thus if 1090 the number of feet sound travels in a second be divided by 256 the number of vibrations in the note C, the result will be very nearly 4¼ feet, the length of a pipe capable of producing that note.

Reciprocally, the velocity if sound may be estimated from the length of a stationery wave in a tube; this length multiplied by the number of vibrations in a second of the sound produced gives the result required. By this means the velocity of sound in the different gases has been ascertained, a subject to which I alluded in the last lecture.

I could refer to a great number of other experiments made since to determine the rate with which sound moves through the air, but none can be so much relied upon as those made in the present century. Greater accuracy in the estimation of the time elapsed, and due attention to the influence of the wind, have been the principal causes of that close agreement between the results of independent observers in England, France, Holland, Germany and India, which is shown in this table.

These observations have been reduced to the zero of temperature or freezing point; and to ascertain the velocity of sound in air of a higher temperature, for every additional degree on Fahrenheit's scale 1.14 foot must be added to the velocity. We shall hereafter see the reasons for this correction.

Comparing together these results as Sir J. Herschel has done we may assume, without the fear of an error exceeding 2 or 3 feet, that sound travels through dry air at the freezing temperature, with a velocity of 1090 feet or 363 yards per second. At the standard temperature of 62° of Fahrenheit, it travels over 9000 feet in 8 seconds, 12¾ miles in a minute or 765 miles in an hour, which is about three fourths of the diurnal velocity of the earth's equator.

The general plan on which the recent experiments have been conducted is as follows. Two stations are chosen and their distance from each other ascertained with sufficient exactness by trigonometrical operation. At each station a gun is placed with a party of observers and each party being provided with well regulated chronometers, it is agreed before hand that a certain number of discharges shall be made from each station at equal intervals from each other, 10' for instance. In the experiment of the French commissioners, the discharges at one station were intermediate between those at the other, and commenced 5' before them; in those of the Dutch philosophers the discharges were simultaneous, both guns being fired at the same precise instant of time. These reciprocal discharges are necessary to eliminate the effect of the wind; to understand how this is effected, let us suppose the wind to blow with uniform velocity directly from one station to the other, the sound which travels in the direction of the wind will be accelerated, and which is transmitted in the contrary direction be retarded by the same quantity. The mean of the two results will consequently give the velocity of sound, unaffected by the wind's motion. The interval between flash and report must be measured by some accurate instrument capable of indicating minute fractions of a second. The French observers employed the stop watch of Bregant, and the chronograph of Rieufrec, a kind of watch having a hand which makes one revolution in a second, which by pressure upon a lever is disengaged from the wheel work and prints with ink a dot upon paper.

The experiments should be performed by night and in clear weather; and careful observations must be made of the temperature of the air, its hygrometric state and the barometer pressure.

VI

In all cases of communicated vibrations and of resonance, which in former lectures I have had occasion to refer to, the newly caused vibrations have always been isochronous with those of the original sounding body; and consequently the resonance and the original sound have always been unisonant.

Page 180

From the experiments which I am about to bring to your notice it will be seen that undulations of one order may give rise in certain cases to undulations of another order, and a sound of one pitch may be reflected as a sound of another and higher pitch.

VI

In the preceding lecture I concluded all I had to say respecting what Savart has properly called normal vibrations because in these cases the motions of the particles are normal or perpendicular to the vibrating surface. I must now call your attention to the tangential vibrations of surfaces, in which the motions of the particles are performed in the plane of the surface itself. It is impossible by any of the usual means to excite immediately the tangential vibrations of surfaces; but they may be excited in a variety of ways by communicated vibrations. Thus if a small glass rod or tube be cemented to a glass plate of any form, and a bow be drawn across the rod in a direction parallel to the plate, the sand distributed over it will assume a nodal figure, as in the case of the normal vibrations of the same plate; but the circumstances of these two kinds of vibration differ very much; in the case of normal vibrations the particles of sand are projected perpendicularly from the surface, in the tangential vibrations they creep or slide along it; the currents of air which are determined on the surface of plates vibrating normally, and give rise to those secondary figures which I referred to in the last lecture, do not occur when the plate vibrates tangentially; and though the same figure which I have here, viz one circular line, may be produced by normal vibrations, yet in this case it occupies a different position and is accompanied by a different sound. When any figure is produced on a surface vibrating normally, no difference can be perceived in it whether the sand be strewed on one surface or the other; and there evidently ought to be no difference; but it is a remarkable peculiarity of surfaces vibrating tangentially that totally different figures are obtained on the two surfaces of the plate so that none of the nodal lines of the upper surface correspond with those of the lower surface. The particles of sand move in the directions of the vibrations of the rod. This is the figure which Savart obtained when he tried this experiment.

It is not necessary that the plate should have any considerable thickness in order to exhibit different figures on its opposite surfaces; the thinnest membrane, gold beater's skin, for instance, will present the same phenomena. I have here a very thin sheet of parchment stretched on a circular frame; the centre of the membrane is traversed by a steel rod which is fixed perpendicularly to a solid block beneath; on drawing the bow across the rod tangential vibrations are excited in the membrane; a few of the figures which simultaneously occur on opposite surfaces are here represented.

These are the most easy and certain methods of obtaining tangential figures, but they are fixed, and a different plate or membrane is required for every new figure.

Tangential vibrations may be excited in surfaces, by communicating to them the vibrations of strings; the figures thus produced are not so well defined as in the former cases, but as the length of the string and its tension may be changed at will, an almost innumerable variety of figures may be produced. Here are two arrangements by which communicated tangential vibrations may be excited; in this the string passes through the centre of the plate, and the bow being drawn across the string the surface is caused to vibrate tangentially; in this other arrangement the plate is supported from the board by a piece of cork, and the string rests on a bridge cemented to the upper surface of the plate. When the string makes its vibration parallel to the surface of the plate, tangential vibrations are communicated to it. By this apparatus normal vibrations may be also communicated to the plate, or any state of vibration intermediate between the tangential and normal, as the string may be made to oscillate in a plane forming any angle with the surface.

Membranes where caused to vibrate by communication through the air become the seat of normal tangential or oblique motions; according to the direction of the exciting vibrations.

Savart has proved by a great variety of experiments that whenever vibrations are communicated to any body or system of bodies, the particles always move in the same direction in which the original vibrations are performed. For the purpose of explaining the phenomena more clearly Savart employs different terms to designate the motions which take place in different directions; thus as in a thin board the motion may be produced tangentially to the surfaces in the direction of its length, he calls this motion longitudinal tangential; as the motion can also be produced tangentially in the direction of its breadth he calls this transversal tangential, and as this motion tangential to the faces may be produced in an infinity of oblique directions, he calls these oblique tangential. He calls that motion normal which takes place perpendicularly to the faces of a plate, and which was formerly designated as transversal vibrations.

To show the gradual transformation of one kind of motion to another Savart has contrived various ingenious arrangements. Holding a thin board perpendicularly to the string of this apparatus the various changes from longitudinal tangential to transversal tangential may be shown, according to the direction in which the string is put into vibration by the bow. <u>This</u> is a more elegant and certain means of exhibiting the same changes; two strings are tuned exactly in unison and the board is held in its position by the tension of the two strings. <u>These</u> are the successive changes in the position of the nodal line which takes place in these circumstances.

<u>This</u> instrument enables us to observe the gradual transformation of longitudinal tangential into normal vibrations. These are the accompanying changes of a nodal figure; when the motion is completely normal the nodal lines correspond on the opposite faces of the board which they do not when the motion is tangential or intermediate.

By this instrument we may trace all the changes from longitudinal tangential to transversal tangential, and from either of these to normal; according to the position in which the sliding box is fixed, and the direction in which the bow is drawn across the string.

These more complicated arrangements were devised for the purpose of showing, that in every case of communicated vibration, the motions of all the particles of the system are parallel to the direction of the vibrations by which they are originally excited. If the vibrations of the string are performed in a plane parallel to the surfaces they all obey tangential motions, and if the plane of the vibrations of the string is perpendicular they all enter into normal vibration.

Having acquired some general idea of the laws of communicated vibrations, we shall now be able to understand the use of the sounding boards of musical instruments, and of the various conditions on which their efficiency depends. All stringed instruments, and these form a very important class, consist of two parts essentially distinct from each other, viz of the original vibrating bodies, which are the strings, and of the sounding boards, which augment or reinforce the original sounds, by entering into simultaneous vibration with them. This participation of the vibrations of an original sounding body is termed resonance.

Some sonorous bodies are capable of producing powerful sounds without accessory means; a vibrating plate, an harmonica glass, and a rod longitudinally excited are instances of this; other sonorous bodies such as insulated strings, or tuning forks, on the contrary, are scarcely audible at a moderate distance from the car, but are capable of considerable augmentation when communicated, to resonant surfaces, as when they are brought in contact with a table, or with the sounding board of a musical instrument.

There are several circumstances which influence the intensity of the resonance of a sounding board. The principal of these is the plane in which the vibrations of the sounding body are made, with respect to the resonant surface. Thus, its vibrations may be so communicated as to be perpendicular, or normal to the surface, in which case the sound is the most augmented; or they may be tangential to, or in the same plane with the surface, when the sound is the most feeble. The first of these cases is easily exemplified by placing a vibrating tuning fork perpendicularly to the surface of a flat board; and the second, by placing it perpendicularly to one of the edges of the board. In intermediate positions, viz when the vibrations are communicated obliquely to the surface – the sound has intermediate degrees of intensity. These facts being understood, the peculiarities of the sounding boards of various musical instruments almost of easy explanation.

The sounding board of the Piano-Forte is better disposed than that of any other stringed instrument, as the planes of the vibrations of the strings, are, on account of the direction in which they are struck by the hammers, always perpendicular to its surface. The difference of intensity when a string is struck this way, and when it is made to vibrate parallel to the surfaces, is very obvious, and may be easily tried by striking it with the finger in these two directions. There is no other instrument now in use, in which the strings make their vibrations perpendicular to the sounding board.

In the Guitar, Lute &c the strings are also parallel to the sounding board, but the vibrations must for the convenience of performance, be made obliquely to it. If the sides of the instrument be of inconsiderable depth and the back be flat, the difference of intensity between the perpendicular and parallel vibrations will be very sensible. But if the sides be deep, very little difference will be perceived, as the vibrations which are tangential to the front sounding board are perpendicular to the sides, which thus enter readily into normal vibrations. This fact may be proved by placing the ear to the side of a Guitar while a string is made to sound with its plane of vibration first parallel to the side, and then perpendicular to it. In some instruments, as the Lute, Mandoline &c the back is polygonal, or curved; by this construction a considerable portion of the resonant surface enters into normal, or nearly normal vibrations when the strings are struck obliquely to the principal sounding board.

In no instruments are the strings placed perpendicular to the sounding board; for in such case, however a string were made to vibrate, its communicated vibrations would be tangential, the least intense of all kinds of vibration. But they are sometimes placed obliquely as in the Harp, and then the same changes of intensity may be observed as when the strings are parallel to the board; for if the plane of their vibrations coincide with the plane in which all the strings lie, the communicated vibrations of the board will be oblique to its surface, and the intensity of the sound will be at its maximum; but if the vibrations be made perpendicular to this plane, the communicated vibrations must then be tangential to the surface, and consequently the intensity will be at its minimum.

These laws are not so immediately applicable to the violin, and other instruments of the same class; an extensive series of experiments will yet be necessary to enable us fully to account for the influence of their various forms and curvatures on the intensity of the sounds they produce.

VI [extra notes]

We shall now be able to understand the use of the sounding boards of musical instruments, and the various conditions on which their efficiency depends.

How far the sounding board may be removed from the original source of sound I showed when explaining in the first lecture the laws of the transmission of sound through solid bodies. I am now about to repeat this experiment but with another end in view, viz to show you the gradual changes.

By this arrangement of wires I shall be able to show that two sounds passing at the same time through the same conductor may be afterwards separated and either them heard alone. The number of instruments and performers which would have been required to make an effective experiment has prevented me this evening from giving to you that the same is true with regard to complete musical performances, and that the sounds of two distinct instruments may be simultaneously transmitted through the same conducting rod or wire, and yet shall so little interfere with each other in their passage that either may be perfectly received by a guitar or other sounding board placed at the end of that rod which is in the plane of the transmitted vibration.

The time will not allow me to point out the intimate analogies which subsist between the transmission of sound through solid conductions, and the propagation of light; suffice it at present to say that sound, when its vibrations are all made in parallel planes is exactly analogous to polarized light, and that when vibrating in planes differently inclined to each other it resembles light in its ordinary state.

In the P.F, Guitar &c the strings are not in immediate contact.
Tuning fork through wood, metal and glass rods.
Experiment from the roof.

Tuning forks are the most convenient instruments for . . .
Theory of the tuning fork.
Vibrations of the sounding board of any stringed instrument may be common original sounds prevented – mode of insulation – square & grand P.F. most convenient – mode of applying the conductor – wooden rod preferable to a wire –
Guitar – Lute – strings not essential – a wooden box – a plane board –
Changes of Normal to Tangential – – Transmission through bent conductors.
Two square P.Fs. or Harps may mutually reciprocate each others sound.
Duet – Echo – Transmission downwards – -
Horizontal transmission; Cabinet P.Fs.
The sounds of an instrument may be transmitted to more than one place.
Transmission of the Violin – Clarinet – Flute –
Orchestra –

In all these experiments I have shown, the conductor has received its impulses from a surface vibrating normally to which it is attached perpendicularly.

It may be interesting to call to recollection the various experiments relating to the conduction of sound which were made previous to those which have just occupied our attention.

The fact of the transmission of sound through solid bodies did not escape the attention of the ancient philosophers, it was observed that when a stick or a metal rod was struck or scratched at one end, and placed to the ear at the other, the sound was far louder than when not so placed to the ear. There is an old experiment of this kind, with which every one is familiar; if a poker be suspended from a string the end of which is held close to the ear, and be struck by some hard body, the sounds which are produced, in consequence of its vibrating as a bar free at both ends, are of far greater intensity than those which would arrive to the ear, did the air and not a solid body form the medium of communication. It was for a long time very erroneously supposed, that an aeriform medium was alone capable of receiving sonorous impressions, and in conformity with this opinion Lord Bacon when noticing these experiments, assumes that the sound is transmitted by spirits contained within the pores of the body.* The first correct observations on this subject appear to have been made by Dr Hooker in 1667, who made an experiment with a distended wire of sufficient length to observe that the same sound was propagated far swifter through the wire than through the air.+ More than a century afterwards Professor Wunsch of Berlin made a similar experiment, substituting for the wire above 1700 feet of wooden lathes joined together, and confirmed Dr Hooke's results. Other experiments of a similar nature were subsequently made by Herhold & Rafn, and Hassenfratz and Gay Lussac. H and GL made their observations in the quarries at Paris; a blow of a hammer against the rock produced two sounds, which separated in their progress; that transmitted through the stone arriving almost instantly while the sound conveyed by the air lagged behind. In all these experiments it was merely observed that sound travelled through iron, wood and stone much quicker than through the air, but no direct estimate of its velocity in these substances was attempted. The only direct observations we have on this subject are those of M. Biot in which he was assisted by Bouvard, Malus and Martin; these experiments were made by transmitting sounds through the sides of the iron conduit pipes of Paris a length of between 3 and 4000 feet was employed, and the mean result of two series of observations made in different ways gave 3459 metres or 11,090 feet per second, for the velocity of sound in cast iron.

Chladni has inferred the velocity of sound in different solid bodies, from the longitudinal vibrations of rods of these substances. The results attained in this way are fully confirmed by calculations from other grounds. His method was founded on Newton's demonstration, that sound travels through space of a given length, filled with air, in the same time that a column of air of the same length, contained in a tube open at both ends makes a single vibration. Chladni's own discovery of the longitudinal vibrations of solid bodies, which are exactly analogous to the ordinary vibrations of columns of air, enabled him to apply this proposition to solid bodies, and to establish the general law, that sound is transmitted through any elastic substance in the same time as a bar or rod of this substance makes one longitudinal vibration. In this manner he ascertained the velocities of sound in a great variety of different substances; through Tin he found it to be 7800 feet in a second; in Silver 9300 feet, in Copper 12,500, in Glass and Steel 17,500, and in various woods from 11,000 to 18,000 feet in a second.

From these determinations we may estimate the time it requires to transmit force, whether by pulling, pushing or by a blow, to any distance by means of metal bars or chairs; for any impulse whatever is transmitted with the same velocity as sound is. Taking Biot's experiments as data, for every 11,090 feet, of distance, the pull, push or blow, will reach its point of action one second after the moment of its emanation from the first mover. In all moderate distances then the interval is utterly insensible. Sir John Herschel has calculated that if the sun and the earth were connected by an iron bar, no less than 1074 days, or nearly three years must elapse before a force at the sun, or a sound produced there, could reach the earth, while the force exerted by their mutual gravity may be proved to acquire no appreciable time for its transmission.

In all the experiments which I have alluded to, as having proved the transmission of sound through solid bodies, the sounds transmitted were either mere noises, such as the blow of a hammer, or as in Herhold & Rafn's experiment, a single musical sound, produced by striking a silver spoon attached to one end of the conducting wire; and in no case were any means employed for the subsequent augmentation of the transmitted sound. I believe that previous to the experiments which I published in 1823, none had been made on the transmission of the modulated sounds of musical instruments; nor had it been shown that sonorous undulations, conveyed through solid linear conductors of considerable length, were capable of exciting, in surfaces brought in contact with them, a quantity of vibratory motion, sufficient to be powerfully audible when communicated through the air.

To extend these experiments much further than has yet been done would be attended with considerable difficulties: but as the velocity of sound is much greater in solid substances than in air, it is not impossible that the transmission of sound through solid conductors may hereafter be applied to useful purposes. Sound travels through the air at the rate of 1142 feet in a second of time; but it is communicated through iron wire, glass cane, or deal wood rods, with the velocity of about 18000 feet per second, so that it would travel the distance of 200 miles in less than a minute.

When sound is allowed to diffuse itself in all directions as from a center, its intensity, according to theory, decreases as the square of the distance increases; but if it be confined to one rectilinear direction, no diminution of intensity ought to take place. But this is on the supposition that the conducting body possesses perfect homogeneity and is uniform in its structure, – conditions which never obtain in our actual experiments. Could any conducting substance be rendered perfectly equal in density and elasticity, so as to allow the undulation to proceed with uniform velocity, without any reflections or interference, it would be as easy to transmit sounds through such conductors from Edinburgh to London, as it is now to establish a communication from one chamber to another. Whether any substance can be rendered thus homogeneous and uniform, remains a problem which time alone can determine.

The last experiment to which I shall call your attention in the present lecture, is one which I have devised for the purpose of proving that in the clarinet or any other reed instrument, the column of air contained within the tube, and the vibrating tongue or reed, mutually influence each other in such a manner, that they not influence each other in such a manner, that they not only produce the same sound with respect to pitch, that is the same number of vibrations, but that the quality of the sound is the same in both cases, which would not be the same vibrations of the reed and of the air were not precisely similar in every respect. To prove this I must separate the sound of the reed from the sound of the column of air; and this I can affect by means of which I am about to describe. Mr Willman whose skill as a performer on the Clarinet is well known to all present, is in the room below; and when I make the signal will commence performing on that instrument. You will not hear any of the sounds which will be communicated to the atmosphere from the column of air vibrating within the instrument, because these will be confined to the room below, and be prevented from reaching your ears by the interposed floor and ceiling. I have connected the lower end of this conducting rod with the fixed end of the clarinet reed, and in accordance with the laws of the transmission of sound through solid bodies which I explained and illustrated in my lecture, the vibrations of the reed will be communicated to this sounding board; and you would find if you were successfully to hear the same performance in the lecture room and in the room below, that one is an exact facsimile of the other in every particular of pitch, quality and intensity. Having told you what to expect you will easily recognise the tones of the clarinet; had I not done so, on account of the strangeness of the sounds of that instrument proceeding from the sounding board of a – --, you might be liable to be mistaken as to the instrument you will hear.

LECTURE VIII

I have now arrived at the conclusion of my course, and it only remains for me to thank you for the attention with which I have been listened to. In these eight lectures I have attempted to give a complete view of the present state of our knowledge concerning the laws of sound and I am not aware that I have omitted any important experiment relating this this subject which has, of late years, been made either here or abroad. It is true that many of the experimental illustrations have been of such nature that they cannot be seen or heard to advantage by an [].

To bestow more attention on a branch of science which has ministered so efficiently to our intellectual wants and pleasures, and which has of late years become so interesting to the philosopher from the strong analogies which its phenomena present with those of light and heat, and which are extending every day by new discoveries.

NOTE

* (Savart had at first in the *Annales de Chimie*, T.14 p 116–119 contradicted what has been said by me respecting the longitudinal vibrations of a rod fixed at one end, and because he could not at that time perform this easy experiment, he represented its production as impossible, and in this false presupposition allowed himself the observation, that I might have therefore concluded principally more from the analogy of the vibrations of the air, than have ascertained it through actually performed experiments (!) But after I had asserted the necessity (in the *Ann de Chimie*. T.20 p 74, and had explained more clearly how the experiment was to be performed, he has likewise produced these vibrations, and has employed himself in further investigations.)

Editorial Headnote

16. David Brewster, *Letters on Natural Magic addressed to Sir Walter Scott* (London: John Murray, 1832), pp. 1–7, 157–243.

David Brewster's *Letters on Natural Magic addressed to Sir Walter Scott*, published in 1832, paid considerable attention to sound and acoustic science, drawing on the very latest sonorous experiments, including Wheatstone's "Kaleidophone". Indeed, Brewster had, himself, invented the Kaleidoscope in 1817, and it was from this that Wheatstone had taken inspiration when naming his own device. Originally intending to enter the Presbyterian Church of Scotland as a minister, Brewster attended the University of Edinburgh between 1794 and 1800, but his fear of public speaking prevented him from pursuing a career as a clergyman. Nevertheless, he remained firmly Calvinistic in his subsequent scientific career, always searching for new knowledge of nature as evidence of God's design. Importantly, Brewster rejected the role of hypothesis in science, believing that to speculate on what God might have ordered, rather than on what He had actually created, risked the sin of pride.[1] Unfortunately, this commitment would see Brewster become increasingly isolated, intellectually, throughout the 1830s and 1840s regarding the question of light. Following French engineer Augustin-Jean Fresnel's (1788–1827) claims that light travelled as a wave, as argued in his 1822 essay, *De la Lumière*, the Newtonian conception of light as a projection of corpuscles came under growing criticism. By the 1830s, Britain's leading mathematicians, such as William Whewell, John Herschel, and George Biddell Airy (1801–1892), were largely in favour of a wave theory of light but Brewster, believing this depended on a propagating medium that was completely speculative, adamantly rejected such mathematical theorizing: he remained committed to Newton's corpuscular theory.[2] Just as a sound wave required a solid body, fluid, or air through which to travel, so Brewster reasoned that any hypothetical light wave would require a similar, as yet unknown, medium. In 1832, however, Brewster concerned himself with the historical role of knowledge of nature, juxtaposing this against modern, scientific learning. In his *Letters on Natural Magic*, Brewster reflected that, at present, philosophical understandings of sound were commonly conscripted within performances of a popular, entertaining character, such as the "Invisible Girl". This consisted of an arrangement of four trumpets, suspended from a frame in which a hidden tube, connected to a woman concealed in a nearby room, projected her voice into the trumpets, which then conveyed the sound out to mystified spectators. For Brewster, this was as a scientifically grounded deception, mobilizing acoustic skill to confuse an audience's senses. Brewster also commentated on the Jew's Harp performances of Eulenstein and Wheatstone's works, including his Kaleidophone. Of all sonorous wonders, however, Brewster believed the principles of harmonics exerted the greatest emotional influence over listeners and related the latest understandings of harmony to Ernst Chladni's observations of nodal lines through his acoustic plate experiments. As Brewster put it, of all "the discoveries of modern science", few rivalled "those which relate to the production of harmonic sounds".[3] Yet the history of sonorous knowledge provided Brewster with a powerful moral lesson, which he was keen to impress on his readers: without understanding the physical laws of sound, it was possible for listeners to be misguided. "The science of *Acoustics*", Brewster explained, "furnished the ancient sorcerers with some of their best deceptions" and seemed to provide them with "supernatural" powers, such as the vocal effects of the statue of Memnon in ancient Egypt.[4] Likewise, the Roman Catholic Church employed music in worship to influence congregations. Knowledge of sound was, Brewster argued, powerful knowledge with the potential to help governing elites control their subjects. Understanding how to mobilize and trust human senses was vital to avoid being the victim of such manipulation.

NOTES

1 Geoffrey N. Cantor, *Optics After Newton: theories of light in Britain and Ireland, 1704–1840* (Manchester: Manchester University Press, 1983), pp. 178 and 181.
2 Geoffrey Cantor, "The reception of the wave theory of light in Britain: a case study illustrating the role of methodology in scientific debate", *Historical Studies in the Physical Sciences* 6 (1975), pp. 109–32; Geoffrey Cantor, "The Changing Role of Young's Ether", *The British Journal for the History of Science* 5, no. 1 (Jun. 1970), pp. 44–62.
3 David Brewster, *Letters on Natural Magic addressed to Sir Walter Scott*, Bart (New York: Harper & Brothers, 1842), pp. 170–78 and 167.
4 Ibid., p. 15.

16

LETTERS ON NATURAL MAGIC ADDRESSED TO SIR WALTER SCOTT

David Brewster

Source: David Brewster, *Letters on Natural Magic addressed to Sir Walter Scott*, Bart (London: John Murray, 1832), pp. 1–7, 157–243.

Letter I

Extent and interest of the subject – Science employed by ancient governments to deceive and enslave their subjects – Influence of the supernatural upon ignorant minds – Means employed by the ancient magicians to establish their authority – Derived from a knowledge of the phenomena of Nature – From the influence of narcotic drugs upon the victims of their delusion – From every branch of science – Acoustics – Hydrostatics – Mechanics – Optics – M. Salverte's work on the occult sciences – Object of the following Letters.

My Dear Sir Walter,

As it was at your suggestion that I undertook to draw up a popular account of those prodigies of the material world which have received the appellation of *Natural Magic*, I have availed myself of the privilege of introducing it under the shelter of your name. Although I cannot hope to produce a volume at all approaching in interest to that which you have contributed to the Family Library, yet the popular character of some of the topics which belong to this branch of Demonology may atone for the defects of the following Letters; and I shall deem it no slight honour if they shall be considered as forming an appropriate supplement to your valuable work.

The subject of Natural Magic is one of great extent as well as of deep interest. In its widest range, it embraces the history of the governments and the superstitions of ancient times, – of the means by which they maintained their influence over the human mind, – of the assistance which they derived from the arts and the sciences, and from a knowledge of the powers and phenomena of nature. When the tyrants of antiquity were unable or unwilling to found their sovereignty on the affections

DOI: 10.4324/b23406-16

and interests of their people, they sought to entrench themselves in the strongholds of supernatural influence, and to rule with the delegated authority of heaven. The prince, the priest, and the sage, were leagued in a dark conspiracy to deceive and enslave their species; and man, who refused his submission to a being like himself, became the obedient slave of a spiritual despotism, and willingly bound himself in chains when they seemed to have been forged by the gods.

This system of imposture was greatly favoured by the ignorance of these early ages. The human mind is at all times fond of the marvellous, and the credulity of the individual may be often measured by his own attachment to the truth. When knowledge was the property of only one caste, it was by no means difficult to employ it in the subjugation of the great mass of society. An acquaintance with the motions of the heavenly bodies, and the variations in the state of the atmosphere, enabled its possessor to predict astronomical and meteorological phenomena with a frequency and an accuracy which could not fail to invest him with a divine character. The power of bringing down fire from the heavens, even at times when the electric influence was itself in a state of repose, could be regarded only as a gift from heaven. The power of rendering the human body insensible to fire was an irresistible instrument of imposture; and in the combinations of chemistry, and the influence of drugs and soporific embrocations on the human frame, the ancient magicians found their most available resources.

The secret use which was thus made of scientific discoveries and of remarkable inventions, has no doubt prevented many of them from reaching the present times; but though we are very ill informed respecting the progress of the ancients in various departments of the physical sciences, yet we have sufficient evidence that almost every branch of knowledge had contributed its wonders to the magician's budget, and we may even obtain some insight into the scientific acquirements of former ages, by a diligent study of their fables and their miracles.

The science of *Acoustics* furnished the ancient sorcerers with some of their best deceptions. The imitation of thunder in their subterranean temples could not fail to indicate the presence of a supernatural agent. The golden virgins whose ravishing voices resounded through the temple of Delphos; – the stone from the river Pactolus, whose trumpet notes scared the robber from the treasure which it guarded; – the speaking head which uttered its oracular responses at Lesbos; – and the vocal statue of Memnon, which began at the break of day to accost the rising sun, – were all deceptions derived from science, and from a diligent observation of the phenomena of nature.

The principles of *Hydrostatics* were equally available in the work of deception. The marvellous fountain which Pliny describes in the Island of Andros as discharging wine for seven days, and water during the rest of the year; – the spring of oil which broke out in Rome to welcome the return of Augustus from the Sicilian war, – the three empty urns which filled themselves with wine at the annual feast of Bacchus in the city of Elis, – the glass tomb of Belus which was full of oil, and which, when once emptied by Xerxes, could not again be filled, – the weeping statues, and the perpetual lamps of the ancients, – were all the obvious effects of the equilibrium and pressure of fluids.

Although we have no direct evidence that the philosophers of antiquity were skilled in *Mechanics*, yet there are indications of their knowledge, by no means equivocal, in the erection of the Egyptian obelisks, and in the transportation of huge masses of stone, and their subsequent elevation to great heights in their temples. The powers which they employed, and the mechanism by which they operated, have been studiously concealed, but their existence may be inferred from results otherwise inexplicable, and the inference derives additional confirmation from the mechanical arrangements which seem to have formed a part of their religious impostures. When in some of the infamous mysteries of ancient Rome, the unfortunate victims were carried off by the gods, there is reason to believe that they were hurried away by the power of machinery; and when Apollonius, conducted by the Indian sages to the temple of their god, felt the earth rising and falling beneath his feet like the agitated sea, he was no doubt placed upon a moving floor capable of imitating the heavings of the waves. The rapid descent of those who consulted the oracle in the cave of Trophonius, – the moving tripods which Apollonius saw in the Indian temples, – the walking statues at Antium, and in the Temple of Hierapolis, – and the wooden pigeon of Archytas, are specimens of the mechanical resources of the ancient magic.

But of all the sciences *Optics* is the most fertile in marvellous expedients. The power of bringing the remotest objects within the very grasp of the observer, and of swelling into gigantic magnitude the almost invisible bodies of the material world, never fails to inspire with astonishment even those who understand the means by which these prodigies are accomplished. The ancients, indeed, were not acquainted with those combinations of lenses and mirrors which constitute the telescope and the microscope, but they must have been familiar with the property of lenses and mirrors to form erect and inverted images of objects. There is reason to think that they employed them to effect the apparition of their gods; and in some of the descriptions of the optical displays which hallowed their ancient temples, we recognize all the transformations of the modern phantasmagoria.

It would be an interesting pursuit to embody the information which history supplies respecting the fables and incantations of the ancient superstitions, and to show how far they can be explained by the scientific knowledge which then prevailed. This task has, to a certain extent, been performed by M. Eusebe Salverte, in a work on the occult sciences, which has recently appeared; but notwithstanding the ingenuity and learning which it displays, the individual facts are too scanty to support the speculations of the author, and the descriptions are too meagre to satisfy the curiosity of the reader.[1]

In the following letters I propose to take a wider range, and to enter into more minute and popular details. The principal phenomena of nature, and the leading combinations of art, which bear the impress of a supernatural character, will pass under our review, and our attention will be particularly called to those singular illusions of sense, by which the most perfect organs either cease to perform their functions, or perform them faithlessly; and where the efforts and the creations of the mind predominate over the direct perceptions of external nature.

In executing this plan, the task of selection is rendered extremely difficult, by the superabundance of materials, as well as from the variety of judgments for which these materials must be prepared. Modern science may be regarded as one vast miracle, whether we view it in relation to the Almighty Being, by whom its objects and its laws were formed, or to the feeble intellect of man, by which its depths have been sounded, and its mysteries explored: – And if the philosopher who is familiarized with its wonders, and who has studied them as necessary results of general laws, never ceases to admire and adore their author, how great should be their effect upon less gifted minds, who must ever view them in the light of inexplicable prodigies. – Man has in all ages sought for a sign from heaven, and yet he has been habitually blind to the millions of wonders with which he is surrounded. If the following pages should contribute to abate this deplorable indifference to all that is grand and sublime in the universe, and if they should inspire the reader with a portion of that enthusiasm of love and gratitude which can alone prepare the mind for its final triumph, the labours of the author will not have been wholly fruitless.

Letter VII

Illusions depending on the ear – Practised by the ancients – Speaking and singing heads of the ancients – Exhibition of the invisible girl described and explained – Illusions arising from the difficulty of determining the direction of sounds – Singular example of this illusion – Nature of ventriloquism – Exhibitions of some of the most celebrated ventriloquists – M. St Gille – Louis Brabant – M. Alexandre – Capt. Lyon's account of Eskimaux ventriloquists.

NEXT to the eye the ear is the most fertile source of our illusions, and the ancient magicians seem to have been very successful in turning to their purposes the doctrines of sound. In the labyrinth of Egypt, which contained twelve palaces and 1500 subterraneous apartments, the gods were made to speak in a voice of thunder; and Pliny, in whose time this singular structure existed, informs us, that some of the palaces were so constructed that their doors could not be opened without permitting the peals of thunder from being heard in the interior. When Darius Hystaspes ascended the throne, and allowed his subjects to prostrate themselves before him as a god, the divinity of his character was impressed upon his worshippers by the bursts of thunder and flashes of lightning which accompanied their devotion. History has of course not informed us how these effects were produced; but it is probable that, in the subterraneous and vaulted apartments of the Egyptian labyrinth, the reverberated sounds arising from the mere opening and shutting of the doors themselves afforded a sufficient imitation of ordinary thunder. In the palace of the Persian king, however, a more artificial imitation is likely to have been employed, and it is not improbable that the method used in our modern theatres was known to the ancients. A thin sheet of iron, three or four feet long, such as that used for German stoves, is held by one corner between the finger and the thumb, and allowed to hang freely by its own weight. The hand is then moved or shaken horizontally, so as to agitate the corner in a direction at right angles to the surface of the sheet. By this simple process a great variety of sounds will be produced, varying from the deep growl of distant thunder to those loud and explosive bursts which rattle in quick succession from clouds immediately over our heads. The operator soon acquires great power over this instrument, so as to be able to produce from it any intensity and character of sound that may be required. The same effect may be produced by sheets of tin-plate, and by thin plates of mica; but on account of their small size, the sound is shorter and more acute. In modern exhibitions an admirable imitation of lightning is produced by throwing the powder of rosin, or the dust of lycopodium, through a flame, and the rattling showers of rain which accompany these meteors are well imitated by a well regulated shower of peas.

The principal pieces of acoustic mechanism used by the ancients were *speaking* or *singing heads*, which were constructed for the purpose of representing the gods, or of uttering oracular responses. Among these, the speaking head of Orpheus, which uttered its responses at Lesbos, is one of the most famous. It was celebrated not only throughout Greece, but even in Persia, and it had the credit of predicting, in the equivocal language of the heathen oracles, the bloody death which terminated the expedition of Cyrus the Great into Scythia. Odin, the mighty magician of the north, who imported into Scandinavia the magical arts of the Past, possessed a speaking head, said to be that of the sage Minos, which he had enchased in gold, and which uttered responses that had all the authority of a divine revelation. The celebrated mechanic Gerbert, who filled the papal chair A. D. 1000, under the name of Sylvester II., constructed a speaking head of brass. Albertus Magnus is said to have executed a head in the thirteenth century, which not only moved but spoke. It was made of earthen-ware, and Thomas Aquinas is said to have been so terrified when he saw it, that he broke it in pieces, upon which the mechanist exclaimed, "There goes the labour of thirty years."

It has been supposed by some authors, that in the ancient speaking machines the deception is effected by means of ventriloquism, the voice issuing from the juggler himself; but it is more probable that the sound was conveyed by pipes from a person in another apartment to the mouth of the figure. Lucian, indeed, expressly informs us, that the impostor Alexander made his figure of Æsculapius speak, by transmitting his voice through the gullet of a crane to the mouth of the statue; and that this method was general, appears from a passage in Theodoretus, who assures us, that in the fourth century, when Bishop Theophilus broke to pieces the statues at Alexandria, he found some which were hollow, and which were so placed against a wall, that the priest could conceal himself behind them, and address the ignorant spectators through their mouths.

Even in modern times, speaking-machines have been constructed on this principle. The figure is frequently a mere head placed upon a hollow pedestal, which, in order to promote the deception, contains a pair of bellows, a sounding-board, a cylinder and pipes supposed to represent the organs of speech. In other cases these are dispensed with, and a simple wooden head utters its sounds through a speaking trumpet. At the court of Charles II. this deception was exhibited with great effect by one Thomas Irson, an Englishman, and when the astonishment had become very general, a popish priest was discovered by one of the pages in an adjoining apartment. The questions had been proposed to the wooden figure by whispering into its ear, and this learned personage had answered them all with great ability, by speaking through a pipe in the same language in which the questions were proposed. Professor Beckmann informs us that children and women were generally concealed either in the juggler's box, or in the adjacent apartment, and that the juggler gave them every assistance by means of signs previously agreed upon. When one of these exhibitions was shown at Gottingen, the Professor was allowed, on the promise of secrecy, to witness the process of deception. He saw the assistant in another room, standing before the pipe with a card in his hand, upon which the signs agreed upon had been marked, and he had been introduced so privately into the house that even the landlady was ignorant of his being there.

An exhibition of the very same kind has been brought forward in our own day, under the name of the *Invisible Girl;* and as the mechanism employed was extremely ingenious, and is well fitted to convey an idea of this class of deceptions, we shall give a detailed description of it.

The machinery, as constructed by M. Charles, is shown in Figure 16.37 in perspective, and a plan of it

Figure 16.37

in Figure 16.38. The four upright posts A, A, A, A, are united at top by a cross rail B, B, and by two similar rails at bottom. Four bent wires *a, a, a, a,* proceeded from the top of these posts, and terminated at *c*. A hollow copper ball M, about a foot in diameter, was suspended from these wires by four slender ribbands *b, b, b, b,* and into the copper ball were fixed the extremities of four trumpets T, T, T, T, with their mouths outwards.

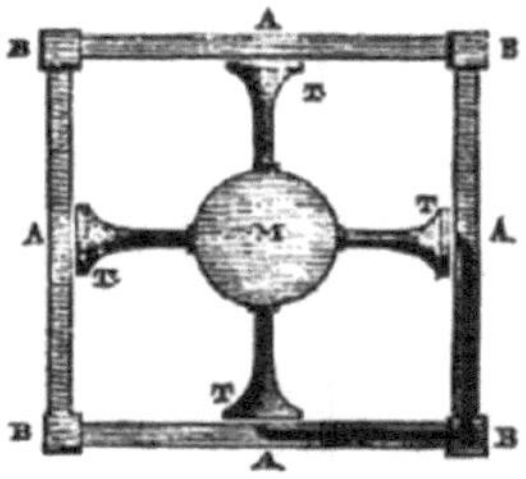

Figure 16.38

The apparatus now described was all that was visible to the spectator; and though fixed in one spot, yet it had the appearance of a piece of separate machinery, which might have occupied any other part of the room. When one of the spectators was requested by the exhibitor to propose some question, he did it by speaking into one of the trumpets at T. An appropriate answer was then returned from all the trumpets, and the sound issued with sufficient intensity to be heard by an ear applied to any of them, and yet it was so weak that it appeared to come from a person of very diminutive size. Hence the sound was supposed to come from an invisible girl, though the real speaker was a full grown woman. The invisible lady conversed in different languages, sang beautifully, and made the most lively and appropriate remarks on the persons in the room.

This exhibition was obviously far more wonderful than the speaking heads which we have described, as the latter invariably communicated with a wall, or with a pedestal through which pipes could be carried into the next apartment. But the ball M and its trumpets communicated with nothing through which sound could be conveyed. The spectator satisfied himself by examination that the ribbands *b, b*, were real ribbands, which concealed nothing, and which could convey no sound, and as he never conceived that the ordinary piece of framework A B, could be of any other use than its apparent one of supporting the sphere M, and defending it from the spectators, he was left in utter amazement respecting the origin of the sound, and his surprise was increased by the difference between the sounds which were uttered and those of ordinary speech.

Though the spectators were thus deceived by their own reasoning, yet the process of deception was a very simple one. In two of the horizontal railings A, A, Figure 16.38, opposite the trumpet mouths T, there was an aperture communicating with a pipe or tube which went to the vertical post B, and descending it, as shown at T A A, Figure 16.39, went beneath the floor *f f* in the direction *p, p*, and entered the apartment N, where the invisible lady sat. On the side of the partition about *h*, there was a small hole, through which the lady saw what was going on in the exhibition room, and communications were no doubt made to her by signals from the person who attended the machine. When one of the spectators asked a question by speaking into one of the trumpets T, the sound was reflected from the mouth of the trumpet back to the aperture at A, in the horizontal rail, Figure 16.38, and

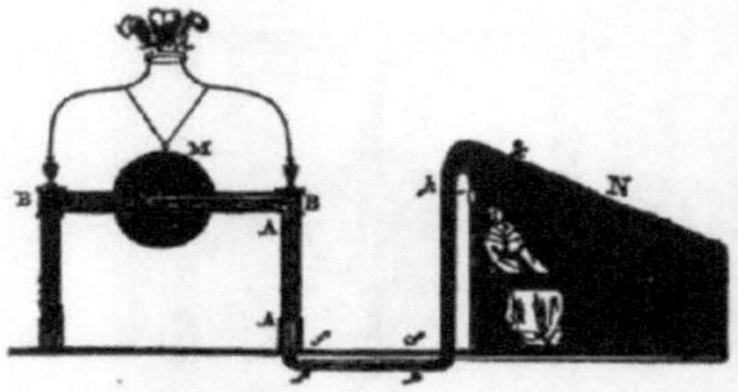

Figure 16.39

was distinctly conveyed along the closed tube into the apartment N. In like manner the answer issued from the aperture A, and being reflected back to the ear of the spectator by the trumpet, he heard the sounds with that change of character which they receive when transmitted through a tube and then reflected to the ear.

The surprise of the auditors was greatly increased by the circumstance, that an answer was returned to questions put in a whisper, and also by the conviction that nobody but a person in the middle of the audience could observe the circumstances to which the invisible figure frequently adverted.

Although the performances of speaking heads were generally effected by the methods now described, yet there is reason to think that the ventriloquist sometimes presided at the exhibition, and deceived the audience by his extraordinary powers of illusion. There is no species of deception more irresistible in its effects than that which arises from the uncertainty with which we judge of the direction and distance of sounds. Every person must have noticed how a sound in their own ears is often mistaken for some loud noise moderated by the distance from which it is supposed to come; and the sportsman must have frequently been surprised at the existence of musical sounds humming remotely in the extended heath, when it was only the wind sounding in the barrel of his gun. The great proportion of apparitions that haunt old castles and apartments associated with death exist only in the sounds which accompany them. The imagination even of the boldest inmate of a place hallowed by superstition, will transfer some trifling sound near his own person to a direction and to a distance very different from the truth, and the sound which otherwise might have no peculiar complexion will derive another character from its new locality. Spurning the idea of a supernatural origin, he determines to unmask the spectre, and grapple with it in its den. All the inmates of the house are found to be asleep – even the quadrupeds are in their lair – there is not a breath of wind to ruffle the lake that reflects through the casement, the reclining crescent of the night; and the massive walls in which he is inclosed forbid the idea that he has been disturbed by the warping of pannelling or the bending of partitions. His search is vain; and he remains master of his own secret, till he has another opportunity of investigation. The same sound again disturbs him, and, modified probably by his own position at the time, it may perhaps appear to come in a direction slightly different from the last. His searches are resumed, and he is again disappointed. If this incident should recur night after night with the same result; – if the sound should appear to depend upon his own motions, or be any how associated with himself, with his present feelings, or with his past history, his personal courage will give way, a superstitious dread, at which he himself perhaps laughs, will seize his mind, and he will rather believe that the sounds have a supernatural origin, than that they could continue to issue from a spot where he knows there is no natural cause for their production.

I have had occasion to have personal knowledge of a case much stronger than that which has now been put. A gentleman, devoid of all superstitious feelings, and living in a house free from any gloomy associations, heard night after night in his bed-room a singular noise, unlike any ordinary sound to which he was accustomed. He had slept in the same room for years without hearing it, and he attributed it at first to some change of circumstances in the roof or in the walls of the room, but after the strictest examination no cause could be found for it. It occurred only once in the night, it was heard almost every night, with few interruptions. It was over in an instant, and it never took place till after the gentleman had gone to bed. It was always distinctly heard by his companion, to whose time of going to bed it had no relation. It depended on the gentleman alone, and it followed him into another apartment with another bed, on the opposite side of the house. Accustomed to such investigations, he made the most diligent but fruitless search into its cause. The consideration that the sound had a special reference to him alone, operated upon his imagination, and he did not scruple to acknowledge that the recurrence of the mysterious sound produced a superstitious feeling at the moment. Many months afterwards it was found that the sound arose from the partial opening of the door of a wardrobe which was within a few feet of the gentleman's head, and which had been taken into the other apartment. This wardrobe was almost always opened before he retired to bed, and the door being a little too tight, it gradually forced itself open with a sort of dull sound, resembling the note of a drum. As the door had only started half an inch out of its place, its change of place never attracted attention. The sound, indeed, seemed to come in a different direction, and from a greater distance.

When sounds so mysterious in their origin are heard by persons predisposed to a belief in the marvellous, their influence over the mind must be very powerful. An inquiry into their origin, if it is made at all, will be made more in the hope of confirming than of removing the original impression, and the unfortunate victim of his own fears will also be the willing dupe of his own judgment.

This uncertainty with respect to the direction of sound is the foundation of the art of ventriloquism. If we place ten men in a row at such a distance from us that they are included in the angle within which we cannot judge of the direction of sound, and if in a calm day each of them speaks in succession, we shall not be able with closed eyes to determine from which of the ten men any of the sounds proceeds, and we shall be incapable of perceiving that there is any difference in the direction of the sounds emitted by the two outermost: If a man and a child are placed within the same angle, and if the man speaks with the accent of a child without any corresponding motion in his mouth or face, we shall necessarily believe that the voice comes from the child: Nay, if the child is so distant from the man that the voice actually appears to us to come from the man, we will still continue in the belief that the child is the speaker; and this conviction would acquire additional strength if the child favoured the deception by accommodating its features and gestures to the words spoken by the man. So powerful, indeed, is the influence of this deception, that if a jack-ass placed near the man were to open its mouth, and shake its head responsive to the words uttered by his neighbour, we would rather believe that the ass spoke than that the sounds proceeded from a person whose mouth was shut, and the muscles of whose face were in perfect repose. If our imagination were even directed to a marble statue or a lump of inanimate matter, as the source from which we were to expect the sounds to issue, we would still be deceived, and would refer the sounds even to these lifeless objects. The illusion would be greatly promoted if the voice were totally different in its tone and character from that of the man from whom it really comes; and if he occasionally speak in his own full and measured voice, the belief will be irresistible that the assumed voice proceeds from the quadruped or from the inanimate object.

When the sounds which are required to proceed from any given object are such as they are actually calculated to yield, the process of deception is extremely easy, and it may be successfully executed even if the angle between the real and the supposed directions of the sound is much greater than the angle of uncertainty. Mr Dugald Stewart has stated some cases in which deceptions of this kind were very perfect. He mentions his having seen a person who, by counterfeiting the gesticulations of a performer on the violin, while he imitated the music by his voice, rivetted the eyes of his audience on the instrument, though every sound they heard proceeded from his own mouth. The late Savile Carey, who imitated the whistling of the wind through a narrow chink, told Mr Stewart that he had frequently practised this deception in the corner of a coffee-house, and that he seldom failed to see some of the company rise to examine the tightness of the windows, while others, more intent on their newspapers, contented themselves with putting on their hats and buttoning their coats. Mr Stewart likewise mentions an exhibition formerly common in some of the continental theatres, where a performer on the stage displayed the dumb show of singing with his lips and eyes and gestures, while another person unseen supplied the music with his voice. The deception in this case he found to be at first so complete as to impose upon the nicest ear and the quickest eye; but in the progress of the entertainment, he came distinctly sensible of the imposition, and sometimes wondered that it should have misled him for a moment. In this case there can be no doubt that the deception was at first the work of the imagination, and was not sustained by the acoustic principle. The real and the mock singer were too distant, and when the influence of the imagination subsided, the true direction of the sound was discovered. This detection of the imposture, however, may have arisen from another cause. If the mock singer happened to change the position of his head, while the real singer made no corresponding change in his voice, the attentive spectator would at once notice this incongruity, and discover the imposition.

In many of the feats of ventriloquism the performer contrives, under some pretence or other, to conceal his face, but ventriloquists of great distinction, such as M. Alexandre, practise their art without any such concealment.

Ventriloquism loses its distinctive character if its imitations are not performed by a voice from the belly. The voice, indeed, does not actually come from that region, but when the ventriloquist utters sounds from the larynx without moving the muscles of his face, he gives them strength by a powerful action of the abdominal muscles. Hence he speaks by means of his belly, although the throat is the real source from which the sounds proceed. Mr Dugald Stewart has doubted the fact, that ventriloquists possess the power of fetching a voice from within: he cannot conceive what aid could be derived from such an extraordinary power; and he considers that the imagination, when seconded by such powers of imitation as some mimics possess, is quite sufficient to account for all the phenomena of ventriloquism which he has heard. This opinion, however, is strongly opposed by the remark made to Mr Stewart himself by a ventriloquist, "that his art would be perfect if it were possible only to speak distinctly without any movement of the lips at all." But, independent of this admission, it is a matter of absolute certainty, that this internal power is exercised by the true ventriloquist. In the account which the Abbé Chapelle has given of the performances of M. St Gille and Louis Brabant, he distinctly states that M. St Gille appeared to be absolutely mute while he was exercising his art, and that no change in his countenance could be discovered. He affirms also that the countenance of Louis Brabant exhibited no change, and that his lips were close and inactive. M. Richerand, who attentively watched the performances of M. Fitz-James, assures us that during his exhibition there was a distension in the epigastric region, and that he could not long continue the exertion without fatigue.

The influence over the human mind which the ventriloquist derives from the skilful practice of his art is greater than that which is exercised by any other species of conjuror. The ordinary magician requires his theatre, his accomplices, and the instruments of his art, and he enjoys but a local sovereignty within the precincts of his own magic circle. The ventriloquist, on the contrary, has the supernatural always at his command. In the open fields, as well as in the crowded city, in the private apartment, as well as in the public hall, he can summon up innumerable spirits; and though the persons of his fictitious dialogue are not visible to the eye, yet they are as unequivocally present to the imagination of his auditors as if they had been shadowed forth in the silence of a spectral form. In order to convey some idea of the influence of this illusion, I shall mention a few well authenticated cases of successful ventriloquism.

M. St Gille, a grocer of St Germain en Lay, whose performances have been recorded by the Abbé de la Chapelle, had occasion to shelter himself from a storm in a neighbouring convent, where the monks were in deep mourning for a much esteemed member of their community who had been recently buried. While lamenting over the tomb of their deceased brother the slight honours which had been paid to his memory, a voice was suddenly heard to issue from the roof of the choir bewailing the condition of the deceased in purgatory, and reproving the brotherhood for their want of zeal. The tidings of this supernatural event brought the whole brotherhood to the church. The voice from above repeated its lamentations and reproaches, and the whole convent fell upon their faces, and vowed to make a reparation of their error. They accordingly chaunted in full choir a *de profundis*, during the intervals of which the spirit of the departed monk expressed his satisfaction at their pious exercises. The prior afterwards inveighed against modern scepticism on the subject of apparitions, and M. St Gille had great difficulty in convincing the fraternity that the whole was a deception.

On another occasion, a commission of the Royal Academy of Sciences at Paris, attended by several persons of the highest rank, met at St Germain en Lay to witness the performances of M. St Gille. The real object of their meeting was purposely withheld from a lady of the party, who was informed that an aërial spirit had lately established itself in the neighbourhood, and that the object of the assembly was to investigate the matter. When the party had sat down to dinner in the open air, the spirit addressed the lady in a voice which seemed to come from above their heads, from the surface of the ground at a great distance, or from a considerable depth under her feet. Having been thus addressed at intervals during two hours the lady was firmly convinced of the existence of the spirit, and could with difficulty be undeceived.

Another ventriloquist, Louis Brabant, who had been valet de chambre to Francis I. turned his powers to a more profitable account. Having fallen in love with a rich and beautiful heiress, he was rejected by her parents as an unsuitable match for their daughter. On the death of her father, Louis paid a visit to the widow, and he had no sooner entered the house than she heard the voice of her deceased husband addressing her from above, "Give my daughter in marriage to Louis Brabant, who is a man of large fortune and excellent character. I endure the inexpressible torments of purgatory for having refused her to him. Obey this admonition, and give everlasting repose to the soul of your poor husband." This awful command could not be resisted, and the widow announced her compliance with it.

As our conjuror, however, required money for the completion of his marriage, he resolved to work upon the fears of one Cornu, an old banker at Lyons, who had amassed immense wealth by usury and extortion. Having obtained an interview with the miser, he introduced the subjects of demons and spectres and the torments of purgatory, and during an interval of silence, the voice of the miser's deceased father was heard complaining of his dreadful situation in purgatory, and calling upon his son to rescue him from his sufferings by enabling Louis Brabant to redeem the Christians that were enslaved by the Turks. The awe-struck miser was also threatened with eternal damnation if he did not thus expiate his own sins; but such was the grasp that the banker took of his gold that the ventriloquist was obliged to pay him another visit. On this occasion, not only his father but all his deceased relations appealed to him in behalf of his own soul and theirs, and such was the loudness of their complaints that the spirit of the banker was subdued, and he gave the ventriloquist ten thousand crowns to liberate the Christian captives.

When the miser was afterwards undeceived, he is said to have been so mortified that he died of vexation.

The ventriloquists of the nineteenth century made great additions to their art, and the performances of M. Fitz-James and M. Alexandre, which must have been seen by many of our countrymen, were far superior to those of their predecessors. Besides the art of speaking by the muscles of the throat and the abdomen, without moving those of the face, these artists had not only studied with great diligence and success the modifications which sounds of all kinds undergo from distance, obstructions and other causes, but had acquired the art of imitating them in the highest perfection. The ventriloquist was therefore able to carry on a dialogue in which the *dramatis voces*, as they may be called, were numerous; and when on the outside of an apartment he could personate a mob with its infinite variety of noise and vociferation. Their influence over an audience was still farther extended by a singular power over the muscles of the body. Mr Fitz-James actually succeeded in making the opposite or corresponding muscles act differently from each other; and while one side of his face was merry and laughing, the other was full of sorrow and in tears. At one moment he was tall, thin, and melancholic, and after passing behind a screen he came out "bloated with obesity and staggering with fulness." M. Alexandre possessed the same power over his face and figure, and so striking was the contrast of two of these forms, that an excellent sculptor, Mr Joseph, has perpetuated them in marble.

This new acquirement of the ventriloquist enabled him, in his own single person and with his own single voice, to represent upon the stage a dramatic composition which would have required the assistance of several actors. Although only one character in the piece could be seen at the same time, yet they all appeared during its performance, and the change of face and figure on the part of the ventriloquist was so perfect that his personal identity could not be recognized in the *dramatis personæ*. This deception was rendered still more complete by a particular construction of the dresses, which enabled the performer to reappear in a new character after an interval so short that the audience necessarily believed that it was another person.

It is a curious circumstance that Captain Lyon found among the Eskimaux of Igloolik ventriloquists of no mean skill. There is much rivalry amongst the professors of the art, who do not expose each other's secrets, and their exhibitions derive great importance from the rarity of their occurrence. The following account of one of them is so interesting that we shall give the whole of it in Captain Lyon's words.

"Amongst our Igloolik acquaintances were two females and a few male wizards, of whom the principal was Toolemak. This personage was cunning and intelligent, and, whether professionally, or from his skill in the chace, but perhaps from both reasons, was considered by all the tribe as a man of importance. As I invariably paid great deference to his opinion on all subjects connected with his calling, he freely communicated to me his superior knowledge, and did not scruple to allow of my being present at his interviews with Tornga, or his patron spirit. In consequence of this, I took an early opportunity of requesting my friend to exhibit his skill in my cabin. His old wife was with him, and by much flattery and an accidental display of a glittering knife and some beads, she assisted me in obtaining my request. All light excluded, our sorcerer began chanting to his wife with great vehemence, and she in return answered by singing the Amna-aya, which was not discontinued during the whole ceremony. As far as I could hear, he afterwards began turning himself rapidly round, and in a loud powerful voice vociferated for Tornga with great impatience, at the same time blowing and snorting like a walrus. His noise, impatience, and agitation, increased every moment, and he at length seated himself on the deck, varying his tones, and making a rustling with his clothes. Suddenly the voice seemed smothered, and was so managed as to sound as if retreating beneath the deck, each moment becoming more distant, and ultimately giving the idea of being many feet below the cabin, when it ceased entirely. His wife now, in answer to my queries, informed me very seriously, that he had dived, and that he would send up Tornga. Accordingly, in about half a minute, a distant blowing was heard very slowly approaching, and a voice, which differed from that at first heard, was at times mingled with the blowing, until at length both sounds became distinct, and the old woman informed me that Tornga was come to answer my questions. I accordingly asked several questions of the sagacious spirit, to each of which inquiries I received an answer by two loud claps on the deck, which I was given to understand were favourable.

"A very hollow, yet powerful voice, certainly much different from the tones of Toolemak, now chanted for some time, and a strange jumble of hisses, groans, shouts, and gabblings like a turkey succeeded in rapid order. The old woman sang with increased energy, and as I took it for granted that this was all intended to astonish the Kabloona, I cried repeatedly that I was very much afraid. This, as I expected, added fuel to the fire, until the poor immortal, exhausted by its own might, asked leave to retire.

"The voice gradually sunk from our hearing as at first, and a very indistinct hissing succeeded; in its advance, it sounded like the tone produced by the wind on the bass chord of an Æolian harp. This was soon changed to a rapid hiss like that of a rocket, and Toolemak with a yell announced his return. I had held my breath at the first distant hissing, and twice exhausted myself, yet our conjuror did not once respire, and even his returning and powerful yell was uttered without a previous stop or inspiration of air.

"Light being admitted, our wizard, as might be expected, was in a profuse perspiration, and certainly much exhausted by his exertions, which had continued for at least half an hour. We now observed a couple of bunches, each consisting of two stripes of white deer-skin and a long piece of sinew, attached to the back of his coat. These we had not seen before, and were informed that they had been sewn on by Tornga while he was below."

Captain Lyon had the good fortune to witness another of Toolemak's exhibitions, and he was much struck with the wonderful steadiness of the wizard throughout the whole performance, which lasted an hour and a half. He did not once appear to move, for he was so close to the skin behind which Captain Lyon sat, that if he had done so he must have perceived it. Captain Lyon did not hear the least rustling of his clothes, or even distinguish his breathing, although his outcries were made with great exertion.

Letter VIII

Musical and harmonic sounds explained – Power of breaking glasses with the voice – Musical sounds from the vibration of a column of air – And of solid bodies – Kaleidophone – Singular acoustic figures produced on sand laid on vibrating plates of glass – and on stretched membranes – Vibration of flat rulers, and cylinders of glass – Production of silence from two sounds – Production of darkness from two lights – Explanation of these singular effects – Acoustic automaton – Droz's bleating sheep – Maillardet's singing bird – Vaucanson's flute player – His pipe and tabor player – Baron Kempelen's talking engine – Kratzenstein's speaking machine – Mr Willis's researches.

AMONG the discoveries of modern science there are few more remarkable than those which relate to the production of harmonic sounds. We are all familiar with the effects of musical instruments, from the deep-toned voice of the organ to the wiry shrill of the Jew's harp. We sit entranced under their magical influence, whether the ear is charmed with the melody of their sounds, or the heart agitated by the sympathies which they rouse. But though we may admire their external form, and the skill of the artist who constructed them, we never think of inquiring into the cause of such extraordinary combinations.

Sounds of all kinds are conveyed to the organ of hearing through the air; and if this element were to be destroyed, all nature would be buried in the deepest silence. Noises of every variety, whether they are musical or discordant, high or low, move through the air of our atmosphere at the surface of the earth with a velocity of 1090 feet in a second, or 765 miles per hour; but in sulphurous acid gas sound moves only through 751 feet in a second, while in hydrogen gas it moves with the great velocity of 3000 feet. Along fluid and solid bodies, its progress is still more rapid. Through water it moves at the rate of 4708 feet in a second, through tin at the rate of 8175 feet, and through iron, glass, and some kinds of wood, at the rate of 18530 feet.

When a number of single and separate sounds follow each other in rapid succession, they produce a continued sound, in the same manner as a continuous circle of light is produced by whirling round a burning stick before the eye. In order that the sound may appear a single one to the ear, nearly sixteen separate sounds must follow one another every second. When these sounds are exactly similar, and recur at equal intervals, they form a musical sound. In order to produce such sounds from the air, it must receive at least sixteen equally distant impulses or strokes in a second. The most common way of producing this effect is by a string or wire A B, Figure 16.40, stretched between the fixed points A, B. If this string is taken by the middle and pulled aside, or if it is suddenly struck, it will vibrate between its two fixed points, as shown in the figure, passing alternately on each side of its axis A B, the vibrations gradually diminishing by the resistance of the air till the string is brought to rest. Its vibrations, however, may be kept up, by drawing a rosined fiddle-bow across it, and while it is vibrating it will give out a sound corresponding to the rapidity of its vibrations, and

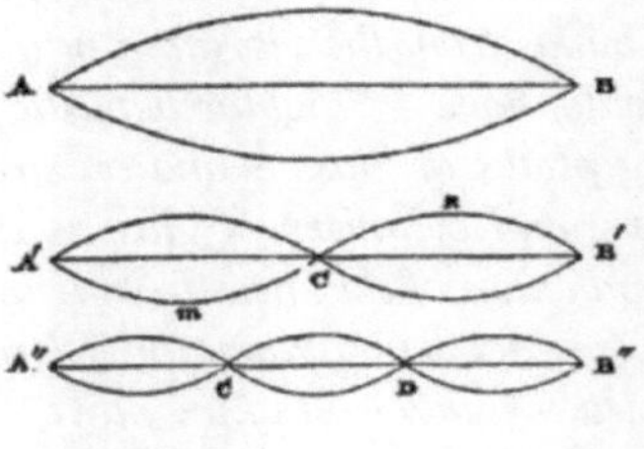

Figure 16.40

arising from the successive blows or impulses given to the air by the string. This sound is called the fundamental sound of the string, and its acuteness or sharpness increases with the number of vibrations which the string performs in a second.

If we now touch the vibrating string A′ B′ lightly with the finger, or with a feather at the middle point C, Figure 16.40, it will give out a more acute but fainter sound than before, and while the extent of its vibrations is diminished, their frequency is doubled. In like manner, if we touch the string A″ B″, Figure 16.40, at a point C, so that A″C is one-third of A″ B″, the note will be still more acute, and correspond to thrice the number of vibrations. All this might have been expected, but the wonderful part of the experiment is, that the vibrating string A′ B′ divides itself at C into two parts A′C, CB′, the part A′C vibrating round A′ and C as fixed points, and the part CB′ round C and B′, but always so that the part A′ C is at the same distance on the one side of the axis A′ B′ as at A *m* C, while the part C B is on the other side, as at C *n* B. Hence the point C, being always pulled by equal and opposite forces, remains at rest as if it were absolutely fixed. This stationary point is called a *node*, and the vibrating portions A′ *m* C, C *n* B′ loops. The very same is true of the string A″ B″, the points C and D being stationary points; and upon the same principle a string may be divided into any number of vibrating portions. In order to prove that the string is actually vibrating in these equal subdivisions, we have only to place a piece of light paper with a notch in it on different parts of the string. At the nodes C and D it will remain perfectly at rest, while at *m* or *n* in the middle of the loops it will be thrown off or violently agitated.

The acute sounds given out by each of the vibrating portions are called *harmonic sounds*, and they accompany the fundamental sound of the string in the very same manner, as we have already seen, that the eye sees the accidental or harmonic colours while it is affected with the fundamental colour.

The subdivision of the string, and consequently the production of harmonic sounds, may be effected without touching the string at all, and by means of a sympathetic action conveyed by the air. If a string A B, for example, Figure 16.40, is at rest, and if a shorter string A″ C, one-third of its length, fixed at the two points A″ and C is set a-vibrating in the same room, the string A B will be set a-vibrating in three loops like A″ B″, giving out the same harmonic sounds as the small string A″ C.

It is owing to this property of sounding bodies that singers with great power of voice are able to break into pieces a large tumbler glass, by singing close to it its proper fundamental note; and it is from the same sympathetic communication of vibrations that two pendulum clocks fixed to the same wall, or two watches lying upon the same table, will take the same rate of going, though they would not agree with one another if placed in separate apartments. Mr Ellicott even observed that the pendulum of the one clock will stop that of the other, and that the stopped pendulum will after a certain time resume its vibrations, and in its turn stop the vibrations of the other pendulum.

The production of musical sounds by the vibrations of a column of air in a pipe is familiar to every person, but the extraordinary mechanism by which it is effected is known principally to philosophers. A column of air in a pipe may be set a-vibrating by blowing over the open end of it, as is done in Pan's pipes, or by blowing over a hole in its side as in the flute, or by blowing through an aperture called a reed, with a flexible tongue, as in the clarionet. In order to understand the nature of this vibration, let A B, Figure 16.41, be a pipe or tube, and let us place in it a spiral spring A B, in which the coil or spire are at equal distances, each end of the spiral being fixed to the end of the tube. This elastic spring may be supposed to represent the air in the pipe, which is of equal density throughout. If we take hold of the spring at *m*, and push the point *m* towards A and towards B in succession, it will give us a good idea of the vibration of an elastic column of air. When *m* is pushed

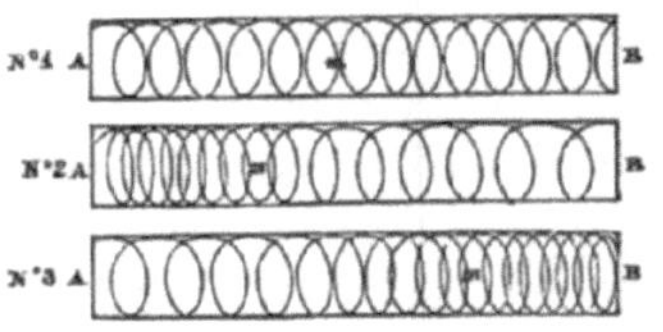

Figure 16.41

towards A, the spiral spring will be compressed or condensed, as shown at *m* A, No. 2, while at the other end, it will be dilated or rarefied, as shown at *m* B, and in the middle of the tube it will have the same degree of compression as in No. 1. When the string is drawn to the other end of the tube B, the spring will be, as in No. 3, condensed at the end B, and dilated at the end A. Now when a column of air vibrates in a pipe A B, the whole of it rushes alternately from B to A, as in No. 2, and from A to B as in No. 3, being condensed at the end A No. 2, and dilated or rarefied at the end B, while in No. 3 it is rarefied at A and condensed at B, preserving its natural density at the middle point between A and B. In the case of the spring the ends A B are alternately pushed outwards and pulled inwards by the spring, the end A being pushed outwards in No. 2, and B pulled inwards, while in No. 3 A is pulled inwards and B pushed outwards.

That the air vibrating in a pipe is actually in the state now described may be shown by boring small holes in the pipe, and putting over them pieces of a fine membrane. The membrane opposite to the middle part between A and B, where the particles of the air have the greatest motion, will be violently agitated, while at points nearer the ends A and B it will be less and less affected.

Let us now suppose two pipes A B, B C, to be joined together as in Figure 16.42, and to be separated by a fixed partition at B; and let a spiral spring be fixed in each. Let the spring AB be now pushed to the end A, while the spring BC is pushed to C, as in No. 1, and back again, as in No. 2,

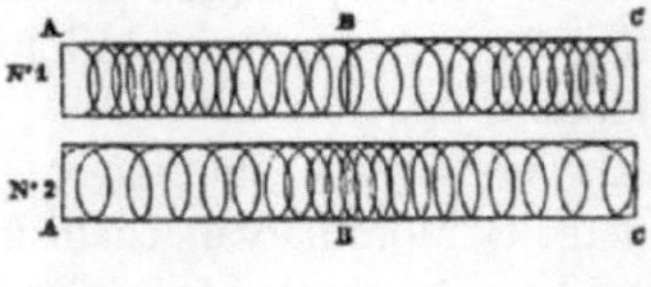

Figure 16.42

but always in opposite directions; then it is obvious that the partition B is in No. 1 drawn in opposite directions towards A and towards C, and always with forces equal to each other, that is, when B is drawn slightly towards A, which it is at the beginning of the motion, it is also drawn slightly towards C, and when it is drawn forcibly towards A, as it is at the end of the motion of the spring, it is also drawn forcibly towards C. If the partition B, therefore, is moveable, it will still remain fixed during the opposite excursions of the spiral springs: Nay, if we remove the partition, and hook the end of one spiral spring to the end of the other, the node or point of junction will remain stationary during the movements of the springs, because at every instant that point is drawn by equal and opposite forces. If *three, four*, or *five*, spiral springs are joined in a similar manner, we may conceive them all vibrating between their nodes in the same manner.

Upon the very same principles we may conceive a long column of air without partitions dividing itself into two, three, or four, smaller columns, each of which will vibrate between its nodes in the same manner as the spiral spring. At the middle point of each small vibrating column, the air will be of its natural density like that of the atmosphere, while at the nodes B, &c. it will be in a state of condensation and rarefaction alternately.

If, when the air is vibrating in one column in the pipe A B, as in Figure 16.41, No 2, 3, we conceive a hole made in the middle, the atmospheric air will not rush in to disturb the vibration, because the air within the pipe and without it has exactly the same density. Nay, if, instead of a single hole, we were to cut a ring out of the pipe at the middle point, the column would vibrate as before. But if we bore a hole between the middle and one of the ends, where the vibrating column must be either in a state of condensation or rarefaction, the air must either rush out or rush in, in order to establish the equilibrium. The air opposite the hole will then be brought to the state of the external air like that in the middle of the pipe, it will become the middle of a vibrating column, and the whole column of air, instead of vibrating as one, will vibrate as two columns, each column vibrating with twice the velocity, and yielding harmonic sounds along with the fundamental sound of the whole columns, in the same manner as we have already explained with regard to vibrating strings. By opening other holes we may subdivide a vibrating column into any number of smaller vibrating columns. The holes in flutes, clarionets, &c. are made for this purpose. When they are all closed up the air vibrates in one column, and by opening and shutting the different holes in succession, the number of vibrating columns is increased or diminished at pleasure, and consequently the harmonic sounds will vary in a similar manner.

Curious as these phenomena are, they are still surpassed by those which are exhibited during the vibration of solid bodies. A rod or bar of metal or glass may be made to vibrate either longitudinally or laterally.

An iron rod will vibrate longitudinally like a column of air if we strike it at one end in the direction of its length, or rub it in the same direction with a wetted finger, and it will emit the same fundamental note as a column of air *ten* or *eleven* times as long, because sound moves as much faster in iron than in air. When the iron rod is thus vibrating along its length, the very same changes which we have shown in Figure 16.41, as produced in a spiral spring or in a column of air, take place in the solid metal. All its particles move alternately towards A and towards B, the metal being in the one case condensed at the end to which the particles move, and expanded at the end from which they move, and retaining its natural density in the middle of the rod. If we now hold this rod in the middle, by the finger and thumb lightly applied, and rub it in the middle either of A B or B C with a piece of cloth sprinkled with powdered rosin, or with a well-rosined fiddle-bow drawn across the rod, it will divide itself into two vibrating portions A B, B C, each of which will vibrate, as shown in Figure 16.42, like the two adjacent columns of air, the section of the rod, or the particles which compose that section at B, being at perfect rest. By holding the rod at any intermediate point between A and B, so that the distance from A to the finger and thumb is one-third, one-fourth, one-fifth, &c. of the whole length A C, and rubbing one of the divisions in the middle, the rod will divide itself into 3, 4, 5, &c. vibrating portions, and give out corresponding harmonic sounds.

A rod of iron may be made to vibrate laterally or transversely by fixing one end of it firmly as in a vice, and leaving the other free, or by having both ends free or both fixed. When a rod, fixed at one end and free at the other, is made to vibrate, its mode of vibrating may be rendered evident to the eye; and for the purpose of doing this Mr Wheatstone has contrived a curious instrument called the *Kaleidophone*, which is shown in Figure 16.43. It consists of a circular base of wood A B, about *nine* inches in diameter and one inch thick, and having four brass sockets firmly fixed into it at C, D, E, and F, Into these sockets are screwed four vertical steel rods C, D, E, and F, about 13 or 14 inches long, one being a square rod, another a bent cylindrical one, and the other two cylindrical ones of different diameters. On the extremities of these rods are fixed

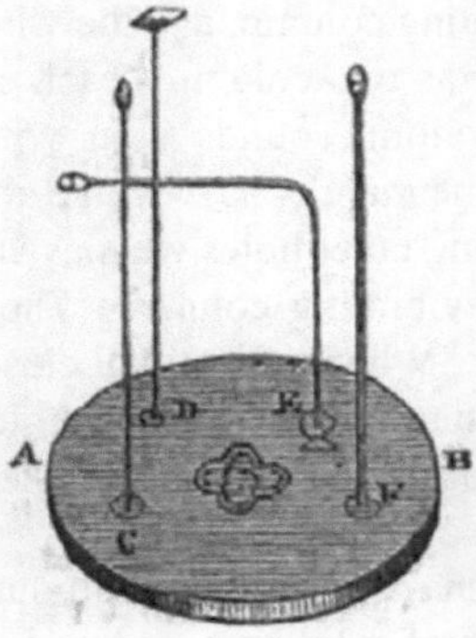

Figure 16.43

small quicksilvered glass beads, either singly or in groups, so that when the instrument is placed in the light of the sun or in that of a lamp, bright images of the sun or candle are seen reflected on each bead. If any of these rods is set a-vibrating, these luminous images will form continuous and returning curve lines in a state of constant variation, each different rod giving curves of different characters.

The melodion, an instrument of great power, embracing five octaves, operates by means of the vibrations of metallic rods of unequal lengths, fixed at one end, and free at the other. A narrow and thin plate of copper is screwed to the free extremity of each rod, and at right angles to its length; and its surface is covered with a small piece of felt impregnated with rosin. This narrow band is placed near the circumference of a revolving cylinder, and by touching the key it is made to descend till it touches the revolving cylinder, and gives out its sound. The sweetness and power of this instrument are unrivalled, and such is the character of its tones, that persons of a nervous temperament are often entirely overpowered by its effects.

The vibrations of plates of metal or glass of various forms exhibit a series of the most extraordinary phenomena which are capable of being shown by very simple means. These phenomena are displayed in an infinite variety of regular figures assumed by sand, or fine lycopodium powder, strewed over the surface of the glass plate. In order to produce these figures, we must pinch or damp the plate at one or more places, and when the sand is strewed upon its surface, it is thrown into vibrations by drawing a fiddle-bow over different parts of its circumference. The method of damping or pinching plates is shown in Figure 16.44 In No. 1, a square plate of glass A B, ground smooth at its edges, is pinched by the finger and thumb. In No. 2 a circular plate is held by the thumb against the top *c* of a perpendicular rod, and damped by the fingers at two different points of its circumference. In No. 3 it is damped at three points of its circumference, *c* and *d* by the thumb and finger, and at *e* by pressing it against a fixed obstacle *a b*. By a means of a clamp like that at No. 4, it may be damped at a greater number of points.

If we take a *square* plate of glass, such as that shown in Figure 16.45, No. 1, and pinching it at its centre, draw the fiddle-bow near one of its angles, the sand will accumulate in the form of a cross, as

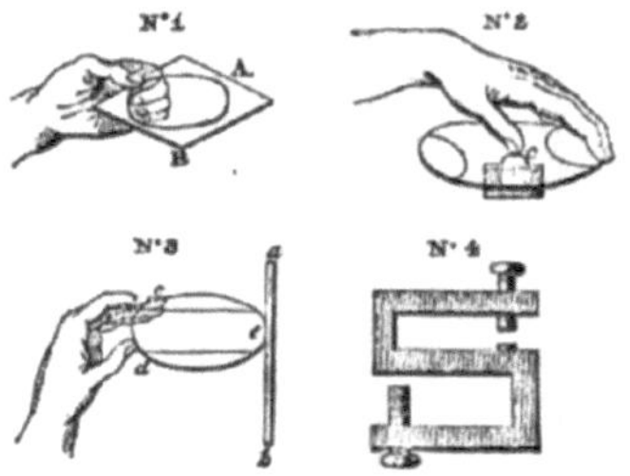

Figure 16.44

shown in the figure, being thrown off the parts of the plate that are in a state of vibration, and settling in the nodes or parts which are at rest. If the bow is drawn across the middle of one of the edges, the sand will accumulate as in No. 2. If the plate is pinched at N, No. 3, and the bow applied at F and perpendicular to A B, the sand will arrange itself in three parallel lines, perpendicular to a fourth passing through F and N. But if the point N where it is pinched, is a little farther from the edge than in No. 3, the parallel lines will change into curves as in No. 4.

If the plate of glass is circular and pinched at its centre, and also at a point of its circumference, and if the bow is applied at a point 45° from the last point, the figure of the sand will be as in Figure 16.46, No. 1. If with the same plate, similarly pinched,

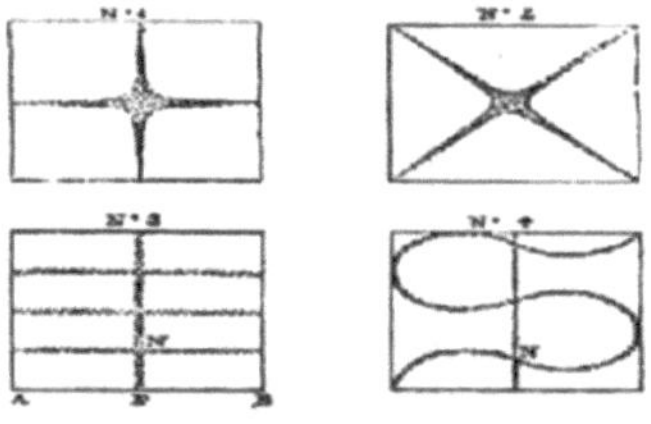

Figure 16.45

the bow is drawn over a part 30° from the pinched point of the circumference, the sand will form six radii as in No. 2. When the centre of the plate is left free, a different set of figures is produced as shown in No. 3 and No. 4. When the plate is pinched near its edge, and the bow applied 45° from the point pinched, a circle of sand will pass through that point, and two diameters of sand at right angles to each other, will be formed, as in No. 3. When a point of the circumference is pressed against a fixed obstacle, and the bow applied 30° from that point, the figure in No. 4 is produced.

If, in place of a solid plate, we strew the sand over a stretched membrane, the sand will form itself

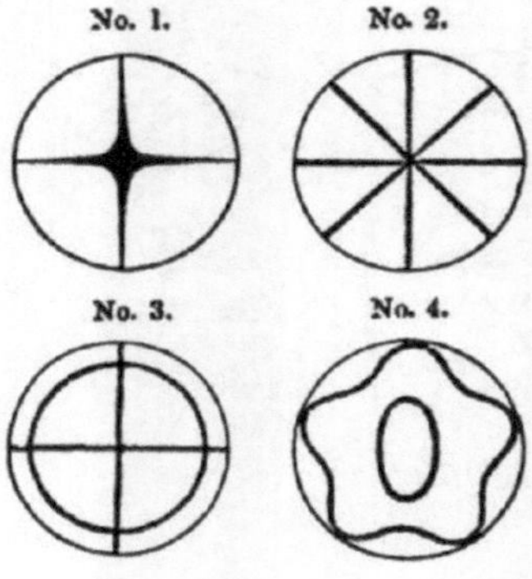

Figure 16.46

into figures, even when the vibrations are communicated to the membrane through the air. In order to make these experiments, we must stretch a thin sheet of wet paper, such as vegetable paper, over the mouth of a tumbler-glass with a footstalk, and fix it to the edges with glue. When the paper is dry, a thin layer of dry sand is strewed upon its surface. If we place this membrane upon a table, and hold immediately above it and parallel to the membrane, a plate of glass vibrating so as to give any of the figures shown in Figure 16.46, the sand upon the membrane will imitate exactly the figure upon the glass. If the glass plate, in place of vibrating horizontally, is made to vibrate in an inclined position, the figures on the membrane will change with the inclination, and the sand will assume the most curious arrangements. The figures thus produced vary with the size of the membrane, with its material, its tension, and its shape. When the same figure occurred several times in succession, a breath upon the paper will change its degree of tension, and produce an entirely new figure, which, as the temporary moisture evaporates, will return to the original figure, through a number of intermediate ones. The pipe of an organ at the distance of a few feet, or the notes of a flute at the distance of half a foot, will arrange the sand on the membrane into figures which perpetually change with the sound that is produced.

The manner in which flat rulers and cylinders of glass perform their vibrations is very remarkable. If a glass plate about twenty-seven inches long, six-tenths of an inch broad, and six-hundredths of an inch thick, is held by the edges between the finger and thumb, and has its lower surface, near either end, rubbed with a piece of wet cloth, sand laid upon its upper surface will arrange itself in parallel lines at right angles to the length of the plate. If the place of these lines is marked with a dot of ink, and the other side of the glass ruler is turned upwards, and the ruler made to vibrate as before, the sand will now accumulate in lines intermediate between the former lines, so that the motions of one-half the thickness of the glass ruler are precisely the reverse of those of the corresponding parts of the other half.

As these singular phenomena have not yet been made available by the scientific conjuror, we must be satisfied with this brief notice of them; but there is still one property of sound, which has its analogy also in light, too remarkable to be passed without notice. This property has more of the marvellous in it than any result within the wide range of the sciences. *Two loud sounds may be made to produce silence*, and *two strong lights may be made to produce darkness!*

If two equal and similar strings, or the columns of air in two equal and similar pipes, perform exactly 100 vibrations in a second, they will produce each equal waves of sound, and these waves will conspire in generating an uninterrupted sound, double of either of the sounds heard separately. If the two strings or the two columns of air are not in unison, but nearly so, as in the case where the one vibrates 100, and the other 101 times in a second, then at the first vibration the two sounds will form one of double the strength of either; but the one will gradually gain upon the other, till at the fiftieth vibration it has gained half a vibration on the other. At this instant the two sounds will *destroy one another*, and an interval of perfect silence will take place. The sound will instantly commence, and gradually increase till it becomes loudest at the hundredth vibration, where the two vibrations conspire in producing a sound double of either. An interval of silence will again occur at the 150th, 250th, 350th vibration, or every second, while a sound of double the strength of either will be heard at the 200dth, 300dth, and 400dth vibration. When the unison is very defective, or when there is a great difference between the number of vibrations which the two strings or columns of air perform in a second, the successive sounds and intervals of silence resemble a rattle. With a powerful organ the effect of this experiment is very fine, the repetition of the sounds *wow – wow – wow*, – representing the doubled sound and the interval of silence which arise from the total extinction of the two separate sounds.

The phenomena corresponding to this in the case of light, is perhaps still more surprising. If a beam of *red* light issues from a luminous point, and falls upon the retina, we shall see distinctly the luminous object from which it proceeds; but if another pencil of red light issues from another luminous point any how situated, provided the difference between its distance and that of the other luminous point from the point of the retina, on which the first beam fell, is the 258th thousandth part of an inch, or exactly *twice, thrice, four* times &c. that distance; and if this second beam falls upon the same point of the retina, the one light will increase the intensity of the other, and the eye will see *twice* as much light as when it received only one of the beams separately. All this is nothing more than what might be expected from our ordinary experience. But if the difference in the distances of the two luminous points is only *one-half* of the 258th thousandth part of an inch, or 1½, 2½, 3½, 4½, times that distance, *the one light will extinguish the other, and produce absolute darkness.* If the two luminous points are so situated, that the difference of their distances from the point of the retina, is intermediate between 1 and 1½, or 2 and 2½, above the 258th thousandth part of an inch, the intensity of the effect which they produce will vary from absolute darkness to double the intensity of either light. At 1¼, 2¼, 3¼, times, &c. the 258th thousandth of an inch, the intensity of the two combined lights will be equal only to one of them acting singly. If the lights, in place of falling upon the retina, fall upon a sheet of white paper, the very same effect will be produced, a black spot being produced in the one case, and a bright white one in the other, and intermediate degrees of brightness in intermediate cases. If the two lights are *violet*, the difference of distances at which the preceding phenomena will be produced, will be the 157th thousandth part of an inch, and it will be intermediate between the 258th and the 157th thousand part of an inch, for the intermediate colours. This curious phenomenon may be easily shown to the eye, by admitting the sun's light into a dark room through a small hole about the 40th or 50th part of an inch in diameter, and receiving the light on a sheet of paper. If we hold a needle or piece of slender wire in this light, and examine its shadow, we shall find that the shadow consists of bright and dark stripes succeeding each other alternately, the stripe in the very middle or axis of the shadow being a bright one. The rays of light which are bent into the shadow, and which meet in the very middle of the shadow, have exactly the same length of path, so that they form a bright fringe of double the intensity of either; but the rays which fall upon a point of the shadow at a certain distance from the middle, have a difference in the length of their paths, corresponding to the difference at which the lights destroy each other, so that a *black* stripe is produced on each side of the middle bright one. At a greater distance from the middle, the difference becomes such as to produce a bright stripe, and so on, a bright and a dark stripe succeeding each other to the margin of the shadow.

The explanation which philosophers have given of these strange phenomena is very satisfactory, and may be easily understood. When a wave is made on the surface of a still pool of water, by plunging a stone into it, the wave advances along the surface, while the water itself is never carried forward, but merely rises into a height and falls into a hollow, each portion of the surface experiencing an elevation and a depression in its turn. If we suppose two waves equal and similar to be produced by two separate stones, and if they reach the same spot at the same time, that is, if the two elevations should exactly coincide, they would unite their effects and produce a wave twice the size of either; but if the one wave should be just so far before the other, that the hollow of the one coincided with the elevation of the other, and the elevation of the one with the hollow of the other, the two waves would obliterate or destroy one another, the elevation as it were of the one filling up half the hollow of the other, and the hollow of the one taking away half the elevation of the other, so as to reduce the surface to a level. These effects will be actually exhibited by throwing two equal stones into a pool of water, and it will be seen that there are certain lines of a hyperbolic form where the water is quite smooth, in consequence of the equal waves obliterating one another, while in other adjacent parts, the water is raised to a height corresponding to both the waves united.

In the tides of the ocean, we have a fine example of the same principle. The two immense waves arising from the action of the sun and moon upon the ocean, produce our spring-tides by their combination, or when the elevations of each coincide, and our neap-tides, when the elevation of the one wave coincides with the depression of the other. If the sun and moon had exerted exactly the same force upon the ocean, or produced tide waves of the same size, then our neap-tides would have disappeared altogether, and the spring-tide would have been a wave double of the wave produced by the sun and moon separately. An example of the effect of the equality of the two waves occurs in the port of Batsha, where the two waves arrive by channels of different lengths, and actually obliterate each other.

Now, as sound is produced by undulations or waves in the air, and as light is supposed to be produced by waves or undulations in an etherial medium, filling all nature, and occupying the pores of transparent bodies, the successive production of sound and silence, by two loud sounds, or of light and darkness by two bright lights, may be explained in the very same manner as we have explained the increase and the obliteration of waves formed on the surface of water. If this theory of light be correct, then the breadth of a wave of *red* light will be the 258th thousandth part of an inch, the breadth of a wave of green light the 207th thousandth part of an inch, and the breadth of a wave of violet light, the 157th thousandth part of an inch.

Among the wonders of modern skill, we must enumerate those beautiful automata by which the motions and actions of man and other animals have been successfully imitated. I shall therefore describe at present some of the most remarkable acoustic automata, in which the production of musical and vocal sounds has been the principal object of the artist.

Many very ingenious pieces of acoustic mechanism have been from time to time exhibited in Europe. The celebrated Swiss mechanist, M. Le Droz, constructed for the King of Spain the figure of a sheep, which imitated in the most perfect manner the bleating of that animal, and likewise the figure of a dog watching a basket of fruit, which, when any of the fruit was taken away, never ceased barking till it was replaced.

The singing-bird of M. Maillardet, which he exhibited in Edinburgh many years ago, is still more wonderful. An oval box, about three inches long, was set upon the table, and in an instant the lid flew up, and a bird of the size of the humming-bird, and of the most beautiful plumage, started from its nest. After fluttering its wings, it opened its bill and performed four different kinds of the most beautiful warbling. It then darted down into its nest, and the lid closed upon it. The moving power in this piece of mechanism is said to have been springs which continued their action only four minutes. As there was no room, within so small a figure, for accommodating pipes to produce the great variety of notes which were warbled, the artist used only one tube, and produced all the variety of sounds by shortening and lengthening it with a moveable piston.

Ingenious as these pieces of mechanism are, they sink into insignificance when compared with the machinery of M. Vaucanson, which had previously astonished all Europe. His two principal automata were the flute-player, and the pipe and tabor-player. The flute-player was completed in 1736, and wherever it was exhibited it produced the greatest sensation. When it came to Paris it was received with great suspicion. The French scavans recollected the story of M. Raisin, the organist of Troyes, who exhibited an automaton player upon the harpsichord, which astonished the French court by the variety of its powers. The curiosity of the King could not be restrained, and in consequence of his insisting upon examining the mechanism, there was found in the figure a pretty little musician five years of age. It was natural, therefore, that a similar piece of mechanism should be received with some distrust; but this feeling was soon removed by M. Vaucanson, who exhibited and explained to a committee of the Academy of Sciences the whole of the mechanism. This learned body was astonished at the ingenuity which it displayed; and they did not hesitate to state, that the machinery employed for producing the sounds of the flute, performed in the most exact manner the very operations of the most expert flute-player, and that the artist had imitated the effects produced, and the means employed by nature, with an accuracy which exceeded all expectation. In 1738, M. Vaucanson published a memoir, approved of by the academy, in which he gave a full description of the machinery employed, and of the principles of its construction. Following this memoir, I shall therefore attempt to give as popular a description of the automaton as can be done without lengthened details and numerous figures.

The body of the flute-player was about 5½ feet high, and was placed upon a piece of rock, surrounding a square pedestal 4½ feet high by three and a half wide. When the panel which formed the front of the pedestal was opened, there was seen on the right a clock movement, which, by the aid of several wheels, gave a rotatory motion to a steel axis about 2½ feet long, having cranks at six equidistant points of its length, but lying in different directions. To each crank was attached a cord, which descended and was fixed by its other end to the upper board of a pair of bellows, 2½ feet long and six inches wide. Six pair of bellows arranged along the bottom of the pedestal were then wrought, or made to blow in succession, by turning the steel axis.

At the upper face of the pedestal, and upon each pair of bellows, is a double pulley, one of whose rims is 3 inches in diameter, and the other 1½. The cord which proceeds from the crank coils round the smallest of these pulleys, and that which is fixed to the upper board of the bellows goes round the larger pulley. By this means the upper board of the bellows is made to rise higher than if the cords went directly from them to the cranks.

Round the larger rims of three of these pulleys, viz. those on the right hand, there are coiled three cords, which, by means of several smaller pulleys, terminate in the upper boards of other three pair of bellows placed on the top of the box.

The tension of each cord when it begins to raise the board of the bellows to which it is attached, gives motion to a lever placed above it between the axis and the double pulley in the middle and lower region of the box. The other end of this lever keeps open the valve in the lower board of the bellows, and allows the air to enter freely, while the upper board is rising to increase the capacity of the bellows. By this means there is not only power gained, in so far as the air gains easier admission through the valve, but the fluttering noise produced by the action of the air upon the valves is entirely avoided, and the nine pair of bellows are wrought with great ease, and without any concussion or noise.

These nine bellows discharge their wind into three different and separate tubes. Each tube receives the wind of three bellows, the upper boards of one of the three pair being loaded with a weight of four pounds, those of the second three pair with a weight of two pounds, and those of the other three pair with no weight at all. These three tubes ascended through the body of the figure, and terminated in three small reservoirs placed in its trunk. These reservoirs were thus united into one, which, ascending into the throat, formed by its enlargement the cavity of the mouth terminated by two small lips, which rested upon the hole of the flute. These lips had the power of opening more or less, and by a particular mechanism, they could advance or recede from the hole in the flute. Within the cavity of the mouth there is a small moveable tongue for opening and shutting the passage for the wind through the lips of the figure.

The motions of the fingers, lips, and tongue, of the figure, were produced by means of a revolving cylinder thirty inches long, and twenty-one in diameter. By means of pegs and brass staples fixed in fifteen different divisions in its circumference, fifteen different levers, similar to those in a barrel-organ, were raised and depressed. Seven of these regulated the motions of the seven fingers for stopping the holes of the flute, which they did by means of steel chains rising through the body and directed by pulleys to the shoulder, elbow, and fingers. Other three of the levers communicating with the valves of the three reservoirs, regulated the ingress of the air, so as to produce a stronger or a weaker tone. Another lever opened the lips so as to give a free passage to the air, and another contracted them for the opposite purpose. A third lever drew them backwards from the orifice of the flute, and a fourth pushed them forward. The remaining lever enabled the tongue to stop up the orifice of the flute.

Such is a very brief view of the general mechanism by which the requisite motions of the flute-player were produced. The airs which it played were probably equal to those executed by a living performer, and its construction, as well as its performances, continued for many years to delight and astonish the philosophers and musicians of Europe.

Encouraged by the success of this machine, M. Vaucanson exhibited in 1741 other automata, which were equally, if not more, admired. One of these was the automaton duck, which performed all the motions of that animal, and not only ate its food, but digested it; and the other was his pipe and tabor-player, a piece of mechanism which required all the resources of his fertile genius. Having begun this machine before he was aware of its peculiar difficulties, he was often about to abandon it in despair, but his patience and his ingenuity combined, enabled him not only to surmount every difficulty, but to construct an automaton which performed complete airs, and greatly excelled the most esteemed peformers on the pipe and tabor.

The figure stands on a pedestal, and is dressed like a dancing shepherd. He holds in one hand a flageolet, and in the other the stick with which he beats the tambourin as an accompaniment to the airs of the flageolet, about twenty of which it is capable of performing. The flageolet has only three holes, and the variety of its tones depends principally on a proper variation of the force of the wind, and on the different degrees with which the orifices are covered. These variations in the force of the wind required to be given with a rapidity which the ear can scarcely follow, and the articulation of the tongue was required for the quickest notes, otherwise the effect was far from agreeable. As the human tongue is not capable of giving the requisite articulations to a rapid succession of notes, and generally slurs over one-half of them, the automaton was thus able to excel the best performers, as it played complete airs with articulations of the tongue at every note.

In constructing this machine M. Vaucanson observed that the flageolet must be a most fatiguing instrument for the human lungs, as the muscles of the chest must make an effort equal to 56 pounds in order to produce the highest notes. A single ounce was sufficient for the lowest notes, so that we may, from this circumstance, form an idea of the variety of intermediate effects required to be produced.

While M. Vaucanson was engaged in the construction of these wonderful machines, his mind was filled with the strange idea of constructing an automaton containing the whole mechanism of the circulation of the blood. From some birds which he made he was satisfied of its practicability; but as the whole vascular system required to be made of elastic gum or caoutchouc, it was supposed that it could only be executed in the country where the caoutchouc tree was indigenous. Louis XVI. took a deep interest in the execution of this machine. It was agreed that a skilful anatomist should proceed to Guyana to superintend the construction of the blood-vessels, and the King had not only approved of, but had given orders for, the voyage. Difficulties, however, were thrown in the way: Vaucanson became disgusted, and the scheme was abandoned.

The two automata which we have described were purchased by Professor Bayreuss of Helmstadt; but we have not been able to learn whether or not they still exist.

Towards the end of the nineteenth century a bold and almost successful attempt was made to construct a *talking automaton*. In the year 1779, the Imperial Academy of Sciences at St Petersburgh proposed as the subject of one of their annual prizes an inquiry into the nature of the vowel sounds, A, E, I, O and U, and the construction of an instrument for artificially imitating them. This prize was gained by M. Kratzenstein, who showed that all the vowels could be distinctly pronounced by blowing through a reed into the lower ends of the pipes of the annexed figures, as shown in Figure 16.47, where the corresponding vowels are marked on the different

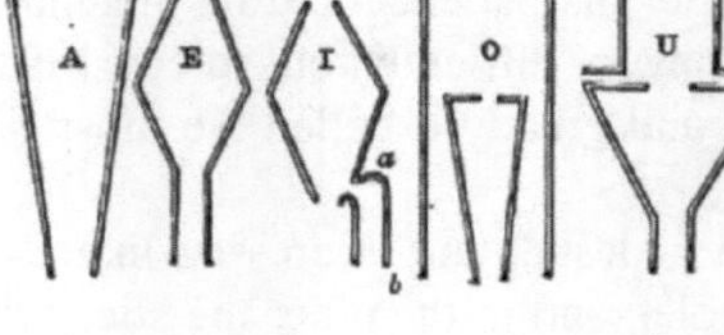

Figure 16.47

pipes. The vowel I is pronounced by merely blowing into the pipe *a b*, of the pipe marked I, without the use of a reed.

About the same time that Kratzenstein was engaged in these researches, M. Kempelen of Vienna, a celebrated mechanician, was occupied with the same subject. In his first attempt he produced the vowel sounds, by adapting a reed R,

Figure 16.48, to the bottom of a funnel-shaped cavity A B, and placing his hand in various positions within the funnel. This contrivance, however, was not fitted for his purpose, but after long study, and a diligent examination of the organs of speech, he contrived a hollow oval box, divided into two portions attached by a hinge as so to resemble jaws. This box received the sound which issued from the tube connected with the reed, and by opening and closing the jaws, he produced the sounds, A, O, O U, and an imperfect E, but no indications of an I. After two years labour he succeeded in obtaining from different jaws

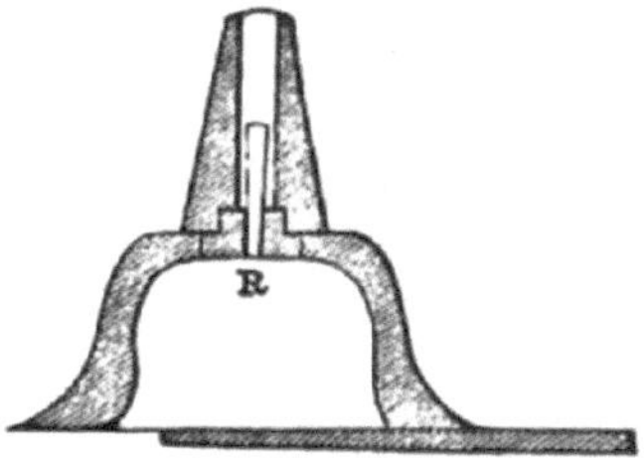

Figure 16.48

the sounds of the consonants P, M, L, and by means of these vowels, and consonants, he could compose syllables and words, such as *mama, papa, aula, lama, mulo*. The sounds of two adjacent letters, however, run into each other, and an aspiration followed some of the consonants, so that instead of *papa* the word sounded *phaa-ph-a;* these difficulties he contrived with much labour to surmount, and he found it necessary to imitate the human organs of speech by having only one mouth and one glottis. The mouth consisted of a funnel or bell-shaped piece of elastic gum, which approximated, by its physical properties, to the softness and flexibility of the human organs. To the mouth-piece was added a nose made of two tin tubes, which communicated with the mouth. When both these tubes were open, and the mouth-piece closed, a perfect M was produced, and when one was closed and the other open, an N was sounded. M. Kempelen could have succeeded in obtaining the four letters D, G, K, T, but by using a P instead of them, and modifying the sound in a particular manner, he contrived to deceive the ear by a tolerable resemblance of these letters.

There seems to be no doubt that he at last was able to produce entire words, and sentences such as, *opera, astronomy, Constantinopolis, vous etés mon ami, je vous aime de tout mon cœur, venez avec moi à Paris, Leopoldus secundus, Romanorum imperator semper Augustus, &c.* but he never fitted up a speaking figure, and probably, from being dissatisfied with the general result of his labours, he exhibited only to his private friends the effects of the apparatus, which was fitted up in the form of a box.

This box was rectangular, and about three feet long, and was placed upon a table and covered with a cloth. When any particular word was mentioned by the company, M. Kempelen caused the machine to pronounce it, by introducing his hands beneath the cloth, and apparently giving motion to some parts of the apparatus. Mr Thomas Collinson, who had seen this machine in London, mentions in a letter to Dr Hutton, that he afterwards saw it at M. Kempelen's own house in Vienna, and that he then gave it the same word to be pronounced, which he gave it in London, viz. the word *Exploitation*, which, he assures us, it again distinctly pronounced with the French accent.

M. Kratzenstein seems to have been equally unsuccessful, for though he assured M. De Lalande, when he saw him in Paris in 1786, that he had made a machine which could speak pretty well, and though he showed him some of the apparatus by which it could sound the vowels, and even such syllables as *papa* and *mama*, yet there is no reason to believe that he had accomplished more than this.

The labours of Kratzenstein and Kempelen have been recently pursued with great success by our ingenious countryman Mr Willis of Cambridge. In repeating Kempelen's experiment shown in Figure 16.48, he used a shallower cavity, such as that in Figure 16.49, and found that he could entirely dispense with the introduction of the hand, and could obtain the

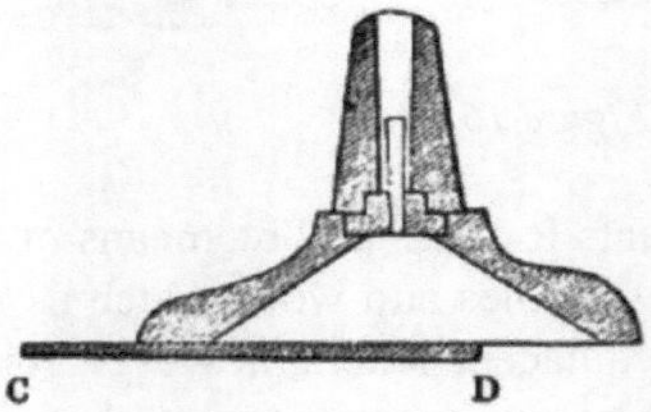

Figure 16.49

whole series of vowels, by sliding a flat board C D over the mouth of the cavity. Mr Willis then conceived the idea of adapting to the reed cylindrical tubes, whose length could be varied by sliding joints. When the tube was greatly less than the length of a stopped pipe in unison with the reed, it sounded I, and by increasing the length of the tube it gave E, A, O and U in succession. But what was very unexpected, when the tube was so much lengthened as to be 1½ times the length of a stopped pipe in unison with the reed, the vowels began to be again sounded in an inverted order, viz. U, O, A, E, and then again in a direct order, I, E, A, O, U, when the length of the tube was equal to twice that of a stopped pipe, in unison with the reed.

Some important discoveries have been recently made by M. Savart respecting the mechanism of the human voice, and we have no doubt that, before another century is completed, a *Talking* and a *Singing machine* will be numbered among the conquests of Science.

Letter IX

Singular effects in nature depending on sound – Permanent character of speech – Influence of great elevations on the character of sounds, and on the powers of speech – Power of sound in throwing down buildings –

Dog killed by sound – Sounds greatly changed under particular circumstances – Great audibility of sounds during the night explained – Sounds deadened in media of different densities – Illustrated in the case of a glass of champagne – and in that of new fallen snow – Remarkable echoes – Reverberations of thunder – Subterranean noises – Remarkable one at the Solfaterra – Echo at the Menai Suspension bridge – Temporary deafness produced in diving bells – Inaudibility of particular sounds to particular ears – Vocal powers of the statue of Memnon – Sounds in granite rocks – Musical mountain of El-Nakous

ALTHOUGH among the phenomena of the material world there is scarcely one, which, when well considered, is not an object of wonder, yet those which we have been accustomed to witness from our infancy lose all their interest from the frequency of their occurrence, while to the natives of other countries they are unceasing objects of astonishment and delight. The inhabitant of a tropical climate is confounded at the sight of falling snow, and he almost discredits the evidence of his senses, when he sees a frozen river carrying loaded waggons on its surface. The diffusion of knowledge by books, as well as by frequent communication between the natives of different quarters of the globe, have deprived this class of local wonders of their influence, and the Indian and the Scandinavian can visit each other's lands without any violent excitement of surprise. Still, however, there are phenomena of rare occurrence, of which no description can convey the idea, and which continue to be as deeply marked with the marvellous as if they had been previously unknown. Among these we may rank the remarkable modifications which sound undergoes in particular situations and under particular circumstances.

In the ordinary intercourse of life, we recognize individuals as much by their voice as by the features of their face and the form of their body. A friend who has been long absent will often stand before us as a stranger, till his voice supplies us with the full power of recognition. The brand imprinted by time on his outer form may have effaced the youthful image which the memory had cherished, but the original character of his voice and its yet remembered tones will remain unimpaired.

An old friend with a new face is not more common in its moral than in its physical acceptation, and though the sagacity of proverbial wisdom has not supplied us with the counterpart in relation to the human voice, yet the influence of its immutability over the mind has been recorded by the poet in some of his most powerful conceptions. When Manfred was unable to recognize in the hectic phantom of Astarte the endeared lineaments of the being whom he loved, the mere utterance of his name recalled "the voice which was his music," and invested her with the desired reality.

Say on, say on –
I live but in the Sound – It is thy voice!

BYRON.

The permanence of character thus impressed upon speech exists only in those regions to whose atmosphere our vocal organs are adapted. If either the speaker or the hearer is placed in air differing greatly in density from that to which they are accustomed, the voice of the one will emit different sounds, or the same sounds will produce a different impression on the ear of the other. But if both parties are placed in this new atmosphere, their tones of communication will suffer the most remarkable change. The two extreme positions, where such effects become sufficiently striking, are in the compressed air of the diving-bell, when it is immersed to a great depth in the sea, or in the rarefied atmosphere which prevails on the summit of the Himalaya or the Andes.

In the region of common life, and even at the stillest hour of night, the ear seldom rests from its toils. When the voice of man and the bustle of his labours have ceased, the sounds of insect life are redoubled, the night breeze awakens among the rustling leaves, and the swell of the distant ocean, and the sounds of the falling cataract or of the murmuring brook, fill the air with their pure and solemn music. The sublimity of deep silence is not to be found even in the steppes of the Volga, or in the forests of the Orinoco. It can be felt only in those lofty regions

> Where the tops of the Andes
> Shoot soaringly forth.

As the traveller rises above the limit of life and motion, and enters the region of habitual solitude, the death-like silence which prevails around him is rendered still more striking by the diminished density of the air which he breathes. The voice of his fellow traveller ceases to be heard even at a moderate distance, and sounds which would stun the ear at a lower level make but a feeble impression. The report of a pistol on the top of Mont Blanc is no louder than that of an Indian cracker. But while the thinness of the air thus subdues the loudest sounds, the voice itself undergoes a singular change: the muscular energy by which we speak experiences a great diminution, and our powers of utterance, as well as our power of hearing, are thus singularly modified. Were the magician, therefore, who is desirous to impress upon his victim or upon his pupil the conviction of his supernatural power, to carry him, under the injunction of silence,

> – – – – – to breathe
> The difficult air of the iced mountain's top,
> Where the birds dare not build, nor insects wing
> Flit o'er the herbless granite,

he would experience little difficulty in arresting his power over the elements, and still less in subsequently communicating the same influence to his companion.

But though the air at the tops of our highest mountains is scarcely capable of transmitting sounds of ordinary intensity, yet sounds of extraordinary power force their way through its most attenuated strata. At elevations where the air is three thousand times more rare than that which we breathe, the explosion of meteors is heard like the sound of cannon on the surface of the earth, and the whole air is often violently agitated by the sound. This fact alone may give us some idea of the tremendous nature of the forces which such explosions create, and it is fortunate for our species, that they are confined to the upper regions of the atmosphere. If the same explosions were to take place in the dense air which rests upon the earth, our habitations and our lives would be exposed to the most imminent peril.

Buildings have often been thrown down by violent concussions of the air, occasioned either by the sound of great guns, or by loud thunder, and the most serious effects upon human and animal life have been produced by the same cause. Most persons have experienced the stunning pain produced in the ear, when placed near a cannon that is discharged. Deafness has frequently been the result of such sudden concussions, and if we may reason from analogy, death itself must often have been the consequence. When peace was proclaimed in London in 1697, two troops of horse were dismounted and drawn up in line in order to fire their vollies. Opposite the centre of the line was the door of a butcher's shop, where there was a large mastiff dog of great courage. This dog was sleeping by the fire, but when the first volley was fired, it immediately started up, ran into another room, and hid itself under a bed. On the firing of the second volley, the dog rose, run several times about the room trembling violently and apparently in great agony. When the third volley was fired, the dog ran about once or twice with great violence, and instantly fell down dead, throwing up blood from his mouth and nose.

Sounds of known character and intensity are often singularly changed even at the surface of the earth according to the state of the ground and the conditions of the clouds. On the extended heath, where there are no solid objects capable of reflecting or modifying sound, the sportsman must frequently have noticed the unaccountable variety of sounds which are produced by the report of his fowling-piece. Sometimes they are flat and prolonged, at other times short and sharp, and sometimes the noise is so strange that it is referred to some mistake in the loading of the gun. These variations, however, arise entirely from the state of the air, and from the nature and proximity of the superjacent clouds. In pure air of uniform density the sound is sharp and soon over, as the undulations of the air advance without any interrupting obstacles. In a foggy atmosphere, or where the vapours produced by heat are seen dancing as it were in the air, the sound is dull and prolonged, and when these clouds are immediately over head, a succession of echoes from them produces a continued or a reverberating sound. When the French astronomers were determining the velocity of sound by firing great guns, they observed that the report was always single and sharp under a perfectly clear sky, but indistinct, and attended by a long continued roll like thunder, when a cloud covered a considerable part of the horizon. It is no doubt owing to the same cause, namely, the reflexion from the clouds, that the thunder rolls through the heavens, as if it were produced by a succession of electric explosions.

The great audibility of sounds during the night is a phenomenon of considerable interest, and one which had been observed even by the ancients. In crowded cities or in their vicinity, the effect was generally ascribed to the rest of animated beings, while in localities where such an explanation was inapplicable, it was supposed to arise from a favourable direction of the prevailing wind. Baron Humboldt was particularly struck with this phenomenon when he first heard the rushing of the great cataracts of the Orinoco in the plain which surrounds the mission of the Apures. These sounds he regarded as three times louder during the night than during the day. Some authors ascribed this fact to the cessation of the humming of insects, the singing of birds, and the action of the wind on the leaves of the trees, but M. Humboldt justly maintains that this cannot be the cause of it on the Orinoco, where the buzz of insects is much louder in the night than in the day, and where the breeze never rises till after sunset. Hence he was led to ascribe the phenomenon to the perfect transparency and uniform density of the air, which can exist only at night after the heat of the ground has been uniformly diffused through the atmosphere. When the rays of the sun have been beating on the ground during the day, currents of hot air of different temperatures, and consequently of different densities, are constantly ascending from the ground and mixing with the cold air above. The air thus ceases to be a homogeneous medium, and every person must have observed the effects of it upon objects seen through it which are very indistinctly visible, and have a tremulous motion, as if they were "dancing in the air." The very same effect is perceived when we look at objects through spirits and water that are not perfectly mixed, or when we view distant objects over a red hot poker or over a flame. In all these cases the light suffers refraction in passing from a medium of one density into a medium of a different density, and the refracted rays are constantly changing their direction as the different currents rise in succession. Analogous effects are produced when sound passes through a mixed medium, whether it consists of two different mediums or of one medium where portions of it have different densities. As sound moves with different velocities through media of different densities, the wave which produces the sound will be partly reflected in passing from one medium to the other, and the direction of the transmitted wave changed; and hence in passing through such media different portions of the wave will reach the ear at different times, and thus destroy the sharpness and distinctness of the sound. This may be proved by many striking facts. If we put a bell in a receiver containing a mixture of hydrogen gas and atmospheric air, the sound of the bell can scarcely be heard. During a shower of rain or of snow, noises are greatly deadened, and when sound is transmitted along an iron wire or an iron pipe of sufficient length, we actually hear two sounds, one transmitted more rapidly through the solid, and the other more slowly through the air. The same property is well illustrated by an elegant and easily repeated experiment of Chladni's. When sparkling champagne is poured into a tall glass till it is half full, the glass loses its power of ringing by a stroke upon its edge, and emits only a disagreeable and puffy sound. This effect will continue while the wine is filled with bubbles of air, or as long as the effervescence lasts; but when the effervescence begins to subside, the sound becomes clearer and clearer, and the glass rings as usual when the air-bubbles have vanished. If we reproduce the effervescence by stirring the champagne with a piece of bread the glass will again cease to ring. The same experiment will succeed with other effervescing fluids.

The difference in the audibility of sounds that pass over homogeneous and over mixed media is sometimes so remarkable as to astonish those who witness it. The following fact is given on the evidence of an officer who observed it: When the British and the American forces were encamped on each side of a river, the outposts were so near that the form of individuals could be easily distinguished. An American drummer made his appearance and began to beat his drum, but though the motion of his arms were distinctly seen, not a single sound reached the ear of the observer. A coating of snow that had newly fallen upon the ground, and the thickness of the atmosphere, had conspired to obstruct the sound. An effect the very reverse of this is produced by a coating of glazed or hardened snow, or by an extended surface of ice or water. Lieutenant Foster was able to carry on a conversation with a sailor across Port Bowen harbour, a distance of no less than a mile and a quarter, and the sound of great guns has been heard at distances varying from 120 to 200 miles. Over hard and dry ground of an uniform character, or where a thin soil rests upon a continuous stratum of rock, the sound is heard at a great distance, and hence it is the practice among many eastern tribes to ascertain the approach of an enemy by applying the ear to the ground.

Many remarkable phenomena in the natural world are produced by the reflexion and concentration of sound. Every person is familiar with the ordinary *Echo* which arises from the reflexion of sound from an even surface, such as the face of a wall, of a house, of a rock, of a hill, or of a cloud. As sound moves at the rate of 1090 feet in a second, and as the sound which returns to the person who emits it has travelled over a space equal to twice his distance from the reflecting surface, the distance in feet of the body which occasions the echo may be readily found by multiplying 545 by the number of seconds which elapse between the emission of the sound and its return in the form of an echo. This kind of echo, where the same person is the speaker and the hearer, never takes place unless when the observer is immediately in front of the reflecting surface, or when a line drawn from his mouth to the flat surface is nearly perpendicular to it, because in this case alone the wave of sound is reflected in the very same direction from the wall in which it reaches it. If the speaker places himself on one side of this line, then the echo will be heard most distinctly by another person as far on the other side of it, because the waves of sound are reflected like light, so that the angle of incidence or the inclination at which the sound falls upon the reflected surface is equal to the angle of reflexion, or the inclination at which the sound is returned from the wall. If two persons, therefore, are placed before the reflecting wall, the one will hear the echo of the sound emitted by the other, and obstacles may intervene between these two persons so that neither of them hears the direct sound emitted by the other; in the same manner as the same persons similarly placed before a looking-glass would see each other distinctly by reflexion, though objects might obstruct their direct view of each other.

Hitherto we have supposed that there is only one reflecting surface, in which case there will be only one echo: but if there are several reflecting surfaces, as is the case in an amphitheatre of mountains, or during a thunder-storm, where there are several strata or masses of clouds; or if there are two parallel or inclined surfaces between which the sound can be repeatedly reflected, or if the surface is curved so that the sound reflected from one part falls upon another part, like the sides of a polygon inscribed in a circle, – in all these cases there will be numerous echoes, which produce a very singular effect. Nothing can be more grand and sublime than the primary and secondary echoes of a piece of ordnance discharged in an amphitheatre of precipitous mountains. The direct or primary echoes from each reflecting surface reach the ear in succession, according to their different distances, and these are either blended with or succeeded by the secondary echoes, which terminate in a prolonged growl ending in absolute silence. Of the same character are the reverberated claps of a thunder-bolt reflected from the surrounding clouds, and dying away in the distance. The echo which is produced by parallel walls is finely illustrated at the Marquis of Simonetta's villa near Milan, which has been described by Addison and Keysler, and which we believe is that described by Mr Southwell in the Philosophical Transactions for 1746. Perpendicular to the main body of this villa there extends two parallel wings about fifty-eight paces distant from each other, and the surfaces of which are unbroken either with doors or windows. The sound of the human voice, or rather a word quickly pronounced, is repeated above forty times, and the report of a pistol from fifty-six to sixty times. The repetitions, however, follow in such rapid succession that it is difficult to reckon them, unless early in the morning before the equal temperature of the atmosphere is disturbed, or in a calm still evening. The echoes appear to be best heard from a window in the main building between the two projecting walls, from which the pistol also is fired. Dr Plot mentions an echo in Woodstock Park which repeats seventeen syllables by day and twenty by night. An echo on the north side of Shipley church in Sussex repeats twenty-one syllables. Sir John Herschel mentions an echo in the Manfroni palace at Venice, where a person standing in the centre of a square room about twenty-five feet high with a concave roof, hears the stamp of his foot repeated a great many times, but as his position deviates from the centre, the echoes become feebler, and at a short distance entirely cease. The same phenomenon, he remarks, occurs in the large room of the library of the museum at Naples. M. Genefay has described as existing near Rouen a curious oblique echo which is not heard by the person who emits the sound. A person who sings hears only his own voice, while those who listen hear only the echo, which sometimes seems to approach, and at other times to recede from the ear; one person hears a single sound, another several sounds, and one hears it on the right and another on the left, the effect always changing, as the hearer changes his position. Dr Birch has described an extraordinary echo at Roseneath in Argyleshire, which certainly does not now exist. When eight or ten notes were played upon a trumpet they were correctly repeated, but on a key a third lower. After a short pause another repetition of the notes was heard in a still lower tone, and after another short interval they were repeated in a still lower tone.

In the same manner as light is always lost by reflexion, so the waves of sound are enfeebled by reflexion from ordinary surfaces, and the echo is in such cases fainter than the original sound. If the reflecting surface, however, is circular, sound may be condensed and rendered stronger in the same manner as light. I have seen a fine example of this, in the circular turn of a garden wall nearly a mile distant from a weir across a river. When the air is pure and homogeneous, the rushing sound of the water is reflected from the hollow surface of the wall, and concentrated in a focus, the place of which the ear can easily discover from the intensity of the sound being there a maximum. A person not acquainted with the locality conceives that the rushing noise is on the other side of the wall.

In Whispering Galleries, or places where the lowest whispers are carried to distances at which the direct sound is inaudible, the sound may be conveyed in two ways, either by repeated reflexions from a curved surface in the direction of the sides of a polygon inscribed in a circle, or where the whisperer is in the focus of one reflecting surface, and the hearer in the focus of another reflecting surface, which is placed so as to receive the reflected sounds. The first of these ways is exemplified in the whispering gallery of St Paul's, and in the octagonal gallery of Gloucester Cathedral, which conveys a whisper 75 feet across the nave, and the second in the baptistery of a church in Pisa, where the architect Giovanni Pisano is said to have constructed the cupola on purpose. The cupola has an elliptical form, and when one person whispers in one focus, it is distinctly heard by the person placed in the other focus, but not by those who are placed between them. The sound first reflected passes across the cupola, and enters the ears of the intermediate persons, but it is too feeble to be heard, till it has been condensed by a second reflexion to the other focus of ellipse. A naval officer, who travelled through Sicily in the year 1824, gives an account of a powerful whispering place in the cathedral of Girgenti, where the slightest whisper is carried with perfect distinctness through a distance of 250 feet, from the great western door to the cornice behind the high altar. By an unfortunate coincidence the focus of one of the reflecting surfaces was chosen for the place of the confessional, and when this was accidentally discovered, the lovers of secrets resorted to the other focus, and thus became acquainted with confessions of the gravest import. This divulgence of scandal continued for a considerable time, till the eager curiosity of one of the dilettanti was punished, by hearing his wife's avowal of her own infidelity. This circumstance gave publicity to the whispering peculiarity of the cathedral, and the confessional was removed to a place of greater secrecy.

An echo of a very peculiar character has been described by Sir John Herschel in his Treatise on Sound, as produced by the suspension bridge across the Menai strait in Wales. "The sound of a blow with a hammer," says he, "on one of the main piers, is returned in succession from each of the cross beams which support the road-way, and from the opposite pier at a distance of 576 feet; and in addition to this, the sound is many times repeated between the water and the road-way. The effect is a series of sounds which may be thus written: the first return

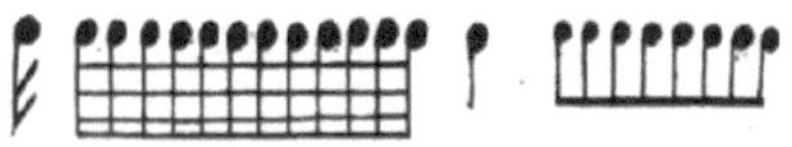

Figure 16.50

is sharp and strong from the road-way overhead; the rattling which succeeds dies away rapidly, but the single repercussion from the opposite pier is very strong, and is succeeded by a faint palpitation repeating the sound at the rate of twenty-eight times in five seconds, and which, therefore, corresponds to a distance of 184 feet, or very nearly the double interval from the road-way to the water. Thus it appears that in the repercussion between the water and road-way that from the latter only affects the ear, the line drawn from the auditor to the water being too oblique for the sound to diverge sufficiently in that direction. Another peculiarity deserves especial notice, namely, that the echo from the opposite pier is best heard when the auditor stands precisely opposite to the middle of the breadth of the pier, and strikes just on that point. As it deviates to one or the other side, the return is proportionally fainter, and is scarcely heard by him when his station is a little beyond the extreme edge of the pier, though another person, stationed (on the same side of the water) at an equal distance from the central point, so as to have the pier between them, hears it well."

A remarkable subterranean echo is often heard when the hoofs of a horse or the wheels of a carriage pass over particular spots of ground. This sound is frequently very similar to that which is produced in passing over an arch or vault, and is commonly attributed to the existence of natural or artificial caves beneath. As such caves have often been constructed in times of war as places of security for persons and property, many unavailing attempts have been made to discover hidden treasures where their locality seemed to be indicated by subterraneous sounds. But though these sounds are sometimes produced by excavations in the ground, yet they generally arise from the nature of the materials of which the ground is composed, and from their manner of combination. If the hollow of a road has been filled up with broken rock, or with large water-worn stones, having hollows either left entirely empty, or filled up with materials of different density, then the sound will be reflected in passing from the loose to the dense materials, and there will arise a great number of echoes reaching the ear in rapid succession, and forming by their union a hollow rumbling sound. This principle has been very successfully applied by Sir John Herschel to explain the subterranean sounds with which every traveller is familiar who has visited the Solfaterra near Naples. When the ground at a particular place is struck violently by throwing a large stone against it, a peculiar hollow sound is distinctly heard. This sound has been ascribed by some geologists to the existence of a great vault communicating with the ancient seat of the volcano, by other writers to a reverberation from the surrounding hills with which it is nearly concentric, and by others to the porosity of the ground. Dr Daubeny, who says that the hollow sound is heard when any part of the Solfaterra is struck, accounts for it by supposing that the hill is not made up of one entire rock, but of a number of detached blocks, which, hanging as it were by each other, form a sort of vault over the abyss within which the volcanic operations are going on. Mr Forbes, who has given the latest and most interesting description of this singular volcano, agrees in opinion with Dr Daubeny, while

Mr Scrope and Sir John Herschel concur in opinion that no such cavities exist. "It seems most probable," says the latter, "that the hollow reverberation is nothing more than an assemblage of partial echoes arising from the reflexion of successive portions of the original sound in its progress through the soil at the innumerable half-coherent surfaces composing it: were the whole soil a mass of sand, these reflexions would be so strong and frequent as to destroy the whole impulse, in too short an interval to allow of a distinguishable after sound. It is a case analogous to that of a strong light, thrown into a milky medium or smoky atmosphere; the whole medium appears to shine with a nebulous undefined light. This is to the eye what such a hollow sound is to the ear."

It has been recently shown by M. Savart, that the human ear is so extremely sensible as to be capable of appreciating sounds which arise from about *twenty-four thousand* vibrations in a second, and consequently, that it can hear a sound which lasts only the twenty-four thousandth part of a second. Vibrations of such frequency afford only a shrill squeak or chirp; and Dr Wollaston has shown that there are many individuals with their sense of hearing entire, who are altogether insensible to such acute sounds, though others are painfully affected by them. Nothing, as Sir John Herschel remarks, can be more surprising than to see two persons, neither of them deaf, the one complaining of the penetrating shrillness of a sound, while the other maintains there is no sound at all. Dr Wollaston has also shown that this is true also of very grave sounds, so that the hearing or not hearing of musical notes at both extremities of the scale seems to depend wholly on the pitch or frequency of vibration constituting the note, and not upon the intensity or loudness of the noise. This affection of the ear sometimes appears in cases of common deafness, where a shrill tone of voice, such as that of women and children, is often better heard than the loud and deeper tone of men.

Dr Wollaston remarked, that when the mouth and nose are shut, the tympanum or drum of the ear may be so exhausted by a forcible attempt to take breath by the expansion of the chest, the pressure of the external air upon the membrane gives it such a tension that the ear becomes insensible to grave tones, without losing in any degree the perception of sharper sounds. Dr Wollaston found, that after he had got into the habit of making the experiment, so as to be able to produce a great degree of exhaustion, his ears were insensible to all sounds below F, marked by the base cliff. "If I strike the table before me," says he, "with the end of my finger, the whole board sounds with a deep dull note. If I strike it with my nail, there is also at the same time a sharp sound produced by quicker vibrations of parts around the point of contact. When the ear is exhausted it hears only the latter sound, without perceiving in any degree the deeper note of the whole table. In the same manner, in listening to the sound of a carriage, the deeper rumbling noise of the body is no longer heard by an exhausted ear; but the rattle of a chain or loose screw remains at least as audible as before exhaustion." Dr Wollaston supposes that this excessive tension of the drum of the ear, when produced by the compressed air in the diving-bell, will also produce a corresponding *deafness to low tones*. This curious experiment has been since made, by Dr Colladon when descending in the diving-bell at Howth in 1820. "We descended," says he, "so slowly that we did not notice the motion of the bell; but as soon as the bell was immersed in water, we felt about the ears and the forehead a sense of pressure, which continued increasing during some minutes. I did not, however, experience any pain in the ears; but my companion suffered so much that we were obliged to stop our descent for a short time. To remedy that inconvenience, the workmen instructed us, after having closed our nostrils and mouth, to endeavour to swallow, and to restrain our respiration for some moments, in order that, by this exertion, the internal air might act on the Eustachian tube. My companion, however, having tried it, found himself very little relieved by this remedy. After some minutes, we resumed our descent. My friend suffered considerably; he was pale; his lips were totally discoloured; his appearance was that of a man on the point of fainting; he was in involuntary low spirits, owing, perhaps, to the violence of the pain, added to that kind of apprehension which our situation unavoidably inspired. This appeared to me the more remarkable, as my case was totally the reverse. I was in a state of excitement resembling the effect of some spirituous liquor. I suffered no pain; I experienced only a strong pressure round my head, as if an iron circle had been bound about it. I spoke with the workmen and had some difficulty in hearing them. This difficulty of hearing rose to such a height, that during three or four minutes I could not hear them speak. I could not, indeed, hear myself speak, though I spoke as loudly as possible; nor did even the great noise caused by the violence of the current against the sides of the bell reach my ears."

The effect thus described by Dr Colladon is different from that anticipated by Dr Wollaston. He was not merely deaf to low tones but to all sounds whatever; and I have found by repeated experiment, that my own ears become perfectly insensible even to the shrill tones of the female voice, and of the voice of a child, when the drum of the ear is thrown into a state of tension by yawning.

With regard to sounds of high pitch at the other extremity of the scale, Dr Wollaston has met with persons, whose hearing was in other respects perfect, who never heard the chirping of the *Gryllus campestris*, which commonly occurs in hedges during a summer's evening, or that of the house-cricket, or the squeak of the bat, or the chirping of the common house-sparrow. The note of the bat is a full octave higher than that of the sparrow; and Dr Wollaston believes that the note of some insects may reach one octave more, as there are sounds decidedly higher than that of a small pipe, one-fourth of an inch in length, which he conceives cannot be far from six octaves above the middle E of the piano-forte. "The suddenness of the transition," says Dr Wollaston, "from perfect hearing to total want of perception, occasions a degree of surprise, which renders an experiment on this subject with a series of small pipes among several persons rather amusing. It is curious to observe the change of feeling manifested by various individuals of the party, in succession, as the sounds approach and pass the limits of their hearing. Those who enjoy a temporary triumph are often compelled in their turn to acknowledge to how short a distance their little superiority extends." In concluding his interesting paper on this subject, Dr Wollaston conjectures that animals, like the grylli, (whose powers of hearing appear to commence nearly where ours terminate,) may have the power of hearing still sharper sounds which at present we do not know to exist, and that there may be other insects having nothing in common with us, but who are endowed with a power of exciting, and a sense of perceiving vibrations which makes no impression upon our organs, while their organs are equally insensible to the slower vibrations to which we are accustomed.

With the view of studying the class of sounds inaudible to certain ears, we would recommend it to the young naturalist to examine the sounds emitted by the insect tribe, both in relation to their effect upon the human ear, and to the mechanism by which they are produced. The Cicadæ or locusts in North America appear, from the observations of Dr Hildreth, to be furnished with a bag-pipe on which they play a variety of notes. "When any one passes," says he, "they make a great noise and screaming with their air bladder or bagpipes. These bags are placed under, and rather behind, the wings in the axilla, something in the manner of using the bagpipes with the bags under the arms, – I could compare them to nothing else; and, indeed, I suspect the first inventor of the instrument borrowed his ideas from some insect of this kind. They play a variety of notes and sounds, one of which nearly imitates the scream of the tree toad."

Among the acoustic wonders of the natural world may be ranked the vocal powers of the statue of Memnon, the son of Aurora, which modern discoveries have withdrawn from among the fables of ancient Egypt. The history of this remarkable statue is involved in much obscurity. Although Strabo affirms that it was overturned by an earthquake, yet as Egypt exhibits no traces of such a convulsion, it has been generally believed that the statue was mutilated by Cambyses. Ph. Casselius, in his dissertation on vocal or speaking stones, quotes the remark of the scholiast in Juvenal, "that, when mutilated by Cambyses, the statue which saluted both the sun and the king, afterwards saluted only the sun." Philostratus, in his life of Apollo, informs us, that the statue looked to the east, and that it spoke as soon as the rays of the rising sun fell upon its mouth. Pausanias, who saw the statue in its dismantled state, says that it is a statue of the sun, that the Egyptians call it Phamenophis, and not Memnon, and *that it emits sounds every morning at sunrise, which can be compared only to that of the breaking of the string of a lyre*. Strabo speaks only of a single sound which he heard; but Juvenal, who had probably heard it often during his stay in Egypt, describes it as if it emitted several sounds.

> Dimidio magicæ resonant ubi Memnone chordæ.
> Where broken Memnon sounds his magic strings.

The simple sounds which issued from the statue were in the progress of time magnified intelligible words, and even into an oracle of seven verses, and this prodigy has been recorded in a Greek inscription on the left leg of the statue. But though this new faculty of the colossus was evidently the contrivance of the Egyptian priests, yet we are not entitled from this to call in question the simple and perfectly credible fact that it emitted sounds. This property, indeed, it seems to possess at the present day; for we learn, that an English traveller, Sir A. Smith, accompanied with a numerous escort, examined the statue, and that at six o'clock in the morning he heard very distinctly the sounds which had been so celebrated in antiquity. He asserts that this sound does not proceed from the statue but from the pedestal; and he expresses his belief that it arises from the impulse of the air upon the stones of the pedestal, which are arranged so as to produce this surprising effect. – This singular description is to a certain extent confirmed by the description of Strabo, who says, that he was quite certain that he heard a sound which proceeded either *from the base*, or from the colossus, or from some one of the assistants. As there were no Egyptian priests in the escort of Sir A. Smith, we may now safely reject this last, and, for many centuries, the most probable hypothesis.

The explanation suggested by Sir A. Smith had been previously given in a more specific form by M. Dussaulx, the translator of Juvenal. "The statue," says he, "being hollow, the heat of the sun heated the air which it contained, and this air, issuing at some crevice, produced the sounds of which the priests gave their own interpretation."

Rejecting this explanation, M. Langles, in his dissertation on the vocal statue of Memnon, and M. Salverte, in his work on the occult sciences, have ascribed the sounds entirely to Egyptian priestcraft, and have even gone so far as to describe the mechanism by which the statue not only emitted sounds, but articulated distinctly the intonations appropriate to the seven Egyptian vowels, and consecrated to the seven planets. M. Langles conceives that the sounds may be produced by a series of hammers, which strike either the granite itself, or sonorous stones like those which have been long used in China for musical instruments. M. Salverte improves this imperfect apparatus, by supposing that there might be adapted to these hammers, a clepsydra or water-clock, or any other instrument fitted to measure time, and so constructed as to put the hammers in motion at sunrise. Not satisfied with this supposition, he conjectures that the spring of all this mechanism was to be found in the art of concentrating the rays of the sun, which was well known to the ancients. Between the lips of the statue, or in some less remarkable part of it concealed from view by its height, he conceives an aperture to be perforated, containing a lens or a mirror capable of condensing the rays of the rising sun upon one or more metallic levers which by their expansion put in motion the seven hammers in succession. Hence he explains why the sounds were emitted only at sunrise, and when the solar rays fell upon the mouth of the statue, and why they were never again heard till the sun returned to the eastern horizon. As a piece of mechanism, this contrivance is defective in not providing for the change in the sun's amplitude, which is very considerable even in Egypt, for as the statue and the lens are both fixed, and as the sounds were heard at all seasons of the year, the same lens which threw the midsummer rays of the sun upon the hammers could not possibly throw upon them his rays in winter. But even if the machinery were perfect, it is obvious that it could not have survived the mutilation of the statue, and could not, short of a miracle, have performed its part in the time of Sir A. Smith.

If we abandon the idea of the whole being a trick of the priesthood, which has been generally done, and which the recent observations of Sir A. Smith authorizes us to do, we must seek some natural cause for the phenomena similar to that suggested by Dussaulx. It is curious to observe how the study of nature gradually dispels the consecrated delusions of ages, and reduces to the level of ordinary facts what time had invested with all the characters of the supernatural: And in the present case it is no less remarkable that the problem of the statue of Memnon should have been first solved by means of an observation made by a solitary traveller wandering on the banks of the Orinoco. "The granitic rock," says Baron Humboldt, "on which we lay, is one of those where travellers on the Orinoco have heard from time to time, towards sunrise, subterraneous sounds resembling those of the organ. The missionaries call these stones *loxas de musica*. 'It is witchcraft,' said our young Indian pilot. We never ourselves heard these mysterious sounds either at Carichana Vieja or in the upper Orinoco; but from information given us by witnesses worthy of belief, the existence of a phenomenon that seems to depend on a certain state of the atmosphere cannot be denied. The shelves of rock are full of very narrow and deep crevices. They are heated during the day to about 50°. I often found their temperature at the surface during the night at 39°, the surrounding atmosphere being at 28°. It may easily be conceived that the difference of temperature between the subterraneous and the external air attains its maximum about sunrise, or at that moment which is at the same time farther from the period of the maximum of the heat of the preceding day. May not these sounds of an organ, then, which are heard when a person sleeps upon the rock, his ear in contact with the stone, be the effect of a current of air that issues out through the crevices? Does not the impulse of the air against the elastic spangles of mica that intercept the crevices contribute to modify the sounds? May we not admit that the ancient inhabitants of Egypt, in passing incessantly up and down the Nile, had made the same observation on some rock of the Thebaid, and that the music of the rocks there led to the jugglery of the priests in the statue of Memnon?"

This curious case of the production of sounds in granite rocks at sunrise might have been regarded as a transatlantic wonder which was not applicable to Egypt; but by a singular coincidence of observation, MM. Jomard, Jollois, and Devilliers, who were travelling in Egypt nearly about the same time that M. Humboldt was traversing the wilds of South America, heard *at sunrise, in a monument of granite*, situated near the centre of the spot on which the palace of Carnac stands, *a noise resembling that of a breaking string*, the very expression by which Pausanias characterizes the sound in the Memnonian granite. The travellers regarded these sounds as arising from the transmission of rarefied air through the crevices of a sonorous stone, and they were of the same opinion with Humboldt, that these sounds might have *suggested* to the Egyptian priests *the juggleries of the Memnonium*. Is it not strange that the Prussian and the French travellers should not have gone a step farther, and solved the problem of two thousand years, by maintaining that the sound of the statue of Memnon was itself a natural phenomenon, or a granite sound elicited at sunrise by the very same causes which operated on the Orinoco and in the Temple of Carnac, in place of regarding it as a trick in imitation of natural sounds? If, as Humboldt supposes, the ancient inhabitants of Egypt had, in passing incessantly up and down the Nile, become familiar with the music of the granite rocks of the Thebaid, how could the imitation of such natural and familiar sounds be regarded by the priests as a means of deceiving the people? There could be nothing marvellous in a colossal statue of granite giving out the very same sounds that were given out at the same time of the day by a granite rock; and in place of reckoning it a supernatural fact, they could regard it in no other light than as the duplicate of a well known natural phenomenon. It is a mere conjecture, however, that such sounds were common in the Thebaid, and it is therefore probable that a granite rock, possessing the property of emitting sounds at sunrise, had been discovered by the priests, who were at the same time the philosophers of Egypt, and that the block had been employed in the formation of the Memnonian statue for the purpose of impressing upon it a supernatural character, and enabling them to maintain their influence over a credulous people.

The inquiries of recent travellers have enabled us to corroborate these views, and to add another remarkable example of the influence of subterraneous sounds over superstitious minds. About three leagues to the north of Tor in Arabia Petræa, is a mountain, within the bosom of which the most singular sounds have been heard. The Arabs of the Desert ascribe these sounds to a convent of monks preserved miraculously under ground and the sound is supposed to be that of the *Nakous*, a long narrow metallic ruler suspended horizontally, which the priest strikes with a hammer for the purpose of assembling the monks to prayer. A Greek was said to have seen the mountain open, and to have descended into the subterranean convent, where he found fine gardens and delicious water; and, in order to give proof of his descent, he produced some fragments of consecrated bread, which he pretended to have brought from the subterranean convent. The inhabitants of Tor likewise declare that the camels are not only frightened but rendered furious when they hear these subterraneous sounds.

M. Seetzen, the first European traveller who visited this extraordinary mountain, set out from Wodyel Nackel on the 17th of June at five o'clock in the morning. He was accompanied by a Greek Christian and some Bedouin Arabs, and after a quarter of an hour's walk they reached the foot of a majestic rock of hard sandstone. The mountain itself was quite bare and entirely composed of it. He found inscribed upon the rock several Greek and Arab names, and also some Koptic characters, which proved that it had been resorted to for centuries. About noon the party reached the foot of the mountains called *Nakous*, where at the foot of a ridge they beheld an insulated peaked rock. This mountain presented upon two of its sides two sandy declivities about 150 feet high, and so inclined that the white and slightly adhering sand which rests upon its surface is scarcely able to support itself; and when the scorching heat of the sun destroys its feeble cohesion, or when it is agitated by the smallest motions, it slides down the two acclivities. These declivities unite behind the insulated rock, forming an acute angle, and, like the adjacent surfaces, they are covered with steep rocks which consist chiefly of a white and friable freestone.

The first sound which greeted the ears of the travellers took place at an hour and a quarter after noon. They had climbed with great difficulty as far as the sandy declivity, a height of seventy or eighty feet, and had rested beneath the rocks where the pilgrims are accustomed to listen to the sounds.

While in the act of climbing, M. Seetzen heard the sound from beneath his knees, and hence he was led to think, that the sliding of the sand was the cause of the sound, and not the effect of the vibration which it occasioned. At three o'clock the sound became louder, and continued six minutes, and after having ceased for ten minutes, it was again heard. The sound appeared to have the greatest resemblance to that of the humming top, rising and falling like that of an Eolian harp. Believing that he had discovered the true origin of the sound, M. Seetzen was anxious to repeat the experiment, and with this view he climbed with the utmost difficulty to the highest rocks, and, sliding down as fast as he could, he endeavoured, with the help of his hands and feet, to set the sand in motion. The effect thus produced far exceeded his expectations, and the sand in rolling beneath him made so loud a noise that the earth seemed to tremble to such a degree that he states he should certainly have been afraid if he had been ignorant of the cause.

M. Seetzen throws out some conjectures respecting the cause of these sounds. Does the rolling layer of sand, says he, act like the fiddle-bow, which on being rubbed upon a plate of glass raises and distributes into regular figures the sand with which the plate is covered? Does the adherent and fixed layer of sand perform here the part of the plate of glass, and the neighbouring rocks that of the sounding body? We cannot pretend to answer these questions, but we trust that some philosopher competent to the task will have an opportunity of examining these interesting phenomena with more attention, and describing them with greater accuracy.

The only person, so far as I can learn, who has visited El Nakous since the time of Seetzen, is Mr Gray of University College Oxford; but he has not added much to the information acquired by his predecessor. During the first visit which he made to the place, he heard at the end of a quarter of an hour a low continuous murmuring sound beneath his feet, which gradually changed into pulsations as it became louder, so as to resemble the striking of a clock, and at the end of five minutes it became so strong as to detach the sand. Returning to the spot next day, he heard the sound still louder than before. He could not observe any crevices by which the externalair could penetrate, and as the sky was serene and the air calm, he was satisfied that the sounds could not arise from this cause.

NOTE

1 We must caution the young reader against some of the views given in M. Salverte's work. In his anxiety to account for every thing miraculous by natural causes, he has ascribed to the same origin some of those events in sacred history which Christians cannot but regard as the result of divine agency.

[illegible] with considerable interest.

[illegible] later to the schools.

[illegible]

[illegible]

[illegible]

[illegible]

Editorial Headnote

17. Felix Savart, "Researches on the elasticity of regularly crystallized bodies", *Edinburgh Journal of Science* I, new series (Edinburgh, Scotland, Apr.–Oct. 1829), pp. 141–46.

In the 1820s, Paris was the seat of a rich, well-established scientific culture, centred on the powerful Académie de sciences. This famed learned society had originally been established in 1666, before the revolutionary government reorganized it into the Institut de France in 1795. When Napoléon Bonaparte fell from power in 1815, the restored Bourbon monarch Louis XVIII reinstated the Académie, although within the confines of the Institut.[1] Around this institution, a network of mathematicians, savants, and natural philosophers secured fame across Europe with their often-revolutionary studies of nature. Most famously, Simon Pierre Laplace's (1749–1827) *Traité de mécanique celeste* and *Système du monde* presented a radical, godless account of the organization of the universe. Across the Channel, in Britain, the likes of Charles Babbage (1791–1871) looked enviously to French science as something worthy of emulation and pressed this point in his claims that English science was in decline.[2] Among Paris's brightest philosophical stars was Felix Savart (1791–1841), a mathematician and specialist in electromagnetic phenomena. Savart also boasted musical credentials, designing a trapezoidal violin and suggesting his name as a unit of measurement for musical intervals, the "Savart". In 1829, Savart presented an account of his extension of Chladni's acoustic plate experiments to the Académie des Sciences. Chladni had, in fact, been prominent in Parisian science during the 1800s, presenting a personal demonstration of his acoustic plates to Emperor Napoléon Bonaparte.[3] Savart took this work much further, using vibrating sand and powders to acoustically examine the structure of crystallized bodies. The French experimentalist cut flat plates from various materials, including crystals; then spread sand or powder over them; caused them to vibrate; and examined the patterns produced. These, Savart asserted, revealed a solid body's structure and offered insights into how it had formed. Savart claimed that analysis "by means of sonorous vibrations" uncovered the forces of cohesion responsible for their formation and suggested that such techniques had the potential to enhance knowledge of industrially produced metals. In 1830, he replicated his crystalline experiments using plates of iron, confident that superior manufacturing practices might result from such analysis. Wheatstone presented a Friday Evening Lecture at the Royal Institution, explaining Savart's work in March 1836. He was confident in "the importance to science of the discovery of the connection of Acoustics with Crystallography".[4] Savart's experiments were of great curiosity to Wheatstone, who also appreciated the French experimentalist's earlier work using powder, rather than sand, in performing Chladni acoustic figures. In Britain, Savart's investigations received considerable interest, which was hardly surprising given the country's industrial prowess, especially in terms of iron production. The *Edinburgh Journal of Science* published a summary of Savart's crystallized body experiments in 1829, as did the *Quarterly Journal of Science, Literature, and Art* in 1830.[5]

NOTES

1 Maurice Crosland, "The French Academy of Sciences in the Nineteenth Century", *Minerva* 16 (1978), pp. 73–102.

2 James A. Secord, *Visions of Science: books and readers at the dawn of the Victorian Age* (Oxford: Oxford University Press, 2014), p. 62.

3 See Viktoria Tkaczyk, "The Making of Acoustics around 1800, or How to Do Science with Words", in Mart Helen Dupree and Sean B. Franzen (eds.), *Performing Knowledge, 1750–1850* (Berlin and Boston: De Gruyter, 2015), pp. 27–56; H. J. Stöckmann, "Chladni meets Napoleon", *European Physical Journal*, Special Topics, 145 (2007), pp. 15–23.

4 (Anon.), "Royal Institution", *The Morning Post*, no. 20371 (22 Mar. 1836, London, England), p. 5.

5 Felix Savart, "Researches on the elasticity of regularly crystallized bodies", *Edinburgh Journal of Science*, I, new series, (Apr.–Oct. 1829, Edinburgh, Scotland), pp. 141–46; (Anon.), "On the structure of metals, by M. Savart", *Quarterly Journal of Science, Literature, and Art* 28 (1830, London, England), pp. 163–67; compare Savart's work to the study of earthquakes by sonorous means, see Edward J. Gillin, "Seismology's acoustic debt: Robert Mallet, Chladni's figures, and the Victorian science of earthquakes", *Sound Studies* 6, no. 1 (Spring 2020), pp. 65–82.

17

RESEARCHES ON THE ELASTICITY OF REGULARLY CRYSTALLIZED BODIES

Felix Savart

Source: Felix Savart, "Researches on the elasticity of regularly crystallized bodies", *Edinburgh Journal of Science* I, new series (Edinburgh, Scotland, Apr.–Oct. 1829), pp. 141–46.

THIS very able and interesting memoir was read before the Academy of Sciences at Paris on the 29th January 1829, and has been printed in the *Ann. de Chimie* for January and February. The object of the author is to determine the distribution of elasticity in solid bodies, and consequently their structure, by cutting plates out of them in various directions, and ascertaining the sound which they emit while vibrating, and the modes in which they divide themselves, as rendered visible by the figures formed by sand or lycopodium strewed on their surface. A condensed abstract of this curious paper is all that our limits will permit us to give.

M. Savart's first experiments were made upon wood, the structure of which he has analyzed by means of sonorous vibrations. The phenomena are here related to three rectangular axes of elasticity, the axis of greatest elasticity corresponding with the axis of the branch, and the other two axes, which are equal, being perpendicular to the annual layers of the wood. When the branch is nearly cylindrical, the elasticity is sensibly uniform in all the diameters of a section perpendicular to the axis of the branch.

The following are some of the leading results obtained by using plates of beech: –

1. When one of the axes of elasticity is in the plane of the plate of wood, one of the nodal figures (viz. those taken by the sand,) is always composed of two straight lines at right angles, one of the lines being in the direction of the axis of elasticity. The other figure is formed by two curves, like the branches of a hyperbola, having their convex summits towards each other, and equi-distant from the centre of the plate.

DOI: 10.4324/b23406-17

2. When the plate does not contain any of the axes of elasticity in its plane, the two nodal figures are always hyperbolic curves.
3. The number of vibrations which accompany each mode of division is generally as much higher as the inclination of the plate to the axis of greatest elasticity becomes less.
4. The plate which emits the most acute sound, or which is susceptible of producing the greatest number of vibrations, is that which contains in its planes both the axis of greatest elasticity, and the axis of mean elasticity.
5. The plate perpendicular to the axis of greatest elasticity is that which emits the gravest sound, or produces the smallest number of vibrations.
6. When one of the axes is in the plane of the plate, and when the elasticity in a direction perpendicular to this axis is equal to that of the axis itself, the two nodal systems are similar. They are each composed of two straight lines at right angles, and the one system is inclined 45° to the other. In bodies with these inequal axes of elasticity, there are only two planes which enjoy this property.

M. Savart next proceeds to the analysis of *rock crystal* by means of sonorous vibrations, and the following are the leading results: –

1. The elasticity of all the diametral lines of any plane whatever perpendicular to the axis of a prism of rock crystal, (the axis of double refraction) is sensibly the same.
2. All planes parallel to the axis *do not possess the same elastic state*; but if we take any *three* of these planes, so that the angles which they form with one another are equal, their elastic state is the same.
   With regard to light, all the planes parallel to the axis possess exactly the same properties, so that the information respecting structure thus given by sonorous vibrations, is not of the same order as that given by means of light. M. Savart is of opinion that the first method indicates more particularly the elastic state, and the force of cohesion in the different directions of all the planes of the integrant particles, whilst the phenomena of light, depending more especially on the form of the particles, and the position which they affect round their centre of gravity, they are to a certain point independent of the mode of junction of the different laminæ of which the crystal is formed.
3. The transformations of the nodal lines of a series of plates cut round one of the edges of the base of the prism, are quite analogous to those which take place in a series of plates cut round the intermediate axis in bodies which possess three unequal and rectangular axes of elasticity.
4. The transformations in a series of plates perpendicular to any one of the three planes which pass through the opposite edges of the hexaedron, are in general analogous to those of a series of plates cut round a line, which divides into two equal parts the plane angle contained between two of the three axes of elasticity, in bodies where these axes are unequal and rectangular.
5. By means of the acoustic figures on a plate cut nearly parallel to the axis, and not parallel to two faces of the hexaedron, we may always distinguish which

are the faces of the pyramid which are susceptible of cleavage. We may also obtain the same results by the arrangement of the modes of division of a plate cut nearly parallel to one of the faces of the pyramid.

6. Whatever be the direction of the plates, the optic axis or its projection on their plane, always occupies there a position which is closely connected with the arrangement of the acoustic lines. For example, in all the plates cut round one of the edges of the base of the prism, the optic axis or its projection corresponds constantly with one of the two straight lines which compose the nodal system formed of two lines at right angles to each other.

Though there is a great analogy between these phenomena and those observed in wood, yet rock crystal cannot be numbered among bodies which have three rectangular and unequal axes of elasticity, and still less in the number of those, all of whose parts are symmetrically arranged round a single straight line. The same phenomena, indeed, are constantly reproduced in three different positions, and it would appear that every thing is related to the different directions of cleavage – to the faces and to the edges of the primitive rhombohedron. Thus all the plates cut parallel to the natural faces of the hexaedron enjoy exactly the same properties, and these properties are very different from those of plates equally parallel to the axis, but which are perpendicular to two opposite faces of the octohedron. Those plates, too, which are parallel to the cleavable faces of the pyramid, emit the same sounds, and produce the same acoustic figures; whilst plates parallel to the three other faces present figures different from those of the preceding. It would appear, therefore, to follow from this identity of phenomena for three distinct positions, that there is in rock crystal three systems of axes or principal lines of elasticity.

By comparing the phenomena in rock crystal with those in wood, M. Savart concludes, that the individual axes of each of these three systems are as follows: The shorter diagonal of each face of the primitive rhombohedron is the axis of greatest elasticity for each system, while the great diagonal of the face of the rhombohedron is the intermediate axis of elasticity. The axis of least elasticity is perpendicular to the axis of intermediate elasticity, and is inclined 57° 40′ 13″ to the axis of greatest elasticity, this angle being the inclination of the rhombohedral face upon the plane passing through the great diagonal of the same face.

"The transparent carbonate of lime, and the carbonate of lime and iron appear to possess elastic properties, in general analogous to those of rock crystal. Like it, they possess three systems of principal lines of elasticity similar to each other; but the extreme facility with which carbonate of lime cleaves, permits us to discover a peculiarity which does not appear in rock crystal."

"It is known that the rhomb of carbonate of lime is *often susceptible of mechanical division, in directions parallel to its diagonal planes;* but as these planes cut one another perpendicularly in pairs, the intersections of each pair with the rhomboidal faces of the crystal form the great and the small diagonal of each face, so that if we imagine a plane which turns round the great diagonal, it ought always to remain perpendicular to the supernumerary joint which passes through the small one. From this it follows, that if we cut a series of plates round this same line, their structure, considered in the direction of their plane, will be different in different directions at right angles to each other, whence arises the production of nodal lines crossed at right angles, as in plates cut round one of the axes of elasticity, in bodies where these axes are rectangular. We may therefore conclude, that *rock crystal possesses, like carbonate of lime, supernumerary planes of cleavage parallel to the diagonal planes of its primitive rhomboid*, and that it is to the existence of these supernumerary joints that we must ascribe the principal peculiarities of the elastic state of this substance.

"The only marked difference which appears to exist between these two minerals is, that in the carbonate the small diagonal of the rhomboidal face is the axis of least elasticity, while in quartz it is the axis of greatest elasticity. This result is curious, as the former is a crystal with *negative* double refraction, and the latter with *positive* double refraction.

"The preceding researches are doubtless far from forming a complete work on the elastic state of rock crystal and carbonate of lime; nevertheless we hope, that they will be sufficient to show, that the mode of experiment we have used may yet become a powerful means of studying the structure of solid bodies, whether regularly or irregularly crystallized. The relation for example, which exists between the modes of division, and the primitive form of crystals, allows us to hope, that by means of sonorous vibrations we may determine the primitive form of certain substances, which do not admit of mechanical division. It is equally natural to suppose, that less imperfect notions than we at present possess of the elastic state and cohesion of crystals, may throw light on many of the peculiarities of crystallization. It is not impossible, for example, that the degrees of elasticity of a given substance may not be exactly the same, and in the same direction relative to the primitive form, whilst from another cause the secondary form is different; and if this is the case, as some facts induce me to suppose, the determination of the elastic state of crystals will lead to the explanation of the most complicated phenomena of the structure of these bodies. It appears to me indeed, that the comparison of the results obtained by means of light respecting the constitution of bodies, and also by means of sonorous vibrations, will necessarily unite in advancing the science of optics as well as that of acoustics."

Editorial Headnote

18. Arthur Trevelyan, "On the vibration of heated metals: By Arthur Trevelyan, Esq.; including a letter on the same subject by Dr W. Knight", *The London and Edinburgh Philosophical Magazine and Journal of Science*, 3rd Series (London, England, Nov. 1833), pp. 321–32.

In November 1830, the Edinburgh chemist David Boswell Reid (1805–1863) received a visit from a local workman named Arthur Trevelyan (1802–1878). Reid was a practical chemistry teacher at the University of Edinburgh with a growing reputation that would, in 1835, see him appointed to manage the construction of a powerful ventilation system for the Palace of Westminster, home to the British Houses of Parliament.[1] Trevelyan was keen to consult Reid regarding a recent observation he had made when spreading pitch with a red-hot iron poker. Trevelyan noticed that when placing his hot poker on a lead rest, a musical note was produced. At the university, Reid and Trevelyan refashioned this accident as an experimental procedure before Reid persuaded Trevelyan to present his findings to the Royal Society of Edinburgh. Trevelyan subsequently delivered two papers to the learned institution during the winter of 1830–31 and then published a paper on his inquiries in 1833. Trevelyan had performed two years of investigations into the production of musical sounds from metals at extreme variations of temperature. A brass bar at high heat on a cold lead rest produced a great sound, as did the cutting of a v-notch into the rest's surface. Trevelyan argued that the key to successfully sustaining a musical pitch was down to the metals employed and the unevenness of their surface. He claimed that the sound resulted from the movement of heat, or "caloric", from the heated metal to the cold. Smooth surfaces, Trevelyan continued, did not produce any sonorous effects because the transfer of caloric occurred at an equal rate.[2] Trevelyan's work impressed several eminent members of Britain's scientific community, including Scottish science writer John Robison (1778–1843) and John Herschel, as well as Michael Faraday. Trevelyan failed to secure further scientific fame, however, with his continuing examination of heat transfer between metals, conducted from 1833 to 1835, attracting little interest.[3] In 1843, Trevelyan wrote to the master of Trinity College Cambridge, William Whewell, with news of his "Theory of Heat", in which he attempted to attribute all natural phenomena to the laws of heat communication. Whewell was unimpressed, providing little encouragement, and Trevelyan's work was subsequently rejected from publication in the *Chemical Gazette*, the *Electrical Magazine*, and the *Phrenological Journal*.[4] Nevertheless, Trevelyan's endeavours between 1829 and 1833 show the potential of sound as a means of examining the nature of heat, as well as the social diversity of those who might become involved in the study of nature.

NOTES

1 On Reid's ventilation, see Edward J. Gillin, *The Victorian Palace of Science: scientific knowledge and the building of the Houses of Parliament* (Cambridge: Cambridge University Press, 2017), pp. 121–83; Henrik Schoenefeldt, *Rebuilding the Houses of Parliament: David Boswell Reid and disruptive environmentalism* (London: Routledge, 2020).

2 Arthur Trevelyan, "On the vibration of heated metals: By Arthur Trevelyan, Esq.; including a letter on the same subject by Dr W. Knight", *The London and Edinburgh Philosophical Magazine and Journal of Science*, 3rd Series (Nov. 1833, London, England), pp. 321–32, 327.

3 Arthur Trevelyan, "Further notice of the vibration of heated metals; with the description of a new and convenient apparatus for experimenting with", *Philosophical Magazine*, Series 3, 6, no. 32 (1835), pp. 85–86.

4 Wren Library, Trinity College Cambridge: Ref.15.b.80.3, Arthur Trevelyan, 'Theory of Heat', (ca. 1843).

18

ON THE VIBRATION OF HEATED METALS: BY ARTHUR TREVELYAN, ESQ.; INCLUDING A LETTER ON THE SAME SUBJECT BY DR. W. KNIGHT

Arthur Trevelyan

Source: Arthur Trevelyan, "On the Vibration of Heated Metals: By Arthur Trevelyan, Esq.; Including a Letter on the Same Subject by Dr W. Knight", *The London and Edinburgh Philosophical Magazine and Journal of Science*, 3rd Series (London, England, Nov. 1833), pp. 321–32.

AS it is always interesting to know the accidental circumstances which lead to the discovery of any new fact in natural science, I shall begin by relating the manner in which I was led to observe the vibrations of heated metals. On the 9th of February 1829, I was about to spread with a plaster-iron some common pitch, when the iron being at too high a temperature, I placed it in a slanting direction against a block of lead, the handle resting on the table: shortly after doing so I heard a shrill high tone, which much resembled a note produced on the chanter of the smaller Northumberland pipes, an instrument played on by my father's gamekeeper. Not knowing the cause of the sound, I thought he might be practising out of doors. The room being on the ground floor, I went outside, when the sound was no longer heard; and seeing no person, returned again to the room, when I was surprised at still hearing the sound as shrill as before. After looking about for a few minutes, I at last approached closer to the iron, perceived the vibration, and thence found the immediate cause of the sound. I was much surprised and pleased; but at that time, not having studied any branch of natural philosophy connected with the subject of heat, I was not aware that it was a fact which had not been before observed (although it struck me so at the time). I came to Edinburgh in November 1830, and attended Dr. D. B. Reid's class of Practical Chemistry. Having mentioned the circumstance and shown him the experiments, he informed me that the fact was new, and recommended me to make further experiments, and prepare a paper on the subject for the Royal Society of Edinburgh. I took his advice, and during

DOI: 10.4324/b23406-18

the winter session had the honour of having two papers read before that Society, which were illustrated by experiments.

Since that time I have made numerous additional experiments, and have succeeded in obtaining vibration with most metals, many of which I failed with previously, from the apparatus not being sufficiently delicate.

On the opposite page is a Table of the different metals in which I have observed vibration.

The experiments were repeated several times before I was satisfied with the results.

I have not yet observed any effect with bismuth, although I have with pale solder, which is an alloy of that metal.

The gold was in too small a piece to try cold, with the hot bar placed on it. The results with it and with platinum would probably have been more numerous, if I could have obtained better-shaped pieces of those metals.

Bars of tin, lead, bismuth, antimony, block tin, solder, tin-solder and pale solder, produced no effect, when heated and placed on cool blocks, or rings, of all the different metals mentioned in the Table.

Cold bars of lead, bell-metal, tin-solder and pale solder, when placed on heated iron or brass, produced vibration and tone. The cold lead block, placed on the heated polished bar of a fire-grate, sounds loudly, and vibrates rapidly.

The vibration continues in the exhausted receiver of an air-pump.

The bars begin to vibrate on lead at a temperature below 212° Fahr., but on harder metals they require a higher temperature.

A cool brass bar, 5 inches long, 2 inches wide and ⅜ths of an inch thick, placed on a cool lead cylinder ½ an inch thick, 2 inches in height and 4¼ inches in diameter, with a spirit-lamp placed under the bar, produced vibration and tone in 6 minutes 15 seconds, and continued sounding for 5 hours and 55 minutes: when I removed the lamp, it ceased vibrating in 6 minutes, but might probably be continued, by the continued application of the lamp, for any length of time. The block had arrived at so high a degree of temperature, that it was too hot to hold with the naked hand: the vibration was only kept up by the brass bar being so superior to lead in its power of conducting caloric. From the small size of the bar, it soon fell to the same temperature as the lead, and then of course the vibration ceased.

When the brass bar at a high temperature is placed on the cool lead, the vibration seldom fails taking place instantly and spontaneously.

A bar 5 inches long, 2 inches wide, and ⅜ths of an inch thick, vibrated when supporting a weight of 12 pounds.

However near the bar and block may be brought to each other, no apparent phænomenon takes place until they are in actual contact.

By balancing the bar in a horizontal position on a narrow lead block (Figure 18.1 p. 336), rounded on the part on which the bar rests, the vibration is well exhibited, the bar moving vertically and laterally at the same time. A rod 10 or

Bars.	*Simple Metals*										*Alloys*							
	1. Platinum.	*2. Gold.*	*3. Silver.*	*4. Copper*	*5. Iron.*	*6. Cast Iron.*	*7. Zinc.*	*8. Tin.*	*9. Antimony.*	*10. Lead.*	*1. Fine Brass.*	*2. Com. Brass.*	*3. Gun Metal.*	*4. Bell Metal.*	*5. Cock Metal.*	*6. Block Tin Solder.*	*7. Tin Solder.*	*8. Pale Solder or Fusible Metal.*
Platinum							+	+		+				+	+	+		
Gold										+	+	+		+	+			
Silver	+				+	+	+	+	+	+	+	+	+	+	+	+	+	+
Copper	+		+	+	+	+	+	+	+	+	+	+	+	+	+	+	+	+
Iron							+	+	+	+				+		+	+	+
Cast Iron								+		+								
Zinc								+	+	+				+			+	+
Fine Brass				+	+	+	+	+	+	+	+	+	+	+	+	+	+	+
Common Brass							+	+		+		+	+	+	+	+	+	
Gun Metal								+		+				+		+	+	
Bell Metal										+							+	
Cock Metal										+				+				

The crosses denote that vibration took place when the heated bar was placed on the cold metal named at the top of the column. The names of metals at the side of the Table are those of the bars which were heated and placed on the cold metal.

more inches long, flattened in the centre to prevent its slipping, with a ball on each end (Figure 18.5), when placed across a heated vibrating bar (Figure 18.6), increases the arc of motion, rendering the movements much more conspicuous.

A thick ring of copper, 5 inches in diameter, when heated, and hung on a lead bar, vibrates backwards and forwards; and when laid on a narrow lead block, upwards and downwards.

The heated bar vibrated on a piece of thin sheet lead, either placed loosely or soldered on brass, and on a lead block burnished with gold-leaf.

A heated copper bar vibrated on the bottom of a glass tumbler. This is an experiment which the late Professor Leslie tried with me, but it is very uncertain and difficult to obtain.

The bars vibrate best when placed on blocks of lead with the surface somewhat rough: both metals also should be kept clean, and free from oxidation, which impairs the vibration.

A bar of heated copper was placed on an iron block rounded on the surface, and being nicely balanced on the centre of the rounded part, showed the vertical motion.

The shape or size of the bars and blocks is of little consequence, except for the more delicate experiments with the hard metals: on lead the hard metal almost of any form will vibrate when heated.

Mr. K. T. Kemp, the skilful electrician and chemist, informed me that after casting some bismuth, whilst it was cooling, after being taken out of the mould, he heard a sound proceed from it; when cool he again heated it, but could not reproduce the sound.

The bars I use to show the experiments are of different sizes: a bar about 5 inches long, 1¼ inch wide, and ⅜ths of an inch thick, will produce considerable tone; a rod of small wire 6 inches long, must be attached at one end, to serve as a handle. A ridge is formed along the centre of one side of the broad part, by its being bevelled off towards each edge; the other side is hollowed out down the centre so as to throw the weight on each side, which assists the vibration: the longitudinal ridge is the part that rests on the block, (Figure 18.3 & 18.4).

A hollow lead cylinder about ½ an inch thick, with the diameter in proportion to the length of the bar, is the best form for showing the vibrations and producing the tone.

An inequality or unequal notch on the lead cylinder increases the sound. When the bar at a high temperature is first placed on the cold lead, the tone is very harsh and undefined; but when the lead has acquired a certain proportion of heat, it becomes clear, full and mellow.

Pressure on the bar alters the note: the greater the pressure the higher is the note. Pressure also applied to the sounding-board or table on which the vibrating bar may happen to be placed, or even walking across the room, by altering the position of the bar, changes the tone.

A common poker heated and placed on a lead block vibrates, producing deep tones. If the vibrating bar be placed on a piano-forte, and certain notes be struck on the instrument, the vibration of the bar, and consequently the tone produced by it, is altered, and sometimes suddenly stopped.

Vibration is prevented by rubbing the surface of the lead with mercury, oil, plaster of Paris, or by oil gilding; also if a piece of thin tissue paper, or of wire-gauze, be interposed between the bar and the block.

The vibrations have not been produced with any substances besides metals, except in the instance of the glass tumbler already mentioned.

To try whether there was any attraction between the hard and soft metals when heated, I had two pieces of brass, about the size of an ounce weight, with a wire attached to them, and then hung on a delicate scale beam: having heated the brass weights, blocks of lead were placed under them and tried at various distances, but without producing the slightest effect.

To try whether electricity caused the phænomenon or made any alteration whilst the bar was vibrating, a hole was drilled in a bar and filled with mercury, a wire was soldered to the lead cylinder on which it was placed, and a galvanic battery of 150 plates, 4 inches square, being charged, the wire from the cylinder was placed in one end of the battery, whilst from the other end a wire was brought to the mercury contained in the cavity of the bar, which was heated, and in motion, previous to completing the circle. We also tried other methods, but with none of them did we succeed in producing any change either in the tone or in the vibration.

A Leyden phial was charged with electricity, and applied to the same bar and block, without producing any effect.

Spirit of wine, or æther, evaporated off the heated bar when in motion, produced no effect.

Air blown with a pair of bellows on the heated bar when vibrating does not affect the tremors.

If the heated bar be ground smooth, on the resting part, and the block of lead be also very smooth, no vibration takes place.

When the bar and block become of the same temperature, the vibration and sound cease.

A cylinder, 5 inches long, with a thick side bevelled off, leaving a flat ridge in the centre, similar to the former bars, when heated and placed on lead, produced no change of tone.

From the above-related experiments it appears:

1st, That in order to produce the vibrations, metals must be employed, either of the same kind or different, for both the bars and blocks; with the single exception of the glass tumbler.

2nd, That the difference of temperature between the two metals must be considerable, although some require a much higher temperature than others: the vibrations on zinc and bell-metal succeed with a lower temperature than others of the hard metals.

3rd, That the surface of the block must have some degree of unevenness, for when rendered quite smooth, the vibration does not take place; but the bar cannot be too smooth.

4th, That the interposition of any matter prevents vibration, with the exception of a burnish of gold-leaf, the thickness of which cannot amount to the 200,000dth part of an inch.

5th, That the air has no share in the production of the vibratory movements, however much its presence is essential to the production of sound.

6th, That it has no connexion with galvanism or electricity, the vibration and tone not being affected by passing a current through the bar when in action. Professor Forbes found no appearance of thermo-electricity after trying a number of experiments.

7th, That all the metals, both simple and alloys, produce vibration and sound, when one is heated and the other cool, on being brought into contact, (with the exception of bismuth,) but with some we find much greater difficulty in obtaining vibration than in others, as it depends on the temperature, but more particularly on the mode of placing the one on the other, which with some requires great nicety.

8th, Although all the metals are not found to vibrate on each other, or on metals of the same kind, – and I have not been able produce vibration with other substances, – yet I do not despair but that hereafter, when this subject is better understood, we shall be able to produce vibration with all the metals, and with all other matter either solid or liquid, when heat is applied to a cool substance, or *vice versâ*.

The following theory, which is partly derived from the late Professor Leslie's mode of accounting for the vibration, and appears the most probable, (and experiments hitherto made strengthen this view,) ascribes the vibratory movements to the usual mechanical changes which caloric occasions in passing from one substance into another, – I mean the expansion and contraction which accompany alternations of temperature.

It appears that some degree of roughness of one metal is essential to the success of the operation. This slight asperity arises from numberless points or ridges projecting from the mass of metal.

When the heated bar is laid upon the cool lead, the caloric passes into these prominences; and as their power of conduction is not great, it does not rapidly diffuse itself through the rest of the mass: of course they instantly expand and elongate, and by that sudden elongation they give an impulse to the incumbent bar. Soon, however, the caloric moves into the adjoining mass, and the prominences contract, and at the same time come into a state ready to admit a renewed accession of caloric from the bar: they receive that caloric, again expand, and give a second impulse to the bar. This goes on incessantly, and though the first impulse be infinitely small, and altogether inadequate to produce any sensible movement of the bar, yet by incessant repetition an accumulation of effect takes place, and the movements gradually reach a sufficient magnitude to become easily discernible.

As soon as the bar and block arrive within a certain limit of difference of temperature, the impulses become feebler and feebler, and at length the bar comes to rest. It has been mentioned that the smoother the bar is so much greater is the effect: I conceive that this smoothness operates by increasing the celerity with which the surface possessed of it communicates the caloric to the projecting points of the block, and thereby the elongation which gives the impulse to the bar is increased both in quickness and extent.

It is obvious that had the bar any considerable degree of asperity, the points of contact between the two metals would be fewer, and the passage of caloric between them more tardy.

When the two surfaces are highly polished, the experiment does not succeed: no tremors occur. This result proceeds probably from the circumstance, that the caloric enters into every part of the surface of the block equally, and is more quickly diffused through the mass, and hence there are none of those partial and sudden expansions which give the tremor-causing impulse.

The rocking on the narrow lead block may be induced by two causes, either by some slight inequality in the weight of the portions of the bar on the two sides of the ridge, or some difference in the condition of the surface of that part of the block which the ridge of the bar touches; and may be thus explained:

1st, If the first-mentioned inequality exist, as soon as the bar receives an upward heave, the greater weight of the one side will cause it to incline to that side; and as soon as that heave ceases, and the contraction succeeds, the bar approaches its original position, but will not remain in it, for the inclination given to the preponderating side will, on its return, of course, cause the bar to incline to the opposite side. The impulses which the bar receives, in this position, from the renewed expansion and elongation, will not only renew the upward heave, but also incline it again to the preponderating side, and thus increase the lateral movement, which, like the vertical, though altogether insensible at first, by incessant frequent repetition accumulates and increases till the rocking becomes conspicuous, and is rendered much more so by the transverse rod.

2ndly, If there exist any difference in regard to asperity in the condition of the surface of that part of the block upon which the ridge of the bar rests, it must necessarily follow, that the impulse given to the bar on that side which is most rough will be greater than on the other, and consequently the upward heave will be so modified as to create an inclination to one side. The bar thus thrown off its balance to the right on ascending, will incline as far to the left on descending, and there receiving the expansive impulse, it will be driven back, and thus the principle of rocking will be created. The sound depends upon the rapidity of the vibrations; for when slow, no tone is heard.

I think from the above-related experiments, and the conclusions and theory drawn from them, that the hitherto unknown causes of many sounds are now accounted for.

The sounds described by Humboldt as heard at sunrise by those who sleep on certain granite rocks on the banks of the Orinoco, also the sounds at sunrise produced by the statue of Memnon, and the twang, like the breaking of a string, heard by the French naturalists, as if proceeding from a mountain at Carnac, are probably caused by the pyrometric expansions and contractions of the heterogeneous material of which the statue and the mountain consist. Similar sounds, and from the same cause, are emitted, when heat is applied to any connected mass of machinery, and the snapping heard in a fire-grate affords a familiar example. We have often heard of a poker producing tone when heated and the point rested on a knot in a fir board in the floor of a room; also the singing in a teakettle is another example. In distilleries, shortly after the fire is put to the cool coppers, a very loud note is given out, and continues until the liquor boils.[1]

John Robison, Esq., Secretary to the Royal Society of Edinburgh, informed me that he once let a heated bar fall from his hand: it alighted on a painted shelf of wood, when he was surprised at hearing sounds; but they soon ceased.

The following theory of the cause of earthquakes and volcanic eruptions strikes me as being not at all unlikely. Earthquakes, and the sounds accompanying them, may be caused by vibration, occasioned by heat generated far below the surface of the earth, in some enormous metallic mass, which being in contact with some cool substance, not a very good conductor of heat, the latter is violently agitated, thus producing the vibration felt in earthquakes. By its intensity chasms are opened on the face of the earth; and below, caverns filled with condensed combustible matter and liquid lava are torn open, and the contents, by their enormous expansion, and having found an egress more easily upwards, rise to a great height above the surface of the earth.

The following interesting remarks on the vibration of heated metals are copied from a letter received by my brother from Dr. W. Knight, Professor of Natural Philosophy in Marischal College, Aberdeen, dated June 8, 1833.

"I regret that I have not written sooner in reply to your letter of the 9th ult., but I waited until I should find a convenient opportunity for repeating many of the experiments which I hinted the general nature of to you in my letter of the 19th of April. These opportunities have unfortunately not occurred yet, from myself being engaged with a daily class here, and my family at a country residence at some distance; nor are they likely to occur soon, so that I must content myself with sending you a shorter account than I could wish of the new trials made on the vibrations of metallic bodies in the end of March and beginning of April last.

"Having frequently failed as well as succeeded with the iron pokers and masses of lead, &c., and not being always able to account for the failures, a simple plan of producing vibrations occurred to me, which in no instance yet fails upon the greater number of metals employed. This consists in placing a quantity of a melted metal (as tin, lead, fusible metal, bismuth, &c.) in a hemispherical, or, better, in a parabolic conoidal cup of copper, or iron, or brass, lying above a piece of lead or other metal. The oscillations of the cup thus resting upon a small surface continue long after it has become solid, indeed until the cup and its contents have fallen to a temperature not much above that of the metal upon which they rest. I have employed in many trials cups of iron, brass and copper, of various sizes, from 2 to 6 inches in diameter at the top; and of melted metals, from an ounce to some pounds, of bismuth, tin, lead, zinc, &c. (In case of pouring the melted metal too rapidly from the ladle into the cups, and thus overturning the latter, it may be steadied by a small pair of forceps for a few seconds.) No vibrations could be produced *above* any other bodies than metals: they are most striking *above* lead, tin, zinc, antimony; more feeble above silver, gold, platina, brass, copper. Above wrought and cast-iron, I could not observe them. They are very distinct above slabs of the metal of reflecting telescopes, and above fusible metal. The smoothness of the ingots above which the vibrations are performed lessens the effect greatly, or prevents it.

"Among the most interesting trials are:

"1st, 3 or 4 ounces of lead, melted and poured into the parabolic cup of copper, resting above the ingot of 1 pound of tin with uneven surface. The tremors loud, quick and large; *a peculiar crackling noise* is heard, which I cannot distinguish in character from the noise of the same kind which ensues on bending a piece of block tin held close to the ear. On compressing the tin ingot with the fingers, the sound does not cease, but assumes a sort of stifled character: on removing the fingers, the former crackling noise is resumed. The fingers feel distinctly the vibration communicated to the ingot by the heated cup.

"2nd, The same placed on an ingot of 1 pound of zinc. The vibrations commenced immediately on pouring out the lead: they were small, rapid and equal, (not of an unequal sort like those above antimony): the sounds louder than above any other metal. They cease on pressing the ingot between the fingers; when resumed, there seemed some difference of tone: the stopping at the end sudden and distinct.

"3rd, The vibrations of the cups above a platina coin were very slow, and did not commence until after the lead had become solid: they soon ceased; – as if connected with the bad conducting power of platina?

"4th, The solidifying of the melted metals in the cup, and their crystallization when made to solidify during their vibrations, offered several singular phænomena, particularly with bismuth, lead and tin. As far as trials have yet gone, the crystallizing property seemed to be more conspicuous when solidifying during the vibrations, than when the same quantities of metals were not vibrating.

"These observations, like those of your brother, seem to connect this subject closely with the arcana of cohesion. On seeing the copy of your brother's paper, which you were so good as to send me, on the 19th of April, I was naturally much interested in the perusal, and cannot but think his theory the true one. An extensive field seems to be opened for examining the passage of heat through metallic bodies, in connexion with the structure of their atoms in cohesion. I could wish much that he or you would repeat some of the above experiments. I do not know when I shall find time to resume them. I found indeed, soon after beginning, that I could not advance without one of the small metallic Breguet's thermometers, with which the temperature of the ingots might be approximated to. I omitted to mention that the cups themselves will vibrate, if heated above a gas flame; but the other mode of pouring from a ladle some melted metal is more convenient, and increases the range of the phænomena; or hot mercury may be poured in, but the fumes are disagreeable.

"P.S. If you try any of the experiments, it will give me great pleasure to hear of your success. I have tried them so very often that I am quite sure you will not lose your trouble."

At one of the meetings of the Royal Society of Edinburgh this spring, Professor Forbes, of the Edinburgh University, read an able paper, and one that showed great research, on the vibration of metals; he also produced tables showing the difference of metals in the conduction of heat and electricity, and for showing the vibrations: there was great similarity in the position of the different metals in the three columns of the table, showing the difference of power in producing the three various effects. – In the Literary Gazette for May 17, 1831, Mr. Faraday is mentioned to have shown the experiment of a curved silver plate vibrating and sounding when placed on cold iron, and to have stated that this effect had long been known to working silversmiths.

DESCRIPTION OF THE FIGURES

Figure 18.1 Is the lead block on the rounded surface of which the bar rests, by that means the vibrations being more distinctly seen: this block is screwed to a flat brass plate, having three small flat knobs to rest on.

Figure 18.1

Figure 18.2 Shows the ring on which the bar is to be placed to produce the tone: it is better to have an unequal notch for the bar to rest on.

Figure 18.2

Figure 18.3 Is the back part of the bar, showing the ridge on which it ought to rest when placed on the lead.

Figure 18.3

Figure 18.4 Is the upper side of the bar, hollowed out in the centre so as to bring the gravity more to each side.

Figure 18.4

Figure 18.5 Is the bar with a ball at each end, to be laid across the vibrating bar, so as to cause it to move through a larger arc.

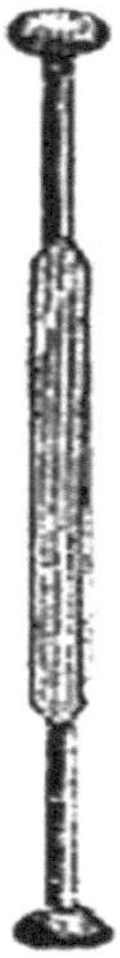

Figure 18.5

Figure 18.6 Shows how the bars ought to be arranged for exhibiting both the vertical and lateral motion.

Figure 18.6

Figure 18.7 Shows the bar as placed for producing the tone.

Figure 18.7

NOTE

1 The sounds from the statue of Memnon and the mountain at Carnac, and those from machinery and a fire-grate, have already been explained, nearly in the above words, by Sir John Herschel, (see Lond. and Edinb. Phil. Mag. vol. i. p. 221,) but without reference, however, to the new phænomena of vibrating metals, with the cause of which, indeed, contrary to Mr. Trevelyan's opinion, we apprehend that they are not essentially connected. The sounds heard on the granite rocks of the Orinoco are attributed by Sir J. Herschel, in the same letter, to sonorous vibrations of the air passing through small orifices, either subterranean or communicating with the atmosphere. The singing in a tea-kettle, and the note given out by stills, are referrible, we conceive, to another order of causes, connected with the rapid condensation of vapour, and quite distinct from all the former as well as from those concerned in the vibration of heated metals. – Ed.

Editorial Headnote

19. Michael Faraday, "Trevelyan's Experiments on the Production of Sound during the Conduction of Heat", in Michael Faraday, *Experimental Researches in Chemistry and Physics* (London: Richard Taylor and William Francis, 1859), pp. 311–14.

Among the readers of Arthur Treveylan's 1833 paper on the production of musical sounds from heated metals, few were of greater scientific repute than Michael Faraday. Having learnt of Trevelyan's discovery, Faraday organized a reconstruction of the hot poker experiments at the Royal Institution in April 1831. He was impressed with sound's potential for examining heat which was unsurprising, given his ongoing work on crispations between February and July of that year. The production of musical notes from heated metals was, for Faraday, an ideal subject for a Friday Evening Lecture. Faraday used a hot brass rod and a cold wooden rest to rework Trevelyan's musical tones.[1] His Friday Evening audience watched on as Faraday moved through an entire octave of notes, all resulting from the dissipation of heat from the brass to the wood. Faraday explained how the sounds were produced by the rapid repeated heating and cooling of the brass as it transmitted its heat to the colder material. By applying pressure, he could increase the rate of this action and alter the tone produced. Here then, was a way of measuring heat's motions by the ear. Scottish natural philosopher James David Forbes (1809–1868) shared Faraday's interest, but he took these experiments much further in his study of heat. Forbes had studied at the University of Edinburgh and had attended Reid's lectures on practical chemistry. On 17 January 1831, he had been at Trevelyan's lecture before Edinburgh's Royal Society, and he subsequently attended Faraday's reworking of these experiments at the Royal Institution. In 1833, Forbes went on to publish his own account of Trevelyan's experiments, agreeing with Faraday that sound occurred due to vibrations resulting from the transition of heat. While Faraday did not pursue these techniques beyond his Friday Evening Lecture, Forbes conducted considerable inquiry into the subject, analysing various combinations of hot and cold metals to determine the relative vibratory power of different materials. Forbes was eager to relate Trevelyan's discovery to electromagnetic phenomena, defining apparent laws that would account for the conductivity of various metals, especially iron, for both heat and electricity. He claimed that "when the best data are collected, the order of conducting powers of the metals for *heat* and for *electricity* is the same".[2] Nevertheless, Forbes's development of Trevelyan's experiments would draw considerable criticism, notably in 1853, when experimentalist John Tyndall (1820–1893) returned to the hot poker arrangement as a subject for a Friday Evening Lecture at the Royal Institution.[3]

NOTES

1 (Anon.), “Friday-Evening Proceedings at the Royal Institution of Great Britain”, *Philosophical Magazine* IX (Jan.–June 1831, London, England), p. 461.
2 James D. Forbes, “Experimental researches regarding certain vibrations which take place between metallic masses having different temperatures”, *Transactions of the Royal Society of Edinburgh* XII, part 2 (1833, London, England), pp. 429–61, 446.
3 “Letter 2704: John Tyndall to Faraday, 24 Jul., 1853”, in Frank A. J. L. James (ed.), *The Correspondence of Michael Faraday, Vol. 4: January 1849–October 1855. Letters 2146–3032* (London: Institution of Electrical Engineers, 1996), pp. 540–41; “Letter 2707: John Tyndall to Faraday, 28 Jul., 1853”, in James (ed.), *The Correspondence of Michael Faraday, Vol. 4*, pp. 544–45

19

TREVELYAN'S EXPERIMENTS ON THE PRODUCTION OF SOUND DURING THE CONDUCTION OF HEAT

Michael Faraday

Source: Michael Faraday, "Trevelyan's Experiments on the Production of Sound during the Conduction of Heat", in Michael Faraday, *Experimental Researches in Chemistry and Physics* (London: Richard Taylor and William Francis, 1859), pp. 311–14.

[Read Friday evening, April 29, 1831.]

MR. TREVELYAN had remarked that when a heated poker was laid down upon a table, so that the knob rested upon it, whilst the hot part was supported by an interposed block of cold lead, regular musical notes were frequently produced. By extending the experiments, he found that a better form than that of a poker might be used for the hot metal: a piece of brass about four inches long, one inch and a quarter broad, and half an inch thick, should have a groove of one-eighth of an inch in width, formed down the middle of one of the broad faces, and then that face bevelled from the edges of the groove on each side. Being now placed with the groove downwards upon a table, and shaken, it rocks to and fro, and is in right condition for the experiment. It is convenient to fasten a brass wire, terminated by a knob, to one end of this rocker, so as to act as a prolongation of an axis: it renders the whole arrangement steady and regular in action. When this piece of metal is used instead of the poker, musical sounds are almost always produced. The surface of the lead upon which it rests should be clean.

The peculiar effects exhibited in these experiments depend upon the occurrence of isochronous vibrations performed by the rocker. When by loading the rocker these are rendered slow, they become visible; but when they occur with sufficient rapidity, they produce the necessary result, a musical note, of higher or lower pitch, as the vibrations or tappings are more or less numerous. It often happens that other and extraneous sounds, as those due to the ringing of the metal, the vibration of the table, or subdivisions of the whole vibrating system, mingle with the true sound produced by the blows of the rocker; these were referred to and illustrated, and a method shown of easily distinguishing the latter from the

DOI: 10.4324/b23406-19

former. It consisted in pressing perpendicularly with a small stick or pointed metal rod on the back of the rocker, exactly over the groove, so as to make the vibrations quicker, but not to disturb their regularity; the true sound of the beats of the rocker immediately rises in pitch, and may be sometimes made to pass through an octave or more at pleasure, falling again as the pressure is removed.

As the sound is evidently due to the rapid blows of the rocker, the only difficulty was to discover the true cause of the sustaining power by which the rocker was continued in motion, whilst any considerable difference of temperature existed between it and the block of lead beneath; this Mr. Faraday referred to the ultimate expansion and contraction, as Professor Leslie and Mr. Trevelyan have done generally; but he gave a minute account of the manner in which, according to his views, such expansion and contraction could produce the effect. When the heated rocker is reposing upon a horizontal ridge of lead, it touches at two points, which are heated and expanded, and form, as it were, two hills; when one side of the rocker is raised, the point relieved from its contact is instantly cooled by the neighbouring portions of lead, the expansion ceases, and the hill falls. When the rocker, therefore, is left free, the raised side descends through a greater space than that through which it was lifted, and also to a lower level than the other side; in consequence of which a momentum is given to it, which carries its centre of gravity beyond the point to which it would pass if there had been no alteration in the heights of the sustaining points. It is this additional force which acts as a maintaining power; and recurs twice in each vibration, *i. e.* once on each side. The force is gained by the whole rocker being lifted bodily by the point on which it is for the time supported, and comes into play by the side of the rocker which is descending, having a greater space to fall through than that which is passed over by the mere force of its momentum during its previous rise. A curious consequence of this action is, that the force which really lifts the rocker is on one side of the centre of gravity, whilst the rising side of the rocker itself is on the other.

This, however, is not the only maintaining cause or mechanical force generated by the alternate expansion and contraction of the lead. If the vertical direction of the forces be put out of consideration for a time, and the two points of support be examined, it will be found that whilst the rocker is quiescent, both points (with their neighbouring parts) being heated, will expand and compress the lateral portions of the lead, until the tension of the latter is equal to their own. When one side of the rocker is raised, the point that it rested upon is instantly cooled, and therefore contracts; but as the neighbouring parts retain their tension, they move towards the contracting part, the other point of support moving with the rest. When the rocker returns in its oscillation, it reheats and re-expands the first point of support, whilst the second, now out of contact, is cooled and contracted, and the first point, therefore, moves towards the second. A necessary consequence of this mutual relation of the points is, that the one under process of heating is always moving towards the other which is under process of cooling, and, consequently, towards a perpendicular from the centre of gravity; but as it is at the same time the supporting point to the rocker, that supporting point is, by irresistible impulse, carried in a direction under and towards the line passing from the centre of gravity towards the earth, at the same instant that the centre of gravity of the rocker is, by the momentum of the latter, moving in the opposite direction: hence a very simple maintaining power, sufficient, whenever the rocker continues to vibrate, to compensate for the loss of force in each half of the vibration which would occur if the rocker and lead were of the same temperature. Mr. Faraday illustrated the sustaining force of the lateral motion of the points of support, by placing a rocker on a piece of lead, and the latter on a board. A pair of sugar-tongs was held tightly by the bend against the edge of the board, so that the line from the tongs towards the rocker was perpendicular to the axis of the latter. On making the limbs of the sugar-tongs vibrate in the manner of a tuning-fork, they communicated longitudinal vibrations of equal duration and number to the board, and through it to the lead and points supporting the rocker; which latter itself immediately acquired vibratory motion isochronous with the vibrations of the tongs, and by successive blows upon the lead produced sound; upon removing the rocker, and repeating the other parts of the experiment, no sound was produced.

Experiments with other metals were then made. A piece of curved silver plate being heated and placed on an iron triblet, rocked and sang in the manner of the others; this is an effect which working silversmiths have long known. The superiority of lead, as the cold metal, was referred to its great expansive force by heat, combined with its deficient conducting power, which is not a fifth of that of copper, silver, or gold; so that the heat accumulates much more at the point of contact in it, than it could do in the latter metals.

Mr. Trevelyan's paper had been read to the Royal Society of Edinburgh, but is not yet published. Mr. Faraday stated that Mr. Trevelyan had very liberally allowed him the use of a written copy.

Editorial Headnote

20. J. P. Marrian, "On Sonorous Phænomena in Electro-Magnets", *Philosophical Magazine*, Series 3, 25, no. 167 (1844), pp. 382–84.

Writing in the *Philosophical Magazine* in 1844, J. P. Marrian published an account of a series of observations on acoustic phenomena in relation to electromagnetism. This represented an example, albeit by an otherwise unknown experimentalist, of the ways in which sound might be employed as a means of examining nature: the ear, as well as the eye, was an organ of philosophical and epistemological value. Marrian claimed to have heard a sound, apparently related to the generation of an electric current. After constructing a helix of copper, wound around an iron bar, Marrian set out to measure the conducting power of various metals. Yet in sustaining a current with the copper and magnet, Marrian reported how "at the instant that the magnetism is imparted *a sound is given from it*", which was repeated on breaking the current. Furthermore, Marrian asserted that he could vary the tone of these sounds by altering the physical dimensions of the metal bar in use. The "tonic of each bar", Marrian maintained, corresponded to that of the bar, which he could imitate "by striking the end of the bar with any metallic substance".[1] Marrian's work did not go unnoticed among Britain's scientific readers. John Percy (1817–1889) drew Faraday's attention to Marrian's article in October 1844.[2] A metallurgist, Percy was convinced Faraday would be keen to hear of Marrian's sonorous observations. Faraday, however, was less than impressed. While sound might have been prominent in his earliest thoughts surrounding electromagnetic induction back in 1831 and 1832, it had long since passed as a serious line of inquiry for the Royal Institution's experimentalist. Back in 1832, Faraday speculated that he was "inclined to compare the diffusion of magnetic forces from a magnetic pole, to the vibrations upon the surface of disturbed water, or those of air in the phenomena of sound", and believed that his future research would unite the "vibratory theory" of sound to the motions of light.[3] Nevertheless, in 1844, Faraday had less enthusiasm for any relating of sound to electromagnetic phenomena. Faraday explained how he had seen the experiment long before Marrian and knew well the sounds produced: "But as I heard it suspect it is merely some conduction of the sound of the snap at the place of making and breaking contact".[4] In fact, Faraday's attention to sound had disappeared almost entirely by the 1840s. Instead, he worked towards a solution to a far more perplexing problem: how to demonstrate that electricity and magnetism, which Ørsted had shown to be connected back in 1820, were also related to light. In September 1845, Faraday noted how he had, by passing a ray of polarized light through a prism positioned on top of a large circular iron bar, surrounded by copper wire, produced an effect on the light. On sustaining an electrical current through this helix, Faraday caused the light ray to bend, confirming the unity of electricity, magnetism, and light. Sometime around 1862, Scottish physicist James Clerk Maxwell (1831–1879), who knew Faraday personally during the 1850s, would confirm this connection when he calculated the speed of an electromagnetic field to be extremely close to the speed of light. He published "A dynamical theory of the electromagnetic field" in 1865, giving an account of this calculation and confirming Faraday's experimental claims.[5]

NOTES

1 J. P. Marrian, "On Sonorous Phænomena in Electro-Magnets", *Philosophical Magazine*, Series 3, 25, no. 167 (1844), pp. 382–84, 384.

2 "Letter 1629: John Percy to Faraday, 23 Oct., 1844", in Frank A. J. L. James (ed.), *The Correspondence of Michael Faraday, Vol. 3: 1841–December 1848. Letters 1334–2145*, (London: Institution of Electrical Engineers, 1996), pp. 262–64.

3 "Letter 557: Faraday to John George Children and sealed note, 12 Mar., 1832", in Frank A. J. L. James (ed.), *The Correspondence of Michael Faraday, Vol. 2: 1832-December 1840. Letters 525–1333* (London: Institution of Electrical Engineers, 1993), pp. 26–27.

4 "Letter 1639: Faraday to William Robert Grove, 31 Oct., 1844", in James (ed.), *The Correspondence of Michael Faraday, Vol. 3*, pp. 273–74, 273.

5 James Clerk Maxwell, "A dynamical theory of the electromagnetic field", *Philosophical Transactions of the Royal Society of London* 155 (1865), pp. 459–512.

20

ON SONOROUS PHÆNOMENA IN ELECTRO-MAGNETS

J. P. Marrian

Source: J. P. Marrian, "On Sonorous Phaenomena in Electro-Magnets", *Philosophical Magazine*, Series 3, 25, no. 167 (1844), pp. 382–84.

To the Editors of the Philosophical Magazine and Journal

Gentlemen,

HAVING noticed a few phænomena resulting from magnetic influence in bars of soft iron, and which I have not before heard of, I am induced to submit an account of some experiments in the hope that they may lead to further investigation, as I am much inclined to think that many extraordinary facts may be gathered as well as important results obtained from a more extended prosecution of the subject. I have already made a great number of experiments, and shall proceed to detail a few of those which I consider to be the most interesting.

Having some considerable time since commenced a series of experiments on the conducting powers of different metals, among others I made use of several bars of soft iron, and found that when a bar of that metal is so circumstanced that it may be converted into an electro-magnet, at the instant that the magnetism is imparted *a sound is given from it*, and another sound is again perceptible when the galvanic current is broken; the two sounds however are not exactly alike, but appear to correspond with such as may be produced by the alternate separation and attraction of *the* particles or *certain* particles composing the material acted upon. In order however that the subject may be more conveniently detailed, I will describe a few experiments.

I constructed a helix of thick copper wire, on the outside of which is another of a much thinner wire, with such an arrangement that they can be easily separated. I inserted a bar of soft iron one inch in diameter and twenty feet long in the compound helix, the primary wire of which I connected with a battery; on breaking the circuit a very audible sound was given from the iron, and again on forming contact. I then took bars of different dimensions, and varying in thickness from a quarter of an inch to two inches, and in length from six inches to twenty feet, and a sound was given by each, varying however in intensity; first, according to the size of the bar; second, to its diameter, filling or not filling the helix; third, the

DOI: 10.4324/b23406-20

power of the battery; fourth, its position in the helix; for if the end were not placed beyond the centre of the primary coil no magnetic effects were produced, *and the bar did not sound*; and lastly, much appeared to depend on the state of the *secondary coil*, as the sounds diminished considerably on the ends being united, and the diminution was greater when the contact was metallic, than when I held the wires in my hands. The sound produced was the tonic of each bar, and I could imitate it by striking the end of the bar with any metallic substance, but could produce no resemblance by means of a blow laterally. I used electrodes of different metals, and tried numerous experiments in order to prove that the sonorous effects were not produced by the jarring of the apparatus; one of these was the partial envelopment of the bar in wet clay and not suffering it to touch the helix, but the effect was still the same. I tried bars of silver, copper, zinc, tin, lead and brass, but no similar effect was produced.

In three bars of equal sizes, one of soft iron, one of hard steel, and the third a permanent magnet, no sensible difference in these effects could be detected.

These experiments tend to prove that the particles of a bar of iron, when changing its electrical state, have a tendency decidedly *motive*, and that too along the bar in the direction of its axes. I have not yet been able to prove that any excitation in the same direction will induce an electric current (which I think is highly probable), but have put together these facts in a hurried manner, in order that we may have an opportunity of hearing of such investigation as we may naturally expect from the immense amount of talent employed in this interesting branch of science. I think also that these facts may serve to elucidate many others which I have not yet heard explained, but must reserve hypotheses for a future opportunity.

Yours, very respectfully,
J. P. Marrian.
Birmingham.

BIBLIOGRAPHY

PRIMARY SOURCES

Manuscripts

King's College London Library: Wheatstone 5, "Lecture IV" in "Various: 8 lectures on sound", (ca. 1835).

Royal Society Archives: HS 18.146, "Wheatstone to John Herschel" (13 May 1835).

Wren Library, Trinity College Cambridge: Ref.15.b.80.3, Arthur Trevelyan, "Theory of Heat" (ca. 1843).

Published

Airy, George Biddell, *On Sound and Atmospheric Vibrations, with the mathematical elements of music. Designed for the use of students of the University* (London: Macmillan and Co, 1868).

Airy, Wilfrid (ed.), *Autobiography of Sir George Biddell Airy* (Cambridge: Cambridge University Press, 1896).

(Anon.), "The New Theatre Royal, Covent-Garden, erected 1809, from the designs of Robert Smirke, esq. jun. Associate of the Royal Academy of London", *The Monthly Magazine*, 28 (1 Nov. 1809, London, England), p. 416.

(Anon.), "The Enchanted Lyre", *The Literary Gazette* 5 (1821, London), p. 586.

(Anon.), "The Enchanted Lyre and the Diaphonicon", *The New Monthly Magazine and Literary Journal*, Vol. 6 (London, 1822), pp. 200–201.

(Anon.), "Enchanted Lyre", *The Circulator of Useful Knowledge, Amusement, Literature, Science, and General Information* (London, 1825), p. 284.

(Anon.), "The Kaleidophone, or phonic Kaleidoscope", *Mechanics' Magazine* 8 (11 Aug. 1827, London, England), pp. 49–52.

(Anon.), "Proceedings of the Royal Institution: February 15th", *The Quarterly Journal of Science, Literature and Art* 25 (1828, London), pp. 173–74.

(Anon.), "Musical Sound", *Arcana of Science and Art* (1830, London, England), pp. 65–67.

(Anon.), "On the structure of metals, by M. Savart", *Quarterly Journal of Science, Literature, and Art* 28 (1830, London, England), pp. 163–67.

(Anon.), "Royal Institution", *The Dublin Literary Gazette, or Weekly Chronicle of Criticism, Belles Lettres, and Fine Arts* 1 (Dublin, Ireland, 1830), pp. 395–96.

(Anon.), "Royal Institution", *The Literary Gazette* (1830, London, England), pp. 223–24.

(Anon.), "Friday-Evening Proceedings at the Royal Institution of Great Britain", *Philosophical Magazine*, Vol. IX (Jan.–June 1831, London, England), p. 461.

(Anon.), "Mr Eulenstein's Jew's Harp", *Liverpool Mercury ect*, no. 1121 (26 Oct. 1832, Liverpool, England).

(Anon.), "Review: Mechanisms of the Heavens", *Monthly Review*, n.s., 1 (Jan. 1832), pp. 133–41.

(Anon.), *A Sketch of the Life of C. Eulenstein, the celebrated performer on the Jews' Harps* (London: James J. Welsh, 1833).

(Anon.), "Eulenstein", *Bristol Mercury*, no. 2242 (6 Apr. 1833, Bristol, England).

(Anon.), "Royal Institution", *The Morning Post*, no. 20371 (22 Mar 1836, London, England).

(Anon.), "The Speaking Machine", *Punch* XI (1846, London, England), p. 83.

(Anon.), "The New House of Lords", *Punch* XII (1847, London, England), p. 184.

(Anon.), "The New House of Lords", *Punch* XII (1847, London, England), p. 206.

(Anon.), "The Peers and the Silent System", *Punch* XVII (1849, London, England), p. 230.

(Anon.), "Opening of the New House of Commons", *The Times*, no. 20503 (31 May 1850, London, England), p. 5.

(Anon.), "Supply – the New House of Commons", Commons Sitting, 2 Aug., 1850, *Hansard*, Third Series, Vol. 113, cc. 726–39.

(Anon.), "The New House of Commons", *The Times*, no. 20546 (20 July 1850, London, England), p. 5.

(Anon.), "An Imperial Pitchfork", *The Spectator*, 28 Aug. 1858 (London, England), pp. 910–11.

(Anon), "English committee on musical pitch", *The Spectator*, 18 June 1859 (London, England), pp. 639–640.

(Anon.), "Tuning-forks and musical pitch", *Chamber's Journal of Popular Literature Science and Arts* 34, no. 346 (Jul.–Dec. 1860, London, England), 18 Aug. 1860, pp. 98–101.

(Anon.), "Orchestral Tuning", *The Musical Standard*, 26 June 1875 (London, England), pp. 409–10.

(Anon.), "Printers' Pension Corporation", *The Times*, no. 29567 (14 May 1879, London, England), p. 7.

(Anon.), "Recent Developments in Acoustical Science", *The Times*, no. 29337 (19 Aug. 1878, London, England), p. 4.

(Anon.), "Telegraphy", *The Times*, no 29160 (24 Jan. 1878, London, England), p. 11.

(Anon.), "The question of a standard musical pitch", *Moonshine* (27 Jun. 1885, London, England), p. 302.

(Anon.), "Royal Academy of Music", *The Musical Times and Singing Class Circular* 37, no. 636 (1 Feb. 1896), pp. 97–98.

Aveling, Thomas, *Recreations, Physical and Mental, lawful and unlawful: a lecture* (London, 1849).

Aveling, Thomas, *Great Sights: A Discourse Delivered in Kingsland Chapel, on the Sabbath Evening Preceding the Opening of the Great Exhibition* (London: Snow, Gurney, and Turner, 1851).

Aveling, Thomas, *"Sorrow On The Sea"; a discourse delivered after the wreck of the royal charter, and the catastrophe on board the great eastern* (London: John Snow, 1859).

Bacon, Francis, *Sylva Sylvarum: or a naturall historie* (London: William Lee, 1635).

Bagenal, Hope and Wood, Alex, *Planning for Good Acoustics* (London: Methuen & Co. Ltd, 1931).

Blackburn, John, "Description of a sounding board in Attercliffe Church," *Philosophical Transactions of the Royal Society*, 1828, (London, England), pp. 361–63.

Blackburn, John, "Description of a parabolic sounding board, erected in Attercliffe Church," *Philosophical Magazine* 6, no. 31 (London, England), pp. 21–27.

Brewster, David, "Review of *On the Connexion of the Physical Sciences*", *Edinburgh Review* 59 (Apr. 1834), pp. 154–71.

Brown, Colin, "Reviews: *A demonstration of the Musical Scale, founded upon the law of vibrations of sound*", *The Musical Times* 14 (1 Apr. 1870, London, England), p. 436.

Brown, Robert, *Elements of Musical Science* (London: Hamilton, Adams, and Co, 1860).

"Builder", "Hearing in the new Houses of Parliament", *The Times*, no. 20572 (20 Aug. 1850, London, England), p. 3.

Busby, Thomas, *A Grammar of Music: to which are prefixed observations explanatory of the properties and powers of music as a science. And of the general scope and object of the work* (London: John Walker, 1818).

Busby, Thomas, "The Acoucryptophone, or Enchanted Lyre", *Concert Room and Orchestra Anecdotes of Music and Musicians, ancient and modern*, Vol. I (London, 1825), pp. 9–10.

Brewster, David, *A Treatise on the Kaleidoscope* (Edinburgh: Archibald Constable & Co, 1819).

Brewster, David, *Letters on Natural Magic addressed to Sir Walter Scott*, Bart (New York: Harper & Brothers, 1842).

Darwin, Charles, *The Descent of Man, and Selection in Relation to Sex*, Vol 2 of 2 (London: John Murray, 1871).

De Morgan, Augustus, "On the beats of imperfect consonances", *Transactions of the Cambridge Philosophical Society* 10 (1864), pp. 129–45.

Ellis, Alexander J., *The History of Musical Pitch* (London: W. Tounce, 1880).

Faraday, Michael, "On a Peculiar Class of Acoustical Figures, and on the Forms of Fluids Vibrating on Elastic Surfaces", *Philosophical Transactions of the Royal Society* 121 (1831, London, England), pp. 299–340.

Faraday, Michael, *Experimental Researches in Chemistry and Physics* (London: Richard Taylor and William Francis, 1859).

Forbes, James D., "Experimental researches regarding certain vibrations which take place between metallic masses having different temperatures", *Transactions of the Royal Society of Edinburgh* XII, pt. 2 (1833, London, England), pp. 429–61.

Galloway, Thomas, "Review of Mechanisms of the Heavens", *Edinburgh Review* 55 (Apr. 1832), pp. 1–25.

Godard, John George, *George Birkbeck, the Pioneer of Popular Education: a memoir and a review* (London: Bemrose & Sons, 1884).

Goldingham, John, "Experiments for ascertaining the velocity of sound, at Madras in the East Indies", *Philosophical Transactions of the Royal Society* (1823, London, England), pp. 96–139.

Goldingham, J., *Madras Observatory Papers* (Madras: College Press of Fort St George, 1827).

Gurney, Edmund, *The Power of Sound* (London: Smith, Elder, & Co, 1880).

Hanslick, Eduard, *On the Musically Beautiful: a contribution towards the revision of the aesthetics of music*, Geoffrey Payzant (Trans.) (Indianapolis, IN: Hackett Publishing Company, 1986).

Helmholtz, Hermann, *On the Sensations of Tone as a Physiological Basis for the Theory of Music*, Alexander J. Ellis (Trans.), (London: Longmans, Green, and Co, 1875).

Helmholtz, Hermann, "Lord Rayleigh's *Theory of Sound*," *Nature* 19, no. 476 (12 Dec. 1878), pp. 117–18.

Herschel, John, *A Preliminary Discourse on the Study of Natural Philosophy*, (London: Longman, Rees, Orme, Brown, and Green, 1830).

Herschel, John, "Light", *Encyclopaedia Metropolitana*, Vol. IV (London, 1830), pp. 341–586.

Herschel, John, "Physical Astronomy", *Encyclopaedia Metropolitana*, Vol. IV (London, 1830), pp. 647–734.

Herschel, John, "Sound", *Encyclopaedia Metropolitana*, Vol. IV (London, 1830), pp. 747–824.

Herschel, John, "Review of Mechanisms of the Heavens", *Quarterly Review* 47 (Jul. 1832), pp. 537–59.

Herschel, John, "Uniform Musical Pitch", *Leeds Mercury*, no. 6985 (2 Aug. 1859, Leeds, England).

Herschel, John, "On Musical Scales", *The Quarterly Journal of Science* XIX (July 1868, London, England), pp. 338–52.

Herschel, John, *Familiar Lectures on Scientific Subjects* (New York: George Routledge & Sons, 1869).

Hegel, G. W. F., *Hegel's Aesthetics: lectures on fine art*, T. N. Knox (Trans.), 2 Vols., (Oxford: Clarendon Press, 1988).

Higgins, William Mullinger, *The Philosophy of Sound, and History of Music* (London: Wm S. Orr and Co, 1838).

Lange, Frederick Albert, *History of Materialism and Criticism of Its Present Importance*, Vol. 3 of 3 (London: Trübner & Co, 1877).

Lunn, Henry Charles, "Musical Pitch", *The Musical Times, and Singing Class Circular* 13, no. 312 (1 Feb. 1869), pp. 663–65.

Lunn, Charles, *Vocal Expression: empirical or scientific? A lecture, delivered at the Royal Academy, Tenterden Street, on Friday, May 17, 1878* (London: Stanley Lucas, Weber, & Co, 1878).

Mallock, Arnulph, "Note on the effect of wind pressure on the pitch of organ pipes," *Philosophical Transactions of the Royal Society* 95, no. 666 (1918), pp. 99–106.

Marrian, J. P., "On sonorous phaenomena in electro-magnets", *Philosophical Magazine*, Series 3, 25, no. 167 (1844), pp. 382–84.

Maxwell, James Clerk, "A dynamical theory of the electromagnetic field", *Philosophical Transactions of the Royal Society of London* 155 (1865), pp. 459–512.

Mersenne, Marin, *Harmonie universelle contenant la théorie et la pratique de la musique* (1636), facsimile, ed. F. Lesure (Paris: Éditions du Centre National de la Recherche Scientifique, 1963).

Moll, Gerard, "On Captain Parry's and Lieutenant Foster's experiments on the velocity of sound", *Philosophical Transactions of the Royal Society* (1828, London, England), pp. 97–104.

Neuffer, "To Eulenstein, Musician on the Jew's Harp", in Leonard Fox (ed.), *The Jew's Harp: a comprehensive anthology* (Lewisburg: Bucknell University Press, 1988), pp. 162–63.

Gore Ouseley, F. A., *A Treatise on Harmony* (Oxford: Clarendon Press, 1868).

Pole, William, "Explanation of the Musical Scale and its Component Intervals", in F. A. Gore Ouseley, *A Treatise on Harmony* (Oxford: Clarendon Press, 1868), pp. 259–63.

Rayleigh, Lord, "Acoustical observations. – IV," *Philosophical Magazine*, 5th Series, no. 82 (1882), pp. 340–47.

Reid, David Boswell, *Illustrations of the Theory and Practice of Ventilation, with remarks on warming, exclusive lighting, and the communication of sound* (London: Longman, Brown, Green & Longmans, 1844).

Report from the Select Committee on weights and measures; together with the proceedings of the committee, minutes of evidence, appendix, and index, House of Commons Paper 411 (1862).

Ross, James Clark, "On the position of the north magnetic pole", *Philosophical Transactions of the Royal Society* 124, no. 1834 (London, England), pp. 47–52.

Ross, John, *Narrative of a Second Voyage in Search of a North-West Passage, and of a Residence in the Arctic Regions, during the years 1829, 1830, 1831, 1832, 1833*, (Brussels: A.D. Wahlen, 1835).

Russell, John Scott, "Elementary considerations of some principles in the construction of buildings designed to accommodate spectators and auditors", *The Edinburgh New Philosophical Journal* XXVII. (Apr.–Oct. 1839, Edinburgh, Scotland), pp. 131–36.

Russell, John Scott, *The Wave of Translation in the Oceans of Water, Air, and Ether* (London: Trübner & Co, 1885).

Saunders, George, *A Treatise on Theatres* (London: I. and J. Taylor, 1790).

Savart, Felix, "Researches on the elasticity of regularly crystallized bodies", *Edinburgh Journal of Science*, I, new series (Apr.–Oct. 1829, Edinburgh, Scotland), pp. 141–46.

Smith, Thomas Roger, *The Acoustics of Public Buildings; or, the principles of the science of sound applied to the purposes of the architect and builder* (London: John Weale, 1861).

Smyth, Charles Piazzi, "On the methods adopted to secure extreme accuracy in the Edinburgh Castle Time-Gun signal", *Transactions of the Royal Scottish Society of Arts* VI (1864, Edinburgh, Scotland), pp. 136–52.

Somerville, Martha, *Personal Recollections from Early Life to Old Age, of Mary Somerville. With selections from her correspondence* (London: John Murray, 1873).

Somerville, Mary, *Mechanism of the Heavens* (London: John Murray, 1831).

Somerville, Mary, *On the Connexion of the Physical Sciences* (London; John Murray, 1834).

Somerville, Mary, *On the Connexion of the Physical Sciences*, 2nd ed. (John Murray: London, 1835).

Spencer, Herbert, "Progress: its Law and Cause", *Westminster Review* 67 (April 1857), pp. 445–85.

Spencer, Herbert, "The Origin and Function of Music", in Herbert Spencer, *Essays Scientific, Political, and Speculative* (London: Longman, Brown, Green, Longmans, and Roberts, 1858), pp. 359–384.

Spencer, Herbert, "Progress: its Law and Cause", in Herbert Spencer, *Essays Scientific, Political, and Speculative*, (London: Longman, Brown, Green, Longmans, and Roberts), pp. 1–54.

Spencer, Herbert, "On the origin of music", *Mind* 16, no. 64 (Oct. 1891), pp. 535–37.

Stone, Edward James, "An experimental determination of the velocity of sound", *Philosophical Transactions of the Royal Society* (1871, London, England), pp. 1–6.

Stratford, Joseph, *Gloucestershire Biographical Notes* (Gloucester: The "Journal" Office, 1887).

Strutt, John William, *The Theory of Sound*, Vol. 1 of 2 (London: Macmillan, 1877).

Strutt, John William, *The Theory of Sound*, Vol. 2 of 2 (London: Macmillan, 1877).

Trevelyan, Arthur, "On the vibration of heated metals: By Arthur Trevelyan, Esq.; including a letter on the same subject by Dr W. Knight", *The London and Edinburgh Philosophical Magazine and Journal of Science*, Series 3 (Nov. 1833, London, England), pp. 321–32.

Trevelyan, Arthur, "Further notice of the vibration of heated metals; with the description of a new and convenient apparatus for experimenting with", *Philosophical Magazine*, Series 3, 6:32, (1835), pp. 85–6.

Tyndall, John, *Sound: a course of eight lectures delivered at the Royal Institution of Great Britain* (London: Longmans, Green, and Co, 1867).

Tyndall, John, *Essays on the Use and Limit of the Imagination in Science* (London: Longmans, Green, and Co, 1870).

Tyndall, John, "On the atmosphere as a vehicle of sound", *Philosophical Transactions of the Royal Society* 164 (1875, London, England), pp. 183–244.

Uniform Musical Pitch. Minutes of a meeting of musicians, amateurs, and others interested in music, held at the house of the Society of Arts, when the Report of the Committee appointed by the council of the society was received and adopted, (Royal Society for the Encouragement of Arts, Manufactures, and Commerce: 1860), published in "Uniform Musical Pitch", *The Journal of the Society of Arts* 8, no. 417 (16 Nov. 1860), pp. 1–8.

Wallaschek, Richard, *Primitive Music: an inquiry into the origin and development of music, songs, instruments, dances, and pantomimes of savage races* (London: Longmans, Green, and Co, 1893).

Wheatstone, Charles, "New Experiments on Sound", *Thomson's Annals of Philosophy* vi (1823, London, England), pp. 81–90.

Wheatstone, Charles, "Description of the Kaleidophone, or Phonic Kaleidoscope", *Quarterly Journal of Science, Literature, and Art* I (1827, London, England), pp. 344–51.

Wheatstone, Charles, "On the resonances, or reciprocated vibrations of columns of air", *Quarterly Journal of Science* III (1828, London, England), pp. 175–83.

Wheatstone, Charles, "On the transmission of musical sounds through solid linear conductors, and on their subsequent reciprocation", *Journal of the Royal Institution of Great Britain*, no. 5, (London: John Murray, Dec. 1831), pp. 223–38.

Wheatstone, Charles, "On the figures obtained by strewing sand on vibrating surfaces, commonly called acoustic figures", *Philosophical Transactions of the Royal Society*123, (1833, London, England), pp. 593–633.

Wheatstone, Charles, *The Scientific Papers of Sir Charles Wheatstone, D.C.L., F.R.S* (London: Taylor and Francis, 1879).

Whewell, William, *Architectural Notes on German Churches, with remarks on the origin of gothic architecture* (J. & J. J. Deighton, 1830).

Whewell, William, *Astronomy and General Physics; considered with reference to natural theology* (London: William Pickering, 1833).

Whewell, William, "On the *Connexion of the Physical Sciences*. by Mrs. Somerville", *Quarterly Review* 51 (1934, London, England), pp. 54–68.

Whewell, William, *The History of the Inductive Sciences, from the earliest to the present time*, Vol. 1 of 3, (London: John W. Parker, 1837).

Whewell, William, *The Philosophy of the Inductive Sciences, founded upon their history*, Vol. 1 of 2, 2nd ed. (London: John W. Parker, 1847).

Whewell, William, *The History of the Inductive Sciences, from the earliest to the present time*, Vol. 2 of 3, 3rd ed. (London: John W. Parker, 1858).

Whewell, William, "Of Certain Analogies between Architecture and the Other Fine Arts", in *Papers read at the Royal Institute of British Architects, Session 1862–63* (London; J. H. James Parker, 1863), pp. 175–89.

Willis, Robert, *An attempt to analyse the automaton chess player, of Mr. De Kempelen*, (J. Booth: London, 1821).

Willis, Robert, "On the vowel sounds, and on reed organ-pipes", *Transactions of the Cambridge Philosophical Society* (Cambridge: Cambridge University Press, 1829), pp. 1–38.

Willis, Robert, *Remarks on the Architecture of the Middle Ages, especially of Italy* (Cambridge: J. & J. J. Deighton, 1835).

Willis, Robert, *Principles of Mechanism, designed for the use of students in the universities, and for engineering students generally* (London: John W. Parker, 1841).

Wordsworth, Christopher, *Sacred Music: a sermon, preached at the anniversary of the Choral Association of the Diocese of Llandaff, in the Cathedral Church of Llandaff, Sept., 2, 1868* (London: Rivingtons, 1868).

Wren, Stephen, *Parentalia: or, memoirs of the family of the Wrens; Viz. of Mathew Bishop of Ely, Christopher Dean of Windsor, &c. but chiefly of Sir Christopher Wren, late surveyor-general of the Royal buildings, president of the Royal Society, &c. &c.* (London: T. Osborn, 1750).

Wyatt, Benjamin, *Observations on the design for the Theatre Royal, Drury Lane, as executed in the year 1812: accompanied by plans, elevation, & sections, of the same* (London: J. Taylor, 1813).

Secondary Sources

Atlas, Allan W., *The Wheatstone English Concertina in Victorian England* (Oxford: Clarendon Press, 1996).

Atlas, Allan W., "Ladies in the Wheatstone Ledgers: the gendered concertina in Victorian England, 1835–1870", *Royal Musical Association Research Chronicle* 39 (2006), pp. 9–10.

Aubin, David, Charlotte Brigg, and H. Otto Sibum (eds.), *The Heavens on Earth: observatories and astronomy in nineteenth-century science and culture* (Durham, NC: Duke University Press, 2010).

Barker, Andrew, *The Science of Harmonics in Classical Greece* (Cambridge: Cambridge University Press, 2007).

Barrett, Paul H. (ed.), *Metaphysics, Materialism & the Evolution of Mind: the early writings of Charles Darwin* (Chicago: University of Chicago Press, 1980).

Barton, Ruth, *The X Club: power and authority in Victorian science* (Chicago: University of Chicago Press, 2018).

Bebbington, David W., *Evangelicalism in Modern Britain: a history from the 1730s to the 1980s*, (London: Unwin Hyman, 1989).

Belteki, Daniel, ""The Grand Strategy of an Observatory": George Airy's vision for the division of astronomical labour among observatories during the nineteenth century", *Notes and Records of the Royal Society*, online version (2021), pp. 1–17.

Beyer, Robert T., *Sounds of Our Times: two hundred years of acoustics* (New York: Springer-Verlag, 1999).

Bijsterveld, Karin, *Mechanical Sound: technology, culture, and public problems of noise in the twentieth century*, (Cambridge, MA: MIT Press, 2008).

Bisset, W. M., "Cape Town's Time-Guns", *Scientia Militaria* 14, no. 4 (1984), pp. 67–71.

Bowers, Brian, *Sir Charles Wheatstone FRS, 1802–1875* (London: Institution of Electrical Engineers, 2001).

Brewer, John, *The Sinews of Power: war, money and the English State, 1688–1783* (London: Unwin Hyman, 1989).

Buchanan, Alexandrina, *Robert Willis (1800–1875): and the foundation of architectural history* (Woodbridge: Boydell Press, 2013).

Buchanan, Alexandrina, James W. P. Campbell, Javier Giron, and Santiago Huerta (eds.), *Robert Willis: science, technology and architecture in the nineteenth century. Proceedings of the international symposium held in Gonville and Caius College, Cambridge UK, 16th–17th September 2016* (Madrid: Instituto Juan De Herrera, 2016).

Bullard, E. C., "Obituary Notices: Charles Galton Darwin", *Quarterly Journal of the Royal Astronomical Society* 4, no. 3 (1963), pp. 316–17.

Cahan, David (ed.), *Hermann Von Helmholtz and the Foundations of Nineteenth-Century Science* (Berkeley: University of California Press, 1993).

Cantor, Geoffrey N., "The changing role of Young's ether", *The British Journal for the History of Science* 5, no. 1 (Jun. 1970), pp. 44–62.

Cantor, Geoffrey N., "The Reception of the Wave Theory of Light in Britain: a case study illustrating the role of methodology in scientific debate", *Historical Studies in the Physical Sciences* 6 (1975), pp. 109–32.

Cantor, Geoffrey N., *Optics After Newton: theories of light in Britain and Ireland, 1704–1840* (Manchester: Manchester University Press, 1983).

Cantor, Geoffrey N., *Michael Faraday: Sandemanian and scientist. A study of science and religion in the nineteenth century* (Basingstoke: Macmillan, 1991).

Cantor, Geoffrey N., *Religion and the Great Exhibition of 1851* (Oxford: Oxford University Press, 2011).

Cantor Geoffrey N., "Science, providence, and progress at the Great Exhibition", *Isis* 103, no. 3, (2012), pp. 439–59.

Cantor, Geoffrey, David Gooding, and Frank A. J. L. James, *Michael Faraday* (New York: Humanity Books, 1991).

Carter, Christopher, *Magnetic Fever: global imperialism and empiricism in the nineteenth century* (Philadelphia: American Philosophical Society, 2009).

Case, Stephen, *Making Stars Physical: the astronomy of Sir John Herschel* (Pittsburgh, PA: University of Pittsburgh Press, 2018).

Chapman, Allan, "Sir George Airy (1801–1892) and the concept of international standards in science, time keeping and navigation", *Vistas in Astronomy* 28 (1985), pp. 321–28.

Charlton, David (ed.), *E. T. A. Hoffmann's Musical Writings: Kreisleriana, the poet and the composer, music criticism* (Cambridge: Cambridge University Press, 1989).

Clarke, Martin V. (ed.), *Music and Theology in Nineteenth-Century Britain* (Farnham: Ashgate, 2012).

Cobb, Aaron D., "Inductivism in practice: experiment in John Herschel's philosophy of science", *Journal of the International Society for the History of Philosophy of Science* 2, no. 1 (Spring 2012), pp. 21–54.

Corbin, Alain, *Village Bells: sound and meaning in the 19th-century French countryside*, Martin Thom (Trans.) (Basingstoke: Macmillan Publishers Ltd, 1999).

Creese, David, *The Monochord in Ancient Greece Harmonic Science* (Cambridge: Cambridge University Press, 2010).

Crosland, Maurice, "The French Academy of Sciences in the nineteenth century", *Minerva* 16 (1978), pp. 73–102.

Davies, James Q., "Instruments of Empire", in James Q. Davies and Ellen Lockhart (eds.), *Sound Knowledge: music and science in London, 1789–1851* (Chicago: University of Chicago Press, 2016), pp. 145–74.

Davies, James Q. and Ellen Lockhart, "Fantasies of total description", in James Q. Davies and Ellen Lockhart (eds.), *Sound Knowledge: music and science in London, 1789–1851* (Chicago: University of Chicago Press, 2016), pp. 1–26.

Davies, James Q. and Ellen Lockhart (eds.), *Sound Knowledge: music and science in London, 1789–1851* (Chicago: University of Chicago Press, 2016).

DeYoung, Ursula, *A Vision of Modern Science: John Tyndall and the role of the scientist in Victorian Culture* (Basingstoke: Palgrave Macmillan, 2011).

Dickson, Melissa, "Charles Wheatstone's Enchanted Lyre and the Spectacle of Sound", in James Q. Davies and Ellen Lockhart (eds.), *Sound Knowledge: music and science in London, 1789–1851* (Chicago: University of Chicago Press, 2016), pp. 125–44.

Dolan, Emily I., *The Orchestral Revolution: Haydn and the technologies of timbre* (Cambridge: Cambridge University Press, 2013).

Drie, Melissa Van, "Hearing through the théâtrophone: sonically constructed spaces and embodied listening in late nineteenth-century French theatre", *Sound Effects – An Interdisciplinary Journal of Sound and Sound Experience* 5, no. 1, (2015), pp. 74–90.

Dunn, Richard and Rebekah Higgitt, *Finding Longitude: how ships, clocks, and stars helped solve the longitude problem* (London: Collins, 2014).

Dupree, Mart Helen and Sean B. Franzen (eds.), *Performing Knowledge, 1750–1850* (Berlin and Boston: De Gruyter, 2015).

Eve, A. S. and C. H. Creasey, *Life and Work of John Tyndall* (London: Macmillan & Co, 1945).

Fisch, Menachem and Simon Schaffer (eds.), *William Whewell: a composite portrait* (Oxford: Clarendon Press, 1991).

Fischer, Sabine von, "Debating volume: architectural *versus* electrical amplification in the League of Nations, 1926–28", *The Journal of Architecture* 23, no. 6 (2018), pp. 904–35.

Flood, Raymond, Adrian Rice, and Robin Wilson (eds.), *Mathematics in Victorian Britain*, (Oxford: Oxford University Press, 2011).

Forsyth, Michael, *Buildings for Music: the architect, the musician, and the listener from the seventeenth century to the present day* (Cambridge: Cambridge University Press, 1985).

Fox, Celina (ed.), *London – World City, 1800–1840* (New Haven, CT: Yale University Press, 1992).

Fox, Leonard (ed.), *The Jew's Harp: a comprehensive anthology* (Lewisburg: Bucknell University Press, 1988).

Francis, Keith A., "Paley to Darwin: natural theology versus science in Victorian sermons", in Keith A. Francis and William Gibson (eds.), *Oxford Handbook of the British Sermon, 1689–1901* (Oxford: Oxford University Press, 2012), pp. 444–62.

Francis, Keith A. and William Gibson (eds.), *Oxford Handbook of the British Sermon, 1689–1901* (Oxford: Oxford University Press, 2012).

Gatens, William J., *Victorian Music in Theory and Practice* (Cambridge: Cambridge University Press, 1986).

Gauld, Alan, "Gurney, Edmund (1847–1888)", *Oxford Dictionary of National Biography*. 25 Mar. 2023. www.oxforddnb.com/view/10.1093/ref:odnb/9780198614128.001.0001/odnb-9780198614128-e-11763.

Gibson, William, "The British Sermon, 1689–1901: quantities, performance, and culture", in Keith A. Francis and William Gibson (eds.), *Oxford Handbook of the British Sermon, 1689–1901*, (Oxford University Press: Oxford, 2012), pp. 3–30.

Gieryn, Thomas F., "Boundary-work and the demarcation of science from non-science: strains and interests in professional ideologies of scientists", *American Sociological Review* 48, no. 6 (Dec. 1983), pp. 781–95.

Gillin, Edward J., "Prophets of progress: authority in the scientific projections and religious realizations of the "Great Eastern" steamship", *Technology and Culture* 56, no. 4 (Oct. 2015), pp. 928–56.

Gillin, Edward J., *The Victorian Palace of Science: scientific knowledge and the building of the Houses of Parliament* (Cambridge: Cambridge University Press, 2017).

Gillin, Edward J., "Seismology's acoustic debt: Robert Mallet, Chladni's figures, and the Victorian science of earthquakes", *Sound Studies* 6, no. 1 (Spring 2020), pp. 65–82.

Gillin, Edward J., *Sound Authorities: scientific and musical knowledge in nineteenth-century Britain* (Chicago: University of Chicago Press, 2022).

Gillin, Edward J., *An Empire of Magnetism: global science and the British Magnetic Survey in the age of imperialism* (Oxford: Oxford University Press, 2024).

Gillin, Edward J. and Fanny Gribenski, "The politics of musical standardization in nineteenth-century France and Britain", *Past & Present*, no. 251 (May 2021), pp. 153–87.

Gillin, Edward J. and Fanny Gribenski, "Science in the church: the sacred spaces of sonorous experiment and the formation of modern acoustics", *Isis* 114, no. 3 (Sep. 2023), pp. 537–558.

Gillin Edward J. and H. Horatio Joyce (eds.), *Experiencing Architecture in the Nineteenth Century: buildings and society in the modern age* (London: Bloomsbury, 2019).

Gitelman, Lisa, *Scripts, Grooves, and Writing Machines: representing technology in the Edison era* (Stanford, CA: Stanford University Press: Stanford, 1999).

Gooday, Graeme J. N., *The Morals of Measurement: accuracy, irony, and trust in late Victorian electrical practice*, (Cambridge University Press: Cambridge, 2004).

Gooday, Graeme, "Architectural Acoustics: Thomas Roger Smith and the science of hearing buildings in nineteenth-century Britain," in Edward Gillin and H. Horatio Joyce (eds.), *Experiencing Architecture in the Nineteenth Century: buildings and society in the modern age* (London: Bloomsbury, 2019), pp. 101–14.

Gooding, David, "'In Nature's School': Faraday as an experimentalist", in David Gooding and Frank A. J. L. James (eds.), *Faraday Rediscovered: essays on the life and work of Michael Faraday, 1791–1867* (Basingstoke: Macmillan Press Ltd, 1985), pp. 105–35.

Gooding, David and Frank A. J. L. James (eds.), *Faraday Rediscovered: essays on the life and work of Michael Faraday, 1791–1867* (Basingstoke: Macmillan Press Ltd, 1985).

Gouk, Penelope M., "Acoustics in the early Royal Society, 1660–1689," *Notes and Records of the Royal Society* 36, no. 2 (1982), pp. 155–75.

Gouk, Penelope M., *Music, Science and Natural Magic in Seventeenth-Century England* (New Haven, CT: Yale University Press, 1999).

Graziano, Amy B. and Johnson, Julene K., "Richard Wallaschek's nineteenth-century contributions to the psychology of music", *Music Perception: an interdisciplinary journal* 23, no. 4 (Apr. 2006), pp. 293–304.

Gribenski, Fanny, "Tuning forks as time travel machines: pitch standardisation and historicism", *Sound Studies* 6, no. 2: Special Issue: Sonic Things: Knowledge Formulation in Flux, (2020), pp. 153–73.

Gribenski, Fanny, *Tuning the World: the rise of 440 hertz in music, science, and politics, 1859–1955* (Chicago: University of Chicago Press, 2024).

Hamilton, James, *Faraday: the life* (London: Harper Collins, 2002).

Hankins, Thomas L. and Robert J. Silverman, *Instruments and the Imagination* (Princeton, NJ: Princeton University Press, 1995).

Harman, P. M. (ed.), *The Scientific Letters and Papers of James Clerk Maxwell*, Vol. 2 of 3 (Cambridge: Cambridge University Press, 1990–2002).

Haynes, Bruce, *History of Performing Pitch: the story of "A"* (Lanham, MD: The Scarecrow Press, 2002).

Herbert, Trevor and Helen Barlow, *Music and the British Military in the Long Nineteenth Century* (Oxford: Oxford University Press, 2013).

Higgitt, Rebekah and Dolan, Graham, "Greenwich, time and 'the line'", *Endeavour* 34, no. 1 (Mar. 2010), pp. 35–39.

Hilton, Boyd, *The Age of Atonement: the influence of evangelicalism on social and economic thought, 1785–1865* (Oxford: Clarendon Press, 1991).

Holmes, Caitlin, "The astronomer royal, the hydrographer and the time ball: collaborations in time signaling 1850–1910", *British Journal for the History of Science* 42, no. 3 (Sep. 2009), pp. 381–406.

Hui, Alexandra, *The Psychophysical Ear: musical experiments, experimental sounds, 1840–1910* (Cambridge, MA: MIT Press, 2013).

Hui, Alexandra, Julia Kursell, and Myles W. Jackson, "Music, sound, and the laboratory from 1750 to 1980", *Osiris* 28, no. 1 (Jan. 2013), pp. 1–11.

Hunt, Bruce J., "The Ohm is where the art is: British telegraph engineers and the development of electrical standards", *Osiris* 9, Instruments (1994), pp. 48–63.

Hunt, Bruce J., "Doing Science in a Global Empire: cable telegraphy and electrical physics in Victorian Britain", in Bernard Lightman (ed.), *Victorian Science in Context* (Chicago: University of Chicago Press, 1997), pp. 312–33.

Jackson, Myles W., *Harmonious Triads: physicists, musicians, and instrument makers in nineteenth-century Germany* (Cambridge, MA: MIT Press, 2006).

Jackson, Myles W., "Charles Wheatstone: musical instrument making, natural philosophy, and acoustics in early nineteenth-century London", in James Q. Davies and Ellen Lockhart (eds.), *Sound Knowledge: music and science in London, 1789–1851* (Chicago: University of Chicago Press, 2016), pp. 101–24.

James, Frank A. J. L. (ed.), *The Correspondence of Michael Faraday, Vol. 2: 1832–December 1840. Letters 525–1333* (London: Institution of Electrical Engineers, 1993).

James, Frank A. J. L. (ed.), *The Correspondence of Michael Faraday, Vol. 3: 1841–December 1848. Letters 1334–2145* (London: Institution of Electrical Engineers, 1996).

James, Frank A. J. L. (ed.), *The Correspondence of Michael Faraday, Vol. 4: January 1849–October 1855. Letters 2146–3032* (London: Institution of Electrical Engineers, 1996).

James, Frank A. J. L., *The Correspondence of Michael Faraday. Vol. 6: Nov., 1860–Aug., 1867, undated letters, additional letters for volumes 1–5. Letters 3874–5053* (London: Institution of Engineering and Technology, 2012).

James, Jamie, *The Music of the Spheres: music, science, and the natural order of the universe*, (London: Little, Brown and Company, 1993).

Kim, Stephen S., *John Tyndall's Transcendental Materialism and the Conflict between Religion and Science in Victorian England* (Lewiston, NY: Mellen University Press, 1996).

Kinns, Roger, "The early history of the Edinburgh Time Ball and Time Gun", *The International Journal for the History of Engineering & Technology* 81, no. 2 (2011), pp. 264–90.

Kivy, Peter, "Charles Darwin on music", *Journal of the American Musicological Society* 12, no. 1 (Spring 1959), pp. 42–48.

Kleinman, Kim, "Darwin and Spencer on the origin of music: is music the food of love?", *Progress in Brain Research* 217 (2015), pp. 3–15.

Kumar, S. Prashant, "The instrumental Brahmin and the "half-caste" computer: astronomy and colonial rule in Madras, 1791–1835", *History of Science*, online version (2022), pp. 1–30.

Kursell, Julia (ed.), *Sounds of Science – Schall im Labor (1800–1930)* (Berlin: Max Planck Institute for the History of Science, 2008).

Kursell, Julia, '"A Gray Box": the phonograph in laboratory experiments and fieldwork, 1900–1920', in Trevor Pinch and Karin Bijsterveld (eds.), *The Oxford Handbook of Sound Studies* (Oxford University Press: Oxford, 2012), pp. 176–97.

Kursell, Julia, "Experiments on tone color in music and acoustics: Helmholtz, Schoenberg, and *Klangfarbenmelodie*", *Osiris* 28 (2013), pp. 191–211.

Kvíčalová, Anna, *Listening and Knowledge in Reformation Europe: hearing, speaking and remembering in Calvin's Geneva* (Cham: Palgrave Macmillan, 2019).

Lambert, Andrew, *Franklin: tragic hero of polar navigation* (London: Faber and Faber, 2009).

Leggett, Donald and Dunn, Richard, (eds.), *Re-inventing the Ship: science, technology and the maritime world, 1800–1918* (Ashgate: Farnham, 2012).

Lenoir, Timothy, 'Helmholtz and the materialities of communication', *Osiris* 9 (1994), pp. 184–207.

Lightman, Bernard (ed.), *Victorian Science in Context* (Chicago: University of Chicago Press, 1997).

Lightman, Bernard, "The visual theology of Victorian popularisers of science: from reverent eye to chemical retina", *Isis* 91, no. 4 (Dec. 2000), pp. 651–80.

Lightman, Bernard, *Victorian Popularizers of Science: designing nature for new audiences*, (Chicago: University of Chicago Press, 2007).

Lightman, Bernard, "Refashioning the Spaces of London Science: elite epistemes in the nineteenth century", in David N. Livingstone and Charles W. J. Withers (eds.), *Geographies of Nineteenth-Century Science* (Chicago: Chicago University Press, 2011), pp. 25–50.

Lightman, Bernard and Bennett Zon (eds.), *Evolution and Victorian Culture* (Cambridge: Cambridge University Press, 2014).

Livingstone, David N. and Charles W. J. Withers (eds.), *Geographies of Nineteenth-Century Science* (Chicago: Chicago University Press, 2011).

MacMahon, M. K. C., "Ellis [formerly Sharpe], Alexander John (1814–1890)", *Oxford Dictionary of National Biography,* 22 Sep. 2022. www.oxforddnb.com/view/10.1093/ref:odnb/9780198614128.001.0001/odnb-9780198614128-e-8683.

Maconie, Robin, "Willis, Speech, Sound, and Music", in Alexandrina Buchanan, James W. P. Campbell, Javier Giron, and Santiago Huerta (eds.), *Robert Willis: science, technology and architecture in the nineteenth century. Proceedings of the international symposium held in Gonville and Caius College, Cambridge UK, 16th–17th September 2016*, (Madrid: Instituto Juan De Herrera, 2016), pp. 75–84.

Mansell, James G., *The Age of Noise in Britain: hearing modernity* (Urbana: University of Illinois Press, 2017).

Marionneau, Ludovic and Josephine Hoegaerts, "Throwing one's voice and speaking for others: performative vocality and transcription in the assemblées of the long nineteenth century", *Journal of Interdisciplinary Voice Studies* 6, no. 1 (2021), pp. 91–108.

Marsden, Ben, ""The progeny of these two "Fellows"": Robert Willis, William Whewell and the sciences of mechanism, mechanics and machinery in early Victorian Britain", *British Journal for the History of Science* 37, no. 4 (Dec. 2004), pp. 401–34.

Marsden, Ben, "Willis, Robert (1800–1875)", *Oxford Dictionary of National Biography* (Oxford: Oxford University Press, 2004); online edition, Oct. 2009. www.oxforddnb.com/view/article/29584, accessed 1 March 2023.

Marsden, Ben and Crosbie Smith, *Engineering Empires: a cultural history of technology in nineteenth-century Britain* (Basingstoke: Palgrave Macmillan, 2005).

Martin, Thomas (ed.), *Faraday's Diary: being various philosophical notes of experimental investigation made by Michael Faraday*, Vol. I: Sep. 1820–11 Jun. 1832 (London: G. Bell and Sons, 1932).

Mawer, Allen, *South by Northwest: the magnetic crusade and the contest for Antarctica* (Edinburgh: Birlinn Limited, 2006).

Metz, Don, "The scientific instruments of Charles Wheatstone and the blending of science, art, and culture", *Interchange* 46 (2015), pp. 19–29.

Meulders, Michel, *Helmholtz: from enlightenment to neuroscience* (Cambridge, MA: MIT Press, 2010).

Miesel, Joseph S., *Public Speech and the Culture of Public Life in the Age of Gladstone* (New York: Columbia University Press, 2001).

Miesen, Leendert van der, "Studying the echo in the early modern period: between the academy and the natural world," in V. Tkaczyk and L. van der Miesen (eds.), "Sonic Things: knowledge formation in flux," *Sound Studies* 6, no. 2 (2020), pp. 196–214.

Mills, Mara, "Deafening: noise and the engineering of communication in the telephone system", *Grey Room* 43 (Spring 2011), pp. 118–43.

Milner, Greg, *Perfecting Sound Forever: the story of recorded music* (London: Granta, 2009).

Morgan, Benjamin, *The Outward Mind: materialist aesthetic in Victorian science and literature*, (Chicago: University of Chicago Press, 2017).

Morus, Iwan, "Seeing and believing science", *Isis* 97, no. 1 (Mar. 2006), pp. 101–10.

Morus, Iwan Rhys, "Worlds of Wonder: sensation and the Victorian scientific performance", *Isis* 101, no. 4 (Dec. 2010), pp. 806–16.

Morus, Iwan, Simon Schaffer, and Jim Secord, "Scientific London", in Celina Fox (ed.), *London – World City, 1800–1840* (New Haven, CT: Yale University Press, 1992), pp. 129–42.

Mowitt, John, "The sound of music in the era of its electronic reproducibility", in Jonathan Stern (ed.), *The Sound Studies Reader* (London: Routledge, 2012), pp. 213–24.

Musselman, Elizabeth Green, *Nervous Conditions: science and the body politic in early industrial Britain* (New York: State University of New York Press, 2006).

Neeley, Kathryn, *Mary Somerville: science, illumination, and the female mind* (Cambridge: Cambridge University Press, 2001).

Olleson, Philip, 'Busby, Thomas (1754–1838)', *Oxford Dictionary of National Biography* (Oxford: Oxford University Press, 2004); online edition May 2011. www.oxforddnb.com/view/article/4158, accessed 17 July 2022.

Otis, Laura, *Networking: communicating with bodies and machines in the nineteenth century*, (Ann Arbor: University of Michigan Press, 2001).

Pantalony, David, *Altered Sensations: Rudolph Koenig's acoustical workshop in nineteenth-century Paris* (New York: Springer, 2009).

Parker, Roger and Susan Rutherford, "Introduction. London Voices, 1820–1840: A "Luminous Guide"", in Roger Parker and Susan Rutherford (eds.), *London Voices, 1820–1840: vocal performers, practices, histories* (Chicago: University of Chicago Press, 2019), pp. 1–14.

Parker, Roger and Susan Rutherford (eds.), *London Voices, 1820–1840: vocal performers, practices, histories* (Chicago: University of Chicago Press, 2019).

Patterson, Elizabeth Chambers, *Mary Somerville and the Cultivation of Science, 1815–1840*, (Boston: Martinus Nijhoff Publishers, 1983).

Pesic, Peter, *Music and the Making of Modern Science* (Cambridge, MA: MIT Press, 2014).

Pearson, Christopher, "Le Corbusier and the acoustical trope: an investigation of its origin", *Journal of the Society of Architectural Historians* 56, no. 2 (June 1997), pp. 168–83.

Picker, John M., *Victorian Soundscapes* (Oxford: Oxford University Press, 2003).

Pinch, Trevor and Bijsterveld, Karin (eds.), *The Oxford Handbook of Sound Studies* (Oxford; Oxford University Press, 2012).

Port, M. H., *600 New Churches: the Church Building Commission, 1818–1856* (Reading: Spire Books Ltd, 2006).

Quinn, Aleta, "William Whewell's philosophy of architecture and the historicization of biology", *Studies in History and Philosophy of Biological and Biomedical Sciences* 59 (2016), pp. 11–19.

Rainbow, Bernarr, *The Choral Revival in the Anglican Church, 1839–1872*, (Woodbridge: Boydell Press, 2001).

Robinson, J., (2015, May 28). "Wyatt, Benjamin Dean (bap. 1775, d. 1855), architect". *Oxford Dictionary of National Biography*. www.oxforddnb.com/view/10.1093/ref:odnb/9780198614128.001.0001/odnb-9780198614128-e-30101, accessed July 17, 2022.

Rosenfeld, Sophia, "On being heard: a case for paying attention to the historical ear", *American Historical Review* 116, no. 2 (Apr. 2011), pp. 316–34.

Secord, James A., *Visions of Science: books and readers at the dawn of the Victorian Age*, (Oxford: Oxford University Press, 2014).

Schaffer, Simon, "Astronomers mark time: discipline and the personal equation", *Science in Context* 2, no. 1 (Mar. 1988), pp. 115–45.

Schaffer, Simon, "The history and geography of the intellectual world: Whewell's politics of language", in Menachem Fisch and Simon Schaffer (eds.), *William Whewell: a composite portrait* (Oxford: Clarendon Press, 1991), pp. 201–31.

Schaffer, Simon, "Metrology, Metrication, and Victorian Values", in Bernard Lightman (ed.), *Victorian Science in Context* (Chicago: Chicago University of Press, 1997), pp. 438–74.

Schaffer, Simon, "Keeping the Books at Parramatta Observatory", in David Aubin, Charlotte Brigg, and H. Otto Sibum (eds.), *The Heavens on Earth: observatories and astronomy in nineteenth-century science and culture* (Durham, NC: Duke University Press, 2010), pp. 118–47.

Schaffer, Simon, "The Bombay case: astronomers, instrument makers and the East India Company", *Journal for the History of Astronomy* xliii (2012), pp. 151–80.

Schickore, Jutta, *The Microscope and the Eye: a history of reflections, 1740–1870* (Chicago: University of Chicago Press, 2007).

Schmidgen, Henning, "Silence in the Laboratory: the history of soundproof rooms", in Julia Kursell (ed.), *Sounds of Science – Schall im Labor (1800–1930)* (Berlin: Max Planck Institute for the History of Science, 2008), pp. 47–61.

Schoenefeldt, Henrik, *Rebuilding the Houses of Parliament: David Boswell Reid and disruptive environmentalism* (London: Routledge, 2020).

Shapin, Steven, *The Scientific Revolution* (Chicago: University of Chicago Press, 1996).

Smith, Crosbie, *The Science of Energy: a cultural history of energy physics in Victorian Britain* (London: The Athlone Press, 1998).

Smith, Crosbie, "'This Great National Undertaking': John Scott Russell, the master shipwrights and the Royal Mail Steam Packet Company", in Donald Leggett and Richard

Dunn, (eds.), *Re-inventing the Ship: science, technology and the maritime world, 1800–1918*, (Farnham: Ashgate, 2012), pp. 25–52.

Smith, Crosbie and M. Norton Wise, *Energy and Empire: a biographical study of Lord Kelvin*, (Cambridge: Cambridge University Press, 1989).

Smith, Mark M., *Sensing the Past: seeing, hearing, smelling, tasting, and touching in history*, (Berkeley: University of California Press, 2007).

Smith, Stanley, "Pole, William (1814–1900)", *Oxford Dictionary of National Biography* (Oxford: Oxford University Press, 2004). www.oxforddnb.com/view/article/22463, accessed 24 Jan. 2023.

Smyth, Fiona, "'A matter of practical emergency': Herbert Baker, Hope Bagenal, and the acoustic legacy of the Assembly Chamber in Imperial Delhi", *Architectural History* 62 (2019), pp. 113–44.

Snyder, Laura J., *Reforming Philosophy: a Victorian debate on science and society* (Chicago: Chicago University Press, 2006).

Stern, Jonathan (ed.), *The Sound Studies Reader* (London: Routledge, 2012).

Strudwick, Vincent, *Christopher Wordsworth: Bishop of Lincoln, 1869–1885* (Lincoln: Honywood Press, 1987).

Steege, Benjamin, *Helmholtz and the Modern Listener* (Cambridge: Cambridge University Press, 2012).

Stephen, Leslie, "Morgan, Augustus De (1806–1871)", rev. I. Grattan-Guinness, *Oxford Dictionary of National Biography* (Oxford University Press, 2004), online edition, May 2006. www.oxforddnb.com/view/article/7470, accessed 4 Jan. 2023.

Sterne, Jonathan, *The Audible Past: cultural origins of sound reproduction* (Durham, NC: Duke University Press, 2003).

Stöckmann, H. J., "Chladni meets Napoleon", *European Physical Journal*, Special Topics, 145 (2007), pp. 15–23.

Strutt, Robert, *Life of John William Strutt: Third Baron Rayleigh, O.M, F.R.S.* (London: Longmans, Green, & Co, 1924).

Sumner, W. L., *Father Henry Willis: organ builder* (London: Musical Opinion Ltd, 1955).

Thompson, Emily, *The Soundscape of Modernity: architectural acoustics and the culture of listening in America, 1900–1933* (Cambridge, MA: MIT Press, 2002).

Tkaczyk, Viktoria, "Listening in circles. Spoken drama and the architects of sound, 1750–1830", *Annals of Science* 71, no. 3 (2014), pp. 299–334.

Tkaczyk, Viktoria, "The Making of Acoustics around 1800, or How to Do Science with Words," in Mart Helen Dupree and Sean B. Franzen (eds.), *Performing Knowledge, 1750–1850* (Berlin and Boston: De Gruyter, 2015), pp. 27–56.

Tkaczyk, V. and L. van der Miesen (eds.), "Sonic things: knowledge formation in flux," *Sound Studies* 6, no. 2 (2020), pp. 105–113.

Tresch, John and Emily I. Dolan, "Towards a new organology: instruments of music and science", *Osiris* 28, no. 1 (Jan. 2013), pp. 278–98.

Trippett, David, *Wagner's Melodies: aesthetics and materialism in German musical identity* (Cambridge: Cambridge University Press, 2013).

Turner, R. Steven, "The OHM-Seebeck dispute, Hermann Von Helmholtz, and the origins of physiological acoustics", *British Journal for the History of Science* 10, no. 34 (1977), pp. 1–24.

Tweney, Ryan D., "Faraday's Discovery of Induction: a cognitive approach", in David Gooding and Frank A. J. L. James (eds.), *Faraday Rediscovered: essays on the life and work of Michael Faraday, 1791–1867* (Basingstoke: Macmillan Press Ltd, 1985), pp. 189–209.

Tweney, Ryan D., "Stopping time: Faraday and the scientific creation of perceptual order", *Physics* 29 (1992), pp. 149–64.

Warner, Brian, *Royal Observatory, Cape of Good Hope, 1820–1831: the founding of a colonial observatory* (Dordrecht: Springer, 1995).

Warwick, Andrew, *Masters of Theory: Cambridge and the rise of mathematical physics* (Chicago: University of Chicago Press, 2003).

Werrett, Simon, "Disciplinary culture: artillery, sound, and science in Woolwich, 1800–1850", *19th Century Music* 39, no. 2 (2015), pp. 87–98.

Winter, Alison, "'Compasses all awry': the iron ship and the ambiguities of cultural authority in Victorian Britain', *Victorian Studies* 38, no. 1 (Autumn 1994), pp. 69–98.

Wise, M. Norton and Smith, Crosbie, "Measurement, work and industry in Lord Kelvin's Britain", *Historical Studies in the Physical and Biological Sciences* 17, no. 1 (1986), pp. 147–73.

Wittje, Roland, *The Age of Electroacoustics: transforming science and sound* (Cambridge, MA: MIT Press, 2016).

Yanni, Carla, "On nature and nomenclature: William Whewell and the production of architectural knowledge in early Victorian Britain", *Architecture History* 40 (1997), pp. 204–21.

Yeo, Richard, *Defining Science: William Whewell, natural knowledge, and public debate in early Victorian Britain* (Cambridge: Cambridge University Press, 1993).

Zon, Bennett, "'Spiritual' selection: Joseph Goddard and the music theology of evolution', in Martin V. Clarke (ed.), *Music and Theology in Nineteenth-Century Britain* (Farnham: Ashgate, 2012), pp. 215–35.

Zon, Bennett, "The 'non-Darwinian' revolution and the Great Chain of Musical Being", in Bernard Lightman and Bennett Zon (eds.), *Evolution and Victorian Culture* (Cambridge: Cambridge University Press, 2014), pp. 196–226.

INDEX